MW00836909

Communications and Control Engineering

Series Editors
A. Isidori • J.H. van Schuppen • E.D. Sontag • M. Thoma • M. Krstic

For further volumes:
www.springer.com/series/61

Shu-Jun Liu · Miroslav Krstic

Stochastic Averaging and Stochastic Extremum Seeking

 Springer

Shu-Jun Liu
Department of Mathematics
Southeast University
Nanjing, People's Republic of China

Miroslav Krstic
Department Mechanical & Aerospace
  Engineering
University of California, San Diego
La Jolla, California, USA

ISSN 0178-5354   Communications and Control Engineering
ISBN 978-1-4471-4086-3   ISBN 978-1-4471-4087-0 (eBook)
DOI 10.1007/978-1-4471-4087-0
Springer London Heidelberg New York Dordrecht

Library of Congress Control Number: 2012941138

Springer is part of Springer Science+Business Media (www.springer.com)

Preface

Inspiration for the Book   This book was inspired by a seemingly non-mathematical question of understanding the biological phenomenon of bacterial chemotaxis, where it is conjectured that a simple extremum seeking-like algorithm, employing stochastic perturbations instead of the conventional sinusoidal probing, enables bacteria to move in space toward areas with higher food concentration by estimating the gradient of the unknown concentration distribution.

While constructing stochastic algorithms that both mimic bacterial motions and are biologically plausible in their simplicity is easy, developing a mathematical theory that supports such algorithms was far from straightforward. The algorithms that perform stochastic extremum seeking violate one or more assumptions of any of the available theorems on stochastic averaging. As a result, we were compelled to develop, from the ground up, stochastic averaging and stability theorems that constitute significant generalizations of the existing stochastic averaging theory developed since the 1960s. This book presents the new theorems on stochastic averaging and then develops the theory and several applications of stochastic extremum seeking, including applications to non-cooperative/Nash games and to robotic vehicles. The new stochastic extremum seeking theory constitutes an alternative to established, sinusoid-based, deterministic extremum seeking.

Stochastic Averaging   The averaging method is a powerful and elegant asymptotic analysis technique for nonlinear time-varying dynamical systems. Its basic idea can be dated back to the late eighteenth century, when in 1788 Lagrange formulated the gravitational three-body problem as a perturbation of the two-body problem. No rigorous proof of its validity was given until Fatou provided the first proof of the asymptotic validity of the method in 1928. After the systematic research conducted by Krylov, Bogolyubov, and Mitropolsky in the 1930s, the averaging method gradually became one of the classical methods in analyzing nonlinear oscillations. In the past three decades, the averaging method has been extensively applied to theoretical research and engineering applications on nonlinear random vibrations.

The stochastic averaging method was first proposed in 1963 by Stratonovich based on physical consideration and later proved mathematically by Khasminskii in 1966. Since then, extensive research interest has developed in stochastic averaging in the fields of mathematics and mechanical engineering.

Stochastic Extremum Seeking   Extremum seeking is a real-time optimization tool and also a method of adaptive control, although it is different from the classical adaptive control in two aspects: (i) extremum seeking does not fit into the classical paradigm or model reference and related schemes, which deal with the problem of stabilization of a known reference trajectory or set point; (ii) extremum seeking is not model based. Extremum seeking is applicable in situations where there is a nonlinearity in the control problem, and the nonlinearity has a local minimum or a maximum. The nonlinearity may be in the plant, as a physical nonlinearity, possibly manifesting itself through an equilibrium map, or it may be in the control objective, added to the system through a cost functional of an optimization problem. Hence, one can use extremum seeking both for tuning a set point to achieve an optimal value of the output, or for tuning parameters of a feedback law.

With many applications of extremum seeking involving mechanical systems and vehicles, which are naturally modeled by nonlinear continuous-time systems, much need exists for continuous-time extremum seeking algorithms and stability theory. Unfortunately, existing stochastic averaging theorems in continuous time are too restrictive to be applicable to extremum seeking algorithms. Such algorithms violate the global Lipschitz assumptions, do not possess an equilibrium at the extremum, the average system is only locally exponentially stable, and the user's interest is in infinite-time behavior (stability) rather than merely in finite-time approximation.

This book develops the framework of stochastic extremum seeking and its applications. In the first part of the book, we develop the theoretical analysis tools of stochastic averaging for general nonlinear systems (Chaps. 3 and 4). In the second part of the book, we develop stochastic extremum seeking algorithms for static maps or dynamical nonlinear systems (Chaps. 5, 8, and 11). In the third part, we investigate the applications of stochastic extremum seeking (Chaps. 6, 7, 9, and 10).

Organization of the Book   Chapter 1 is a basic introduction to the deterministic/stochastic averaging theory. Chapter 2 provides a brief review of developments in extremum seeking in the last 15 years and presents a basic idea of stochastic extremum seeking. Chapter 3 presents stochastic averaging theory for locally Lipschitz systems that maintain an equilibrium in the presence of a stochastic perturbation. Chapter 4 presents stochastic averaging theory developed to analyze the algorithms where equilibrium is not preserved and practical stability is achieved. Chapter 5 presents single-input stochastic extremum seeking algorithm and its convergence analysis. Chapter 6 presents an application of single-parameter stochastic extremum seeking to stochastic source seeking by nonholonomic vehicles with tuning angular velocity. Chapter 7 presents stochastic source seeking with tuning forward velocity. Chapter 8 presents multi-parameter stochastic extremum seeking and slope seeking. Chapter 9 presents the application of multi-parameter stochastic extremum seeking to Nash equilibrium seeking for games with general nonlinear payoffs. Chapter 10 presents some special cases of Chap. 9: seeking of Nash equilibria for games with quadratic payoffs and applications to oligopoly economic markets and to planar multi-vehicle deployment. Chapter 11 introduces a Newton-based stochastic extremum seeking algorithm, which allows the user to achieve an arbitrary convergence rate, even in multivariable problems, despite the unknown Hessian of the cost function.

Acknowledgments   The results on which this book is based were principally developed while the first author was a postdoctoral fellow at University of California, San Diego, and has continued after she has assumed the position of a faculty member in the Department of Mathematics at Southeast University, Nanjing, China.

In the course of this research, we have benefited from interaction with Professors Tamer Basar, Ruth J. Williams, and Gang George Yin, and from the encouragement by Professors P.R. Kumar and Sanjoy Mitter. We thank our collaborators in related efforts that have inspired some of the chapters of the book—Jennie Cochran, Paul Frihauf, Azad Ghaffari, Nima Ghods, Chris Manzie, and Dragan Nesic. Our deep gratitude goes to Petar Kokotovic for his continuous support and inspiration.

We gratefully acknowledge the support that we have received at various stages in conducting this research from the National Natural Science Foundation of China, Cymer Corporation, U.S. National Science Foundation, Office of Naval Research, Los Alamos National Laboratory, and Air Force Office of Scientific Research.

Shu-Jun Liu appreciates her husband Ze-Chun's enduring support and was blessed with the presence and birth of her daughter Han-Wen in the course of writing this book. Miroslav Krstic warmly thanks his daughters Victoria and Alexandra, and his wife Angela, for their support.

Nanjing, China   Shu-Jun Liu
La Jolla, California, USA   Miroslav Krstic

Contents

Chapter 1
Introduction to Averaging

The basic idea of averaging theory—either deterministic or stochastic—is to approximate the original system (time-varying and periodic, almost periodic, or randomly perturbed) by a simpler (average) system (time-invariant, deterministic) or some approximating diffusion system (a stochastic system simpler than the original one). Starting with considerations driven by applications, the averaging method has been developed as a practical tool in mechanics/dynamics [22, 73, 117, 120, 130, 151, 152] as well as a theoretical tool in mathematics [21, 39, 51, 52, 56, 57, 59, 129], both for deterministic dynamics [22, 56, 120, 121] and for stochastic dynamics [39, 59, 78, 145]. Stochastic averaging has been the cornerstone of many control and optimization methods, such as in stochastic approximation and adaptive algorithms [17, 79, 92, 131, 132].

In this chapter, we introduce some basic results about deterministic averaging and stochastic averaging.

1.1 Averaging for Ordinary Differential Equations

1.1.1 Averaging for Globally Lipschitz Systems

1.1.1.1 Simple Case

Consider the system

$$\frac{dZ_t^\varepsilon}{dt} = \varepsilon f\left(Z_t^\varepsilon, \xi_t\right), \quad Z_0^\varepsilon = x, \tag{1.1}$$

where $Z_t^\varepsilon \in \mathbb{R}^n$, ξ_t is a function from $\mathbb{R}_+ \cup \{0\}$ to $\mathbb{R}^l$, ε is a small parameter, and $f(x, y) = [f_1(x, y), \ldots, f_n(x, y)]^T$.

If the functions $f_i(x, y)$, $i = 1, \ldots, n$, do not increase too fast, then the solution of system (1.1) converges to $Z_t^0 \equiv x$ as $\varepsilon \to 0$, uniformly on every finite time interval $[0, T]$. However, the behavior of Z_t^0 on arbitrarily long time intervals or infinite time intervals is more interesting since as time goes on far enough, significant

S.-J. Liu, M. Krstic, *Stochastic Averaging and Stochastic Extremum Seeking*, Communications and Control Engineering, DOI 10.1007/978-1-4471-4087-0_1, © Springer-Verlag London 2012

changes—such as exit from the neighborhood of an equilibrium position or of a periodic trajectory—may take place in system (1.1). Usually, time intervals of order ε^{-1} or larger are considered [39].

Let

$$X_t^\varepsilon = Z_{t/\varepsilon}^\varepsilon. \tag{1.2}$$

Then the equation for X_t^ε assumes the form

$$\frac{dX_t^\varepsilon}{dt} = f\left(X_t^\varepsilon, \xi_{t/\varepsilon}\right), \quad X_0^\varepsilon = x. \tag{1.3}$$

Thus the study of this system on a finite time interval is equivalent to the study of system (1.1) on time intervals of order ε^{-1}.

We assume that

- $f(x, y)$ is bounded, continuous in x and y, and satisfies a global Lipschitz condition in x uniformly in y: for any $x_1, x_2 \in \mathbb{R}^n$, $\forall y \in \mathbb{R}^l$, there exits a constant K (independent of y) such that

$$\left| f(x_1, y) - f(x_2, y) \right| \le K |x_1 - x_2|. \tag{1.4}$$

- The following limit holds

$$\lim_{T \to \infty} \frac{1}{T} \int_0^T f(x, \xi_s) \, ds = \bar{f}(x) \tag{1.5}$$

uniformly in $x \in \mathbb{R}^n$.

It can be shown that under the above assumptions, the function $\bar{f}(x)$ is bounded and satisfies a global Lipschitz condition with the same constant K as in (1.4). Condition (1.5) can be satisfied if ξ_t is periodic or is a sum of periodic functions.

Thus we obtain a simpler system, i.e., the *average system* of the original system (1.1):

$$\frac{d\bar{X}_t}{dt} = \bar{f}(\bar{X}_t), \quad \bar{X}_0 = x. \tag{1.6}$$

Now we consider the error between the solution of the original system (1.1) and that of its average system (1.6). By (1.1), (1.6), and (1.4), we have $\forall t \in [0, T]$

$$
\begin{aligned}
\left| X_t^\varepsilon - \bar{X}_t \right| &= \left| \int_0^t \left[f(X_s^\varepsilon, \xi_{s/\varepsilon}) - \bar{f}(\bar{X}_s) \right] ds \right| \\
&= \left| \int_0^t \left[f(X_s^\varepsilon, \xi_{s/\varepsilon}) - f(\bar{X}_s, \xi_{s/\varepsilon}) \right] ds + \int_0^t \left[f(\bar{X}_s, \xi_{s/\varepsilon}) - \bar{f}(\bar{X}_s) \right] ds \right| \\
&\le \int_0^t \left| f(X_s^\varepsilon, \xi_{s/\varepsilon}) - f(\bar{X}_s, \xi_{s/\varepsilon}) \right| ds + \left| \int_0^t \left[f(\bar{X}_s, \xi_{s/\varepsilon}) - \bar{f}(\bar{X}_s) \right] ds \right| \\
&\le K \int_0^t \left| X_s^\varepsilon - \bar{X}_s \right| ds + \alpha(\varepsilon),
\end{aligned}
\tag{1.7}
$$

where

$$\alpha(\varepsilon) \triangleq \sup_{0 \le t \le T} \left| \int_0^t \left[f(\bar{X}_s, \xi_{s/\varepsilon}) - \bar{f}(\bar{X}_s) \right] ds \right|. \tag{1.8}$$

By Gronwall's inequality, from (1.7) we obtain that

$$\sup_{0 \le t \le T} \left| X_t^\varepsilon - \bar{X}_t \right| \le \alpha(\varepsilon) e^{KT}. \tag{1.9}$$

Since

$$\alpha(\varepsilon) = \sup_{0 \le t \le T} \left| t \frac{\varepsilon}{t} \int_0^{t/\varepsilon} \left[f(\bar{X}_{u\varepsilon}, \xi_u) - \bar{f}(\bar{X}_{u\varepsilon}) \right] du \right| \to 0, \quad \text{as } \varepsilon \to 0, \tag{1.10}$$

we have

$$\lim_{\varepsilon \to 0} \sup_{0 \le t \le T} \left| X_t^\varepsilon - \bar{X}_t \right| = 0. \tag{1.11}$$

From this we obtain a proof of the fact that the trajectory X_t^ε converges to the solution of Eq. (1.6), uniformly on every finite time interval as $\varepsilon \to 0$.

The assertion that the trajectory X_t^ε is close to $\bar{X}_t$ is called the averaging principle [39]. The averaging principle supplies a kind of approximation relation between the original system and its average system. In some problems, analyzing a solution property of the original system by that of its average system is of main interest.

1.1.1.2 General Case

Consider the system

$$\dot{X}_t^\varepsilon = f_1\left(X_t^\varepsilon, \xi_t^\varepsilon\right), \quad X_0^\varepsilon = x, \tag{1.12}$$

$$\dot{\xi}_t^\varepsilon = \varepsilon^{-1} f_2\left(X_t^\varepsilon, \xi_t^\varepsilon\right), \quad \xi_0^\varepsilon = \xi, \tag{1.13}$$

where $X_t^\varepsilon \in \mathbb{R}^n$, $\xi_t^\varepsilon \in \mathbb{R}^l$, and $f_1 : \mathbb{R}^n \times \mathbb{R}^l \to \mathbb{R}^n$, $f_2 : \mathbb{R}^n \times \mathbb{R}^l \to \mathbb{R}^l$. The velocity of the motion of the variables ξ_t^ε has order ε^{-1} as $\varepsilon \to 0$. Therefore, ξ_t^ε is called the fast variable, and X_t^ε is called the slow variable. For Eq. (1.3), the role of fast motion is played by $\xi_t^\varepsilon = \xi_{t/\varepsilon}$. In this case, the velocity of fast motion does not depend on the slow variable.

We consider the fast motion $\xi_t(x)$ for fixed slow variables $x \in \mathbb{R}^n$:

$$\frac{d\xi_t(x)}{dt} = f_2\left(x, \xi_t(x)\right), \quad \xi_0(x) = \xi, \tag{1.14}$$

and assume that

- Both f_1 and f_2 are bounded and continuously differentiable functions.
- The limit

$$\lim_{T \to \infty} \frac{1}{T} \int_0^T f_1\left(x, \xi_s(x)\right) ds = \bar{f}_1(x) \tag{1.15}$$

exists independently of the initial point ξ of the trajectory $\xi_t(x)$.

The averaging principle for system (1.12)–(1.13) is the assertion that, under certain assumptions, the trajectory of the slow motion can be approximated by the trajectory of the average system

$$\frac{d\bar{X}_t}{dt} = \bar{f}_1(\bar{X}_t), \quad \bar{X}_0 = x. \tag{1.16}$$

Although the averaging principle has long been applied to problems of celestial mechanics, oscillation theory and radiophysics, a mathematically rigorous justification remained unavailable for a long time. The brief history of the development of mathematically rigorous theory of averaging is as follows [39]:

- The first general and rigorous proof of the averaging theory was obtained by N.N. Bogolyubov [22] who proved that if the limit (1.5) exists uniformly in x, then the solution X_t^ε of Eq. (1.3) converges to the solution of the average system (1.6), uniformly on every finite time interval.
- In another work [23] (see also [22]), Bogolyubov extended the above results to some cases of systems in the form (1.12)–(1.13), such as systems in which the fast motion is one-dimensional and the equation for ξ_t^ε has the form $\dot{\xi}_t^\varepsilon = \varepsilon^{-1} f_2(X_t^\varepsilon)$, as well as to some more general systems.
- V.M. Volosov [139] obtained a series of results concerning the general case of system (1.12)–(1.13). Nevertheless, in the case of multidimensional fast motions, the requirement of uniform convergence to the limit in (1.15), which is usually imposed, excludes a series of interesting problems, for example, problems arising in perturbations of Hamiltonian systems.
- In [5], it is proved that for every $T > 0$ and $\rho > 0$, the Lebesgue measure of the set F_ρ^ε of those initial conditions in problem (1.12)–(1.13) for which $\sup_{0 \le t \le T} |X_t^\varepsilon - \bar{X}_t| > \rho$ converges to zero with ε. This result was later sharpened for systems of a special form [106, 107].

1.1.2 Averaging for Locally Lipschitz Systems

In Sect. 1.1.1, the averaging principle is formulated for globally Lipschitz systems. In this section, we introduce averaging results for locally Lipschitz systems, which can be used to analyze the convergence or stability of the deterministic extremum seeking algorithm or control. Details can be found in [56, Chap. 10].

1.1.2.1 Averaging in the Periodic Case

Consider the system

$$\frac{dX_t^\varepsilon}{dt} = \varepsilon f\left(t, X_t^\varepsilon, \varepsilon\right), \tag{1.17}$$

where $X_t^\varepsilon \in \mathbb{R}^n$, and f and its partial derivatives with respect to the second and third argument up to the second order are continuous and bounded for $(t, x, \varepsilon) \in$

$[0, \infty) \times D_0 \times [0, \varepsilon_0]$, for every compact set $D_0 \subset D$, where $D \subset \mathbb{R}^n$ is a domain. Moreover, $f(t, x, \varepsilon)$ is T-periodic in t for some $T > 0$ and ε is a positive parameter. We associate with (1.17) an autonomous average system

$$\frac{d\bar{X}_t}{dt} = \varepsilon f_{\mathrm{av}}(\bar{X}_t), \tag{1.18}$$

where

$$f_{\mathrm{av}}(x) = \frac{1}{T} \int_0^T f(r, x, 0) \, dr. \tag{1.19}$$

The basic problem in the averaging method is to determine in what sense the behavior of the autonomous system (1.18) approximates the behavior of the nonautonomous system (1.17). In fact, via a change of variables, the nonautonomous system (1.17) can be represented as a perturbation of the autonomous system (1.18). For details, the reader is referred to [56].

The main result is given next.

Theorem 1.1 ([56, Theorem 10.4]) *Let $f(t, x, \varepsilon)$ and its partial derivatives with respect to (x, ε) up to the second order be continuous and bounded for $(t, x, \varepsilon) \in [0, \infty) \times D_0 \times [0, \varepsilon_0]$, for every compact set $D_0 \subset D$, where $D \subset \mathbb{R}^n$ is a domain. Suppose f is T-periodic in t for some $T > 0$ and ε is a positive parameter. Let X_t^ε and $\bar{X}_t$ denote the solutions of (1.17) and (1.18), respectively.*

1. *If $\bar{X}_{t\varepsilon} \in D$, $\forall t \in [0, b/\varepsilon]$ and $X_0^\varepsilon - \bar{X}_0 = O(\varepsilon)$, then there exists $\varepsilon^* > 0$ such that, for all $0 < \varepsilon < \varepsilon^*$, X_t^ε is defined and*

$$X_t^\varepsilon - \bar{X}_{t\varepsilon} = O(\varepsilon) \quad on \ [0, b/\varepsilon]. \tag{1.20}$$

2. *If the origin $x = 0 \in D$ is an exponentially stable equilibrium point of the average system (1.18), $\Omega \subset D$ is a compact subset of its region of attraction, $\bar{X}_0 \in \Omega$, and $X_0^\varepsilon - \bar{X}_0 = O(\varepsilon)$, then there exists $\varepsilon^* > 0$ such that, for all $0 < \varepsilon < \varepsilon^*$, X_t^ε is defined and*

$$X_t^\varepsilon - \bar{X}_{t\varepsilon} = O(\varepsilon) \quad for \ all \ t \in [0, \infty). \tag{1.21}$$

3. *If the origin $x = 0 \in D$ is an exponentially stable equilibrium point of the average system (1.18), then there exist positive constants ε^* and k such that, for all $0 < \varepsilon < \varepsilon^*$, (1.17) has a unique, exponentially stable, T-periodic solution $X_t^{T,\varepsilon}$ with the property $\|X_t^{T,\varepsilon}\| \le k\varepsilon$.*

If $f(t, 0, \varepsilon) = 0$ for all $(t, \varepsilon) \in [0, \infty) \times [0, \varepsilon_0]$, the origin is an equilibrium point of (1.17). By the uniqueness of the T-periodic solution $X_t^{T,\varepsilon}$, it follows that $X_t^{T,\varepsilon}$ is the trivial solution $x = 0$. In this case, the theorem ensures that the origin is an exponentially stable equilibrium point of (1.17).

1.1.2.2 Averaging in the General Case

Consider the system

$$\frac{dX_t^\varepsilon}{dt} = \varepsilon f\left(t, X_t^\varepsilon, \varepsilon\right), \tag{1.22}$$

where f and its partial derivatives with respect to (x, ε) up to the second order are continuous and bounded for $(t, x, \varepsilon) \in [0, \infty) \times D_0 \times [0, \varepsilon_0]$, for every compact set $D_0 \subset D$, where $D \subset \mathbb{R}^n$ is a domain. The parameter ε is positive.

The average of nonlinear function $f(t, x, \varepsilon)$ is given by the following definition.

Definition 1.1 ([56, Definition 10.2]) A continuous, bounded function $g : [0, \infty) \times D \to \mathbb{R}^n$ is said to have an average $g_{av}(x)$ if the limit

$$g_{av}(x) = \lim_{T \to \infty} \frac{1}{T} \int_t^{t+T} g(r, x)\, dr \tag{1.23}$$

exists and

$$\left\| \frac{1}{T} \int_t^{t+T} g(r, x)\, dr - g_{av}(x) \right\| \le k\sigma(T) \quad \forall (t, x) \in [0, \infty) \times D_0 \tag{1.24}$$

for every compact set $D_0 \subset D$, where k is a positive constant (possibly dependent on D_0) and $\sigma : [0, \infty) \to [0, \infty)$ is a strictly decreasing, continuous, bounded function such that $\sigma(T) \to 0$ as $T \to \infty$. The function σ is called the convergence function.

By this definition, we obtain the average system of (1.22):

$$\frac{d\bar{X}_t}{dt} = \varepsilon f_{av}(\bar{X}_t). \tag{1.25}$$

For the convenience of stating the general averaging theorem, we list some details of the deduction of the theorem.

Let

$$h(t, x) = f(t, x, 0) - f_{av}(x) \tag{1.26}$$

and denote

$$w(t, x, \eta) = \int_0^t h(r, x) e^{-\eta(t-r)}\, dr \tag{1.27}$$

for some positive constant η.

It can be shown that there is a class $\mathcal{K}$ function α such that

$$\eta \left\| w(t, x, \eta) \right\| \le k\alpha(\eta) \quad \forall (t, x) \in [0, \infty) \times D_0, \tag{1.28}$$

$$\eta \left\| \frac{\partial w(t, x, \eta)}{\partial x} \right\| \le k\alpha(\eta) \quad \forall (t, x) \in [0, \infty) \times D_0. \tag{1.29}$$

The main result for the general averaging is as follows:

Theorem 1.2 ([56, Theorem 10.5]) *Let $f(t, x, \varepsilon)$ and its partial derivatives with respect to (x, ε) up to the second order be continuous and bounded for $(t, x, \varepsilon) \in [0, \infty) \times D_0 \times [0, \varepsilon_0]$, for every compact set $D_0 \subset D$, where $\varepsilon > 0$ and $D \subset \mathbb{R}^n$ is a domain. Suppose $f(t, x, 0)$ has the average function $f_{av}(x)$ on $[0, \infty) \times D$ and the Jacobian of $h(t, x) = f(t, x, 0) - f_{av}(x)$ has zero average with the same convergence function as f. Let X_t^ε and $\bar{X}_{\varepsilon t}$ denote the solutions of (1.22) and (1.25), respectively, and α be the class $\mathcal{K}$ function appearing in the estimates of (1.28) and (1.29).*

1. *If $\bar{X}_{\varepsilon t} \in D \ \forall t \in [0, b/\varepsilon]$ and $X_0^\varepsilon - \bar{X}_0 = O(\alpha(\varepsilon))$, then there exists $\varepsilon^* > 0$ such that, for all $0 < \varepsilon < \varepsilon^*$, X_t^ε is defined and*

$$X_t^\varepsilon - \bar{X}_{\varepsilon t} = O(\alpha(\varepsilon)) \quad \text{on } [0, b/\varepsilon]. \tag{1.30}$$

2. *If the origin $x = 0 \in D$ is an exponentially stable equilibrium point of the average system (1.25), $\Omega \subset D$ is a compact subset of its region of attraction, $x_{av}(0) \in \Omega$, and $X_0^\varepsilon - \bar{X}_0 = O(\alpha(\varepsilon))$, then there exists $\varepsilon^* > 0$ such that, for all $0 < \varepsilon < \varepsilon^*$, $x(t, \varepsilon)$ is defined and*

$$X_t^\varepsilon - \bar{X}_{\varepsilon t} = O(\alpha(\varepsilon)) \quad \text{for all } t \in [0, \infty). \tag{1.31}$$

3. *If the origin $x = 0 \in D$ is an exponentially stable equilibrium point of the average system (1.25) and $f(t, 0, \varepsilon) = 0$ for all $(t, \varepsilon) \in [0, \infty) \times [0, \varepsilon_0]$, then there exists ε^* such that, for all $0 < \varepsilon < \varepsilon^*$, the origin is an exponentially stable equilibrium point of the original system (1.22).*

1.2 Stochastic Averaging

Compared with mature theoretical results for the deterministic averaging principle, stochastic averaging offers a much broader spectrum of possibilities for developing averaging theorems (due to several notions of convergence and stability, as well as many possibilities for noise processes), which are far from being exhausted. On finite time intervals, in which case one does not study stability but only approximation accuracy, there have been many averaging theorems about weak convergence [39, 60, 86, 129], convergence in probability [39, 87], and almost sure convergence [52, 86]. However, the study of stochastic averaging on the infinite time interval is not complete compared to complete results for the deterministic case [56, 121].

1.2.1 Averaging for Stochastic Perturbation Process

Consider the system

$$\dot{X}_t^\varepsilon = f(X_t^\varepsilon, \xi_{t/\varepsilon}), \quad X_0^\varepsilon = x, \tag{1.32}$$

where $X_t^\varepsilon \in \mathbb{R}^n$, ξ_t, $t \geq 0$, is a stochastic process with values in $\mathbb{R}^l$.

We assume that

- The function $f(x, y)$ satisfies a global Lipschitz condition: for $x_i \in \mathbb{R}^n$, $y_i \in \mathbb{R}^l$, $i = 1, 2$, there exists a constant K such that

$$\left| f(x_1, y_1) - f(x_2, y_2) \right| \leq K \left(|x_1 - x_2| + |y_1 - y_2| \right). \tag{1.33}$$

- The trajectories of the process $(\xi_t, t \geq 0)$ are continuous with probability one, or on every finite time interval they have a finite number of discontinuities of the first kind and there are no discontinuities of the second kind.

Under these assumptions, the solution of equation (1.32) exists with probability one for any initial condition and it is defined uniquely for all $t \geq 0$ [39].

Compared with the deterministic condition (1.5), the stochastic averaging principle has different types of convergence condition since there are different convergence notions in the stochastic case. In general, if less stringent assumptions are imposed concerning the type of convergence in (1.5), then a weaker result holds. Here we just list two cases (convergence with probability one and convergence in probability):

(i) If condition (1.5) is satisfied with probability one uniformly in $x \in \mathbb{R}^n$, then the ordinary averaging principle implies that with probability one the trajectory of X_t^ε converges to the solution of Eq. (1.6), uniformly on every finite interval ($\bar{f}(x)$ and $\bar{X}_t$ may depend on the sample trajectory ω in general).
(ii) Assume that there exists a vector field $\bar{f}(x)$ in $\mathbb{R}^n$ such that, for any $\delta > 0$ and $x \in \mathbb{R}^n$,

$$\lim_{T \to \infty} P \left\{ \left| \frac{1}{T} \int_t^{t+T} f(x, \xi_s) \, ds - \bar{f}(x) \right| > \delta \right\} = 0, \tag{1.34}$$

uniformly in $t > 0$. It follows from (1.34) that $\bar{f}(x)$ satisfies a global Lipschitz condition (with the same constant as $f(x, y)$). Therefore, there exists a unique solution of the problem

$$\frac{d\bar{X}_t}{dt} = \bar{f}(\bar{X}_t), \quad \bar{X}_0 = x. \tag{1.35}$$

The stochastic process X_t^ε can be considered as a result of stochastic perturbations of the dynamical system (1.35), small on the average.

Theorem 1.3 ([39, Theorem 7.2.1]) *Suppose that condition (1.34) is satisfied and* $\sup_t E|f(x, \xi_t)|^2 < \infty$. *Then for any* $T > 0$ *and* $\delta > 0$,

$$\lim_{\varepsilon \to 0} P \left\{ \sup_{0 \leq t \leq T} \left| X_t^\varepsilon - \bar{X}_t \right| > \delta \right\} = 0. \tag{1.36}$$

1.2.2 Averaging for Stochastic Differential Equations

Consider the system of differential equations

$$dX_t^\varepsilon = f\left(X_t^\varepsilon, Y_t^\varepsilon\right) dt + g\left(X_t^\varepsilon, Y_t^\varepsilon\right) dW_t, \quad X_0^\varepsilon = x, \tag{1.37}$$

$$dY_t^\varepsilon = \varepsilon^{-1} B\left(X_t^\varepsilon, Y_t^\varepsilon\right) dt + \varepsilon^{-1/2} C\left(X_t^\varepsilon, Y_t^\varepsilon\right) dW_t, \quad Y_0^\varepsilon = y, \tag{1.38}$$

where $X_t^\varepsilon \in \mathbb{R}^n$, $Y_t^\varepsilon \in \mathbb{R}^l$, $f(x,y) = (f_1(x,y), \ldots, f_n(x,y))$, $B(x,y) = (B_1(x,y), \ldots, B_l(x,y))$, W_t is an r-dimensional Wiener process and $g(x,y) = (g_{ij})_{n \times r}$, $C(x,y) = (C_{ij}(x,y))_{l \times r}$.

We introduce a stochastic process Y_t^{xy}, $x \in \mathbb{R}^n$, $y \in \mathbb{R}^l$, which is defined by the stochastic differential equation

$$dY_t^{xy} = B\left(x, Y_t^{xy}\right) dt + C\left(x, Y_t^{xy}\right) dW_t, \quad Y_0^{xy} = y. \tag{1.39}$$

The solution of this equation forms a Markov process in $\mathbb{R}^l$, depending on $x \in \mathbb{R}^n$ as a parameter.

We assume that

- The functions $f_i(x,y)$, $B_i(x,y)$, $g_{ij}(x,y)$, $C_{ij}(x,y)$ are bounded and satisfy a global Lipschitz condition.
- There exists a function $\bar{f}(x) = (\bar{f}_1(x), \ldots, \bar{f}_n(x))$, $x \in \mathbb{R}^n$, such that, for any $t \geq 0$, $x \in \mathbb{R}^n$, $y \in \mathbb{R}^l$, we have

$$E \left| \frac{1}{T} \int_t^{t+T} f\left(x, Y_r^{xy}\right) dr - \bar{f}(x) \right| < \kappa(T), \tag{1.40}$$

where $\kappa(T) \to 0$ as $T \to \infty$.

Theorem 1.4 ([39, Theorem 7.9.1]) *Let the entries of $g(x,y) = g(x)$ be independent of y and let condition (1.40) be satisfied. Denote by $\bar{X}_t \in \mathbb{R}^n$ the stochastic process governed by the differential equation*

$$\frac{d\bar{X}_t}{dt} = \bar{f}(\bar{X}_t) dt + g(\bar{X}_t) dW_t, \quad \bar{X}_0 = x. \tag{1.41}$$

Then for any $T > 0$, $\delta > 0$, $x \in \mathbb{R}^n$,

$$\lim_{\varepsilon \to 0} P\left\{ \sup_{0 \leq t \leq T} \left| X_t^\varepsilon - \bar{X}_t \right| > \delta \right\} = 0. \tag{1.42}$$

In general, the averaging principle on infinite time interval is considered under the stability condition of average systems or diffusion approximation. The stability of a stochastic system with wide-band noise disturbances under diffusion approximation conditions is stated in [21]. The stability of dynamic systems with Markov perturbations under the stability condition of the average system is studied in [65]. Under a condition on a diffusion approximation of a dynamical system with Markov perturbations, the problem of stability is solved in [66]. Under conditions of averaging and diffusion approximation, the stability of dynamic systems in semi-Markov medium was studied in [67]. All these results are established under all or almost all of the following conditions:

- The average system or approximating diffusion system is globally exponentially stable;

- The nonlinear vector field of the original system has bounded derivative or is dominated by some form of Lyapunov function of the average system;
- The nonlinear vector field of the original system vanishes at the origin for any value of perturbation process (equilibrium condition);
- The state space of the perturbation process is a compact space.

These conditions largely limit the application of existing stochastic averaging theorems.

In Chaps. 3 and 4, we remove or weaken several restrictions in these existing results and develop more general averaging for our stochastic extremum seeking problems.

Chapter 2
Introduction to Extremum Seeking

In this chapter, we review the motivation behind extremum seeking methodology and the advances in the field of extremum seeking of the last 15 years. Then we present a basic introduction to stochastic extremum seeking, including how it relates to standard deterministic extremum seeking with periodic perturbations and what ideas are behind the study of stability of the resulting stochastic nonlinear system.

2.1 Motivation and Recent Revival

Extremum seeking is a non-model based real-time optimization approach for dynamic problems where only limited knowledge of a system is available, such as when the system has a nonlinear equilibrium map which has a local minimum or maximum. Popular in applications around the middle of the twentieth century, extremum seeking was nearly dormant for several decades until the emergence of a proof of its stability [71], with a subsequent resurgence of interest in extremum seeking for further theoretical developments and applications.

The increasing complexity of engineering systems, including feedback systems, has led to many optimization challenges since analytic solutions to optimization problems for multi-agent, nonlinear, and infinite-dimensional systems are difficult, if not impossible, to obtain. This difficulty arises for many reasons, including the presence of competing or adversarial goals, the high-dimensionality of the system, and the inherent system uncertainty. Moreover, if a model-based solution is obtained for these complicated optimization problems, it is likely to be conservative due to modeling deficiencies. Hence, non-model based extremum seeking methods are an attractive option to solve these problems.

Many works have focused on optimization/learning methods for unknown systems in a wide variety of fields. In games, most algorithms designed to achieve convergence to Nash equilibria require modeling information for the game and assume the players can observe the actions of the other players. The fictitious play strategy is one such strategy (employed in finite games) where a player devises a best

S.-J. Liu, M. Krstic, *Stochastic Averaging and Stochastic Extremum Seeking*,
Communications and Control Engineering,
DOI 10.1007/978-1-4471-4087-0_2, © Springer-Verlag London 2012

response based on the history of the other players actions. A dynamic version of fictitious play and gradient response is developed in [126] and shown to converge to a mixed-strategy Nash equilibrium in cases where previously developed algorithms did not converge. In [38], regret testing with random strategy switches is proved to converge to the Nash equilibrium in finite two-player games where each player measures only its own payoffs. In [150], a synchronous distributed learning algorithm, where players remember their own actions and utility values from the previous two times steps, is shown to converge in probability to a set of restricted Nash equilibria. In [8, 48, 128], games with a continuum of traders are analyzed. Additional results on learning in games can be found in [26, 37, 42, 54, 83, 127].

The extremum seeking (ES) method has seen significant theoretical advances during the past decade, including the proof of local convergence [6, 27, 119, 140], PID tuning [61], slope seeking [7], performance improvement and limitations in ES control [70], extension to semi-global convergence [137], development of scalar Newton-like algorithms [101, 102, 108], inclusion of measurement noise [135], extremum seeking with partial modeling information [1, 2, 33, 36, 50], and learning in noncooperative games [40, 136].

ES has also been used in many diverse applications with unknown/uncertain systems, such as steering vehicles toward a source in GPS-denied environments [28, 30, 146], active flow control [15, 16, 24, 53, 63, 64], aeropropulsion [104, 144], colling systems [82, 84] wind energy [32], photovoltaics [81], human exercise machines [148], optimizing the control of nonisothermal valve actuator [113], controlling Tokamak plasmas [25], and enhancing mixing in magnetohydrodynamic channel flows [96], timing control of HCCI engine combustion [62], formation flight optimization [20], control of aircraft endurance based on atmospheric turbulence [69], beam matching adaptive control [123], optimizing bioreactors [141], control of combustion instability [9], control of nonisothermal continuous stirred reactors [49], control of swirl-stabilized spray combustion [105], optimal control of current profile in the DIII-D tokamak [110], laser pulse shaping [115], control of beam envelope in particle accelerators [122], control of an axial-flow compressor [142], and stabilization of neoclassical tearing modes in tokamak fusion plasmas [143].

2.2 Why Stochastic Extremum Seeking?

In existing perturbation-based extremum seeking algorithms, periodic (sinusoidal) excitation signals are primarily used to probe the nonlinearity and estimate its gradient. Biological systems (such as bacterial chemotaxis) do not use periodic probing in climbing food or light gradients. In man-made source seeking systems, the nearly random motion of the stochastic seeker has its advantage in applications where the seeker itself may be pursued by another pursuer. A seeker, which successfully performs the source finding task but with an unpredictable, nearly random trajectory, is a more challenging target, and is hence less vulnerable, than a deterministic seeker. Furthermore, if the system has high dimensionality, the orthogonality requirements

Fig. 2.1 A quartic static map with local minimum $f(-1) = 1$ and global minimum $f(1) = -3$

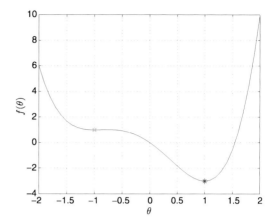

on the elements of the periodic perturbation vector pose an implementation challenge. Thus there is merit in investigating the use of stochastic perturbations within the ES architecture. The first results in that direction were achieved in the discrete-time case [98], using the existing theory of stochastic averaging in the discrete-time case. Source seeking results employing deterministic perturbations in the presence of stochastic noise have been reported in [133, 134], also in discrete time.

Stochastic extremum seeking and its stability analysis have some ideas and techniques in common with classical methods of annealing, stochastic approximation, and stochastic adaptive control [14, 43, 44, 75, 93–95].

2.3 A Brief Introduction to Stochastic Extremum Seeking

In this section, we present the basic idea of stochastic extremum seeking, make a comparison with deterministic (periodically perturbed) extremum seeking, and discuss a heuristic idea of stochastic averaging as a way of studying stability of a stochastic extremum seeking algorithm.

While extremum seeking is applicable to plants with dynamics (plants modeled by ordinary differential equations), in this section we introduce extremum seeking on the simplest possible problem—the optimization of a static map $f(\theta)$. Without loss of generality, we assume that f has a minimum at $\theta = \theta^*$ and we seek that minimum.

For the purpose of illustration, we use the following quartic map

$$f(\theta) = \theta^4 + \theta^3 - 2\theta^2 - 3\theta, \tag{2.1}$$

which is depicted in Fig. 2.1 and has a local minimum $f(-1) = 1$ and a global minimum $f(1) = -3$. The second derivatives at the two minima are $f''(-1) = 2 < 14 = f''(1)$, which is consistent with the global minimum at $\theta = 1$ being much "deeper" and "sharper" than the local minimum at $\theta = -1$.

Fig. 2.2 Block diagram for *deterministic* extremum seeking scheme for a static map

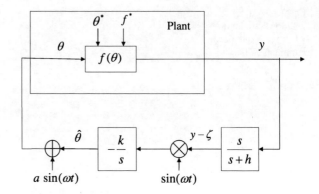

2.3.1 A Basic Deterministic ES Scheme

Let us consider first the deterministic ES scheme shown in Fig. 2.2. The scheme employs a sinusoidal perturbation $\sin(\omega t)$, which additively enters the map $f(\theta)$. The measured output $y = f(\theta)$ is then passed through a washout filter and multiplied by the same perturbation signal, $\sin(\omega t)$, generating an estimate of the derivative (scalar gradient) $f'(\theta)$ at the input of the integrator. The integrator then updates the estimate $\hat{\theta}(t)$ in the direction of driving the gradient to zero. For $k > 0$ the ES scheme drives $\hat{\theta}(t)$ toward the nearest local minimum of $f(\theta)$, whereas for $k < 0$ the scheme converges toward the nearest maximum. The washout filter $\frac{s}{s+h}$ is not required but it helps performance somewhat. The logic behind the use of the washout filter is to kill the DC component of the map, $f(\theta^*)$, although the multiplication of the output y with the zero-mean perturbation $\sin(\omega t)$ also performs that role. The washout filter is just more effective in eliminating the DC component of y, without requiring that the perturbation frequency ω be relatively high.

The scheme in Fig. 2.2 has four design parameters, a, k, ω, and h. The amplitude a provides a trade-off between asymptotic performance and the region of attraction of the algorithm. The smaller the a, the smaller the residual error at the minimum achieved, but also the larger the possibility of getting stuck at a local minimum. Conversely, the larger the a, the larger both the residual error and the possibility of reaching the global minimum. The gain parameter k controls the speed of convergence, jointly with a which also influences the speed of convergence. The perturbation frequency ω controls the separation between the time scale of the estimation process, conducted by the integrator, and of the gradient estimation process, performed by the additive and multiplicative perturbation. The higher the frequency ω, the cleaner the estimate of the gradient and the smaller the effect of the perturbations introduced by the higher-order harmonics and of the DC component of y. The washout filter frequency h should be smaller than ω, so that the filter eliminates the DC component in y without corrupting the estimation of the gradient $f'(\theta)$.

Figure 2.3 shows the time response of a discrete-time version of the deterministic ES algorithm in Fig. 2.2. Even though the algorithm starts from the local minimum $\theta = -1$, it does not remain stuck at the local minimum and converges to the global

Fig. 2.3 Time response of a
discrete-time version of the
deterministic extremum
seeking algorithm in Fig. 2.2,
starting from the local
minimum, $\hat{\theta}(0) = -1$. The
parameters are chosen as
$\omega = 5, a = 0.4, k = 1$

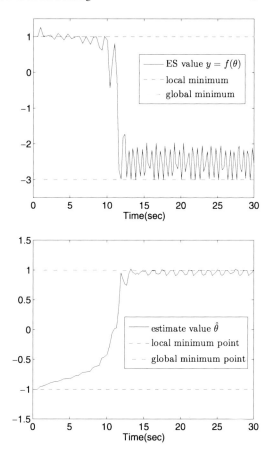

minimum $\theta = 1$. However, if the amplitude a and the gain k were chosen smaller,
the algorithm would be unable to overcome the "hump" between $\theta = -1$ and $\theta = 1$
and it would remain stuck at the local minimum.

2.3.2 A Basic Stochastic ES Scheme

Limitations of the deterministic ES scheme include the fact that the perturbation
is uniformly bounded (by a), which may highly restrict the algorithm's region of
attraction, and the fact that learning using a single-frequency sinusoidal perturbation is rather simple-minded and rare in probing-based learning and optimization
approaches encountered in biological systems.

To overcome such limitations of deterministic probing signals, we consider using
stochastic probing signals. Sinusoidal signals have two properties that are crucial for
extremum seeking: (i) their mean is zero and (ii) when squared, the mean is positive.

Fig. 2.4 Block diagram for *stochastic* extremum seeking scheme with unbounded perturbations for a static map

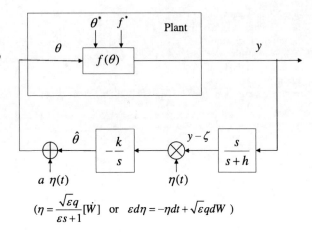

$$\left(\eta = \frac{\sqrt{\varepsilon q}}{\varepsilon s+1}[\dot{W}] \quad \text{or} \quad \varepsilon d\eta = -\eta dt + \sqrt{\varepsilon q}dW \right)$$

Such properties are similar to the properties of Gaussian white noise signals, namely, zero expectation and positive variance.

Hence, we consider replacing the signals $\sin(\omega t)$ in Fig. 2.2 by white noise $\dot{W}(t)$, where $W(t)$ is a standard Brownian motion process (also referred to as the Wiener process). However, such a perturbation is overly aggressive and makes the mathematical analysis intractable because it enters the differential equation in a nonlinear manner (it gives rise to quadratic and other functions of $\dot{W}$).

To soften the impact of the white noise perturbations, while introducing randomness and making the resulting dynamics mathematically tractable, we replace the signals $\sin(\omega t)$ in Fig. 2.2 by the signal $\eta(t)$ obtained by passing white noise $\dot{W}(t)$ through a low-pass filter $\frac{\sqrt{\varepsilon q}}{\varepsilon s+1}$ for some positive constants ε and q, or, given in terms of an Ito differential equation, we employ the perturbation $\eta(t)$ governed by

$$\varepsilon \, d\eta = -\eta \, dt + \sqrt{\varepsilon q} \, dW. \tag{2.2}$$

The resulting stochastic ES scheme is shown in Fig. 2.4.

Figure 2.5 shows the time response of a discrete-time version of the stochastic ES algorithm in Fig. 2.4. Starting from the local minimum $\theta = -1$, the algorithm converges to the global minimum $\theta = 1$.

2.3.3 A Heuristic Analysis of a Simple Stochastic ES Algorithm

To provide the reader with some intuition and motivation, in this section we provide a preliminary and completely informal analysis of the extremum seeking algorithm in Fig. 2.4. We present a series of calculations which, though not reflective of the rigorous methods pursued in the book, do illustrate heuristically the basic ideas behind establishing stability and quantifying the convergence rates of ES schemes.

To simplify our analysis, we eliminate the washout filter from the ES scheme, namely, we replace $\frac{s}{s+h}$ in Fig. 2.4 by a unity gain block. This approximation is

Fig. 2.5 Time response of a
discrete-time version of the
stochastic extremum seeking
algorithm in Fig. 2.4, starting
from the local minimum,
$\hat{\theta}(0) = -1$. The parameters
are chosen as $q = 1$,
$\varepsilon = 0.25$, $a = 0.8$, $k = 10$

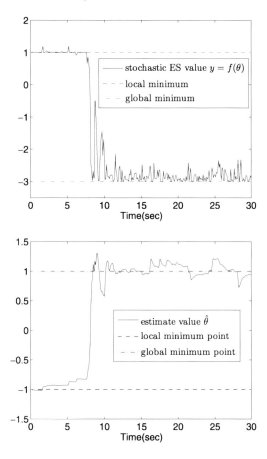

certainly justified for h that is small relative to other parameters, particularly relative
to k. The elimination of the washout filter results in a first-order system, whose sole
state is the state $\hat{\theta}$ of the integrator in Fig. 2.4, and which is driven by another first-
order linear stochastic system with state η. Despite the low order, the analysis of the
closed-loop system is not trivial because the system is nonlinear, time-varying, and
stochastic.

We start by introducing notation to describe the system in Fig. 2.4. We note that

$$\theta(t) = \hat{\theta}(t) + a\eta(t) \tag{2.3}$$

and denote the estimation error as

$$\tilde{\theta}(t) = \theta^* - \hat{\theta}(t). \tag{2.4}$$

Combining (2.3) and (2.4), we get

$$\theta(t) = a\eta(t) - \tilde{\theta}(t). \tag{2.5}$$

Then, from the integrator block we observe that the estimation error is governed
by

$$\dot{\tilde{\theta}}(t) = -\dot{\hat{\theta}}(t)$$
$$= k\eta(t)y(t)$$
$$= k\eta(t)f(\theta(t)). \tag{2.6}$$

Using (2.5) and applying the Taylor expansion to $f(\theta)$ around θ^* up to second order, we get

$$f(\theta) = f(a\eta - \tilde{\theta})$$
$$\approx f(\theta^*) + f'(\theta^*)(a\eta - \tilde{\theta}) + \frac{1}{2}f''(\theta^*)(a\eta - \tilde{\theta})^2. \tag{2.7}$$

Given the assumption that the map $f(\theta)$ has a minimum at θ^*, it follows that $f'(\theta^*) = 0$, which yields

$$f(\theta) \approx f(\theta^*) + \frac{1}{2}f''(\theta^*)(a\eta - \tilde{\theta})^2$$
$$= f(\theta^*) + \frac{1}{2}f''(\theta^*)[a^2\eta^2 - 2a\eta\tilde{\theta} + \tilde{\theta}^2]. \tag{2.8}$$

Substituting (2.8) into (2.6), we get

$$\dot{\tilde{\theta}} \approx k\eta\left\{f(\theta^*) + \frac{1}{2}f''(\theta^*)[a^2\eta^2 - 2a\eta\tilde{\theta} + \tilde{\theta}^2]\right\}$$
$$= k\eta\left[f(\theta^*) + \frac{a^2}{2}f''(\theta^*)\eta^2 - af''(\theta^*)\eta\tilde{\theta} + \frac{1}{2}f''(\theta^*)\tilde{\theta}^2\right]. \tag{2.9}$$

Grouping the terms in powers of η, we obtain

$$\dot{\tilde{\theta}}(t) \approx k\left\{\eta(t)\left[f(\theta^*) + \frac{1}{2}f''(\theta^*)\tilde{\theta}^2(t)\right]\right.$$
$$- \eta^2(t)af''(\theta^*)\tilde{\theta}(t)$$
$$\left. + \eta^3(t)\frac{a^2}{2}f''(\theta^*)\right\}. \tag{2.10}$$

The signal $\eta(t)$ is a stochastic perturbation governed by the stochastic linear differential equation (2.2), where $W(t)$ is the Wiener process. With small ε, the signal η is a close approximation of white noise. Using elementary Ito calculus, it is easy to calculate the expectations of the three powers of η appearing in (2.10). These expectations have the properties that

$$\lim_{t\to\infty} E\{\eta(t)\} = 0, \tag{2.11}$$

$$\lim_{t\to\infty} E\{\eta^2(t)\} = \frac{q^2}{2}, \tag{2.12}$$

$$\lim_{t\to\infty} E\{\eta^3(t)\} = 0. \tag{2.13}$$

Fig. 2.6 Block diagram for *stochastic* extremum seeking scheme with bounded perturbations for a static map

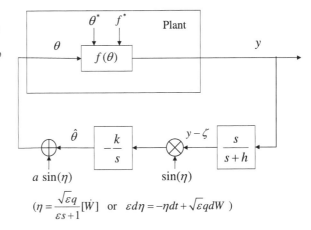

$$\left(\eta = \frac{\sqrt{\varepsilon}q}{\varepsilon s + 1}[\dot{W}] \quad \text{or} \quad \varepsilon d\eta = -\eta dt + \sqrt{\varepsilon}q dW\right)$$

To illustrate how these relations are obtained, we consider the case of η^2, namely, (2.12), which is obtained by applying Ito's differentiation rule to η^2 with the help of (2.2), which yields the ODE

$$\frac{\varepsilon}{2}\frac{dE\{\eta^2\}}{dt} = -E\{\eta^2\} + \frac{q^2}{2} \tag{2.14}$$

The solution of the linear ODE (2.14) is

$$E\{\eta^2(t)\} = e^{-2t/\varepsilon}E\{\eta^2(0)\} + \frac{q^2}{2}\left(1 - e^{-2t/\varepsilon}\right) \tag{2.15}$$

$$\rightarrow \frac{q^2}{2} \quad \text{as } t \rightarrow \infty. \tag{2.16}$$

When ε is small, it is clear from (2.15) that the convergence in time t is very fast. This is the case with the convergence rates of all three expectations given in (2.11), (2.12), and (2.13).

Approximating now the η-terms in (2.10) by their respective expectations, after a short transient whose length is $O(\varepsilon)$, the estimation error is governed by

$$\dot{\tilde{\theta}}(t) \approx -\frac{kaq^2}{2}f''(\theta^*)\tilde{\theta}(t). \tag{2.17}$$

This completes our heuristic preliminary study of stability of the stochastic ES scheme in Fig. 2.4. Local stability is expected, in a suitable probabilistic sense, provided $ka > 0$ and provided the map has a minimum at θ^*. Moreover, the convergence speed is governed by the values of the parameters k, a, q, and also by the value of $f''(\theta^*) > 0$. The "flatter" the extremum, the slower the convergence and, conversely, the "sharper" the extremum, the faster the convergence toward it.

Rigorous stability analysis of stochastic ES algorithms is presented in Chaps. 5 and 6. However, the scheme in Fig. 2.4 with the unbounded stochastic perturbation $\eta(t)$ is not amenable to rigorous analysis. To make analysis feasible, using averaging theorems that we develop in Chap. 4, we replace η in the algorithm in Fig. 2.4 by a bounded stochastic perturbation $\sin(\eta)$, obtaining the algorithm in Fig. 2.6.

Algorithms in Figs. 2.4 and 2.6 have very similar local convergence properties. The convergence speeds of the two algorithms are related as

$$\frac{\text{speed}_{\sin(\eta)}}{\text{speed}_\eta} = \frac{(1 - e^{-q^2})}{q^2}. \tag{2.18}$$

Chapter 3
Stochastic Averaging for Asymptotic Stability

In this chapter, we remove or weaken the restrictions in the existing averaging theory and develop stochastic averaging theorems for studying the stability of a general class of nonlinear systems with a stochastic perturbation. This chapter focuses on the asymptotic stability because the original system considered here is required to satisfy an equilibrium condition. When such a condition does not hold for the original system, practical stability is studied in Chap. 4.

In this chapter, if the perturbation process satisfies a uniform strong ergodic condition and the equilibrium of the average system is exponentially stable, we show that the original system is exponentially practically stable in probability. Under the condition that the equilibrium of the average system is exponentially stable, if the perturbation process is ϕ-mixing with exponential mixing rate and exponentially ergodic, and the original system satisfies an equilibrium condition, we show that the equilibrium of the original system is asymptotically stable in probability. For the case where the average system is globally exponentially stable and all the other assumptions are valid globally, a global result is obtained for the original system.

The chapter is organized as follows. Section 3.1 describes the investigated problem. Section 3.2 presents results for two cases: uniform strong ergodic perturbation process, and exponentially ϕ-mixing and exponentially ergodic perturbation process, respectively. In Sect. 3.3, we give the detailed proofs for the results in Sect. 3.2. In Sect. 3.4, we give three examples. Section 3.5 contains some notes and references.

3.1 Problem Formulation

Consider the system

$$\frac{dX_t^\varepsilon}{dt} = a\left(X_t^\varepsilon, Y_{t/\varepsilon}\right), \quad X_0^\varepsilon = x, \tag{3.1}$$

where $X_t^\varepsilon \in \mathbb{R}^n$, and the stochastic perturbation $Y_t \in \mathbb{R}^m$ is a time homogeneous continuous Markov process defined on a complete probability space $(\Omega, \mathcal{F}, P)$, where

S.-J. Liu, M. Krstic, *Stochastic Averaging and Stochastic Extremum Seeking*,
Communications and Control Engineering,
DOI 10.1007/978-1-4471-4087-0_3, © Springer-Verlag London 2012

Ω is the sample space, $\mathcal{F}$ is a σ-field, P is a probability measure, and ε is a small positive parameter, where $\varepsilon \in (0, \varepsilon_0)$ for some fixed $\varepsilon_0 > 0$.

The average system corresponding to system (3.1) can be defined in various ways, depending on assumptions on the perturbation process $(Y_t, t \geq 0)$. For example, the average system of (3.1) can be defined as

$$\frac{d\bar{X}_t}{dt} = \bar{a}(\bar{X}_t), \quad \bar{X}_0 = x, \tag{3.2}$$

where $\bar{a}(x)$ is a function such that (1.34) holds, i.e., for any $\delta > 0$ and $x \in \mathbb{R}^n$,

$$\lim_{T \to \infty} P\left\{ \left| \frac{1}{T} \int_t^{t+T} a(x, Y_s)\, ds - \bar{a}(x) \right| > \delta \right\} = 0 \tag{3.3}$$

uniformly in $t \geq 0$.

From Theorem 1.3, we know that on a finite time interval $[0, T]$, under certain conditions, the solution of the original system (3.1) can be approximated in probability by the solution of the average system (3.2) as the small parameter ε goes to zero.

In this chapter, we explore the averaging principle when t belongs to the infinite time interval $[0, \infty)$. First, in the case where the original stochastic system may not have an equilibrium, but the average system has an exponentially stable equilibrium at the origin, a stability-like property of the original system is established for ε sufficiently small. Second, when $a(0, y) \equiv 0$, namely, when the original system (3.1) maintains an equilibrium at the origin, despite the presence of noise, we establish stability of this equilibrium for sufficiently small ε.

3.2 Main Theorems

3.2.1 Uniform Strong Ergodic Perturbation Process

In the time scale $s = t/\varepsilon$, define $Z_s^\varepsilon = X_{\varepsilon s}^\varepsilon = X_t^\varepsilon$, $Y_s = Y_{t/\varepsilon}$. Then we transform system (3.1) into

$$\frac{dZ_s^\varepsilon}{ds} = \varepsilon a\left(Z_s^\varepsilon, Y_s\right), \tag{3.4}$$

with the initial value $Z_0^\varepsilon = x$. Let S_Y be the living space of the perturbation process $(Y_t, t \geq 0)$. Notice that S_Y may be a proper (e.g., compact) subset of $\mathbb{R}^m$.

Assumption 3.1 The vector field $a(x, y)$ is separable, i.e., it can be written as

$$a(x, y) = \sum_{i=1}^{l} a_i(x) b_i(y), \tag{3.5}$$

where the functions $b_i : O_Y \to \mathbb{R}$, $i = 1, \dots, l$, are continuous (the set O_Y, which contains S_Y, is an open subset of $\mathbb{R}^n$) and bounded on S_Y; the functions $a_i : D \to \mathbb{R}^n$,

$i = 1, \ldots, l$, and their partial derivatives up to the second order are continuous on some domain (open connected set) $D \subset \mathbb{R}^n$.

Assumption 3.2 For $i = 1, \ldots, l$, there exists a constant $\bar{b}_i$ such that

$$\lim_{T \to \infty} \frac{1}{T} \int_t^{t+T} b_i(Y_s)\, ds = \bar{b}_i \quad \text{a.s.} \tag{3.6}$$

uniformly in $t \in [0, \infty)$.

By Assumption 3.2, we obtain the average system of (3.4) as

$$\frac{d\bar{Z}_s^\varepsilon}{ds} = \varepsilon \bar{a}(\bar{Z}_s^\varepsilon), \tag{3.7}$$

with the initial value $\bar{Z}_0^\varepsilon = x$, where

$$\bar{a}(x) = \sum_{i=1}^l a_i(x)\bar{b}_i. \tag{3.8}$$

Theorem 3.1 *Suppose that Assumptions 3.1 and 3.2 hold. Let $Z_s^\varepsilon(\omega)$ and $\bar{Z}_s^\varepsilon$ denote the solutions of system (3.4) and the average system (3.7), respectively. If the origin $\bar{Z}_s^\varepsilon \equiv 0$ is an exponentially stable equilibrium point of the average system, $K \subset D$ is a compact subset of its region of attraction, and $\bar{Z}_0^\varepsilon = x \in K$, then, for any $\varsigma \in (0, 1)$, there exist a measurable set $\Omega_\varsigma \subset \Omega$ with $P(\Omega_\varsigma) > 1 - \varsigma$, a class $\mathcal{K}$ function α_ς, and a constant $\varepsilon^*(\varsigma) > 0$ such that if $Z_0^\varepsilon - \bar{Z}_0^\varepsilon = O(\alpha_\varsigma)$, then, for all $0 < \varepsilon < \varepsilon^*(\varsigma)$,*

$$Z_s^\varepsilon(\omega) - \bar{Z}_s^\varepsilon = O(\alpha_\varsigma(\varepsilon)) \quad \text{for all } s \in [0, \infty) \tag{3.9}$$

uniformly in $\omega \in \Omega_\varsigma$, which implies

$$P\left\{ \sup_{s \in [0,\infty)} \left| Z_s^\varepsilon(\omega) - \bar{Z}_s^\varepsilon \right| = O(\alpha_\varsigma(\varepsilon)) \right\} > 1 - \varsigma. \tag{3.10}$$

Next we extend the finite-time result (1.36) of [39, Theorem 7.2.1] to infinite time.

Theorem 3.2 *Suppose that Assumptions 3.1 and 3.2 hold. Let $Z_s^\varepsilon(\omega)$ and $\bar{Z}_s^\varepsilon$ denote the solutions of system (3.4) and the average system (3.7), respectively. If the origin $\bar{Z}_s^\varepsilon \equiv 0$ is an exponentially stable equilibrium of the average system, $K \subset D$ is a compact subset of its region of attraction, and $\bar{Z}_0^\varepsilon = Z_0^\varepsilon = x \in K$, then, for any $\delta > 0$,*

$$\lim_{\varepsilon \to 0} P\left\{ \sup_{s \in [0,\infty)} \left| Z_s^\varepsilon(\omega) - \bar{Z}_s^\varepsilon \right| > \delta \right\} = 0, \tag{3.11}$$

i.e., $\sup_{s \in [0,\infty)} |Z_s^\varepsilon(\omega) - \bar{Z}_s^\varepsilon|$ converges to 0 in probability as $\varepsilon \to 0$.

The above two theorems are about systems in the time scale $s = t/\varepsilon$. Now we turn to the X-system (3.1) and its average system (3.2), where $\bar{X}_t = \bar{Z}^\varepsilon_{t/\varepsilon}$, and $X^\varepsilon_t = Z^\varepsilon_{t/\varepsilon}$. Theorems 3.1 and 3.2 yield the following corollaries.

Corollary 3.1 *Suppose that Assumptions 3.1 and 3.2 hold. If the origin $\bar{X}_t = 0$ is an exponentially stable equilibrium point of the average system (3.2), $K \subset D$ is a compact subset of its region of attraction, $\bar{X}_0 = x \in K$, then, for any $\varsigma \in (0, 1)$, there exists a class $\mathcal{K}$ function α_ς and a constant $\varepsilon^*(\varsigma) > 0$ such that if $X^\varepsilon_0 - \bar{X}_0 = O(\alpha_\varsigma)$, then, for all $0 < \varepsilon < \varepsilon^*(\varsigma)$,*

$$P\left\{ \sup_{t\in[0,\infty)} \left| X^\varepsilon_t(\omega) - \bar{X}_t \right| = O(\alpha_\varsigma(\varepsilon)) \right\} > 1 - \varsigma. \tag{3.12}$$

Corollary 3.2 *Suppose that Assumptions 3.1 and 3.2 hold. If the origin $\bar{X}_t = 0$ is an exponentially stable equilibrium point of the average system (3.2), $K \subset D$ is a compact subset of its region of attraction, and $X^\varepsilon_0 = \bar{X}_0 = x \in K$, then, for any $\delta > 0$,*

$$\lim_{\varepsilon \to 0} P\left\{ \sup_{t\in[0,\infty)} \left| X^\varepsilon_t(\omega) - \bar{X}_t \right| > \delta \right\} = 0. \tag{3.13}$$

From Theorem 3.1 and the definition of exponential stability of deterministic systems, we obtain the following stability result.

Theorem 3.3 *Suppose that Assumptions 3.1 and 3.2 hold. If the origin $\bar{X}_t \equiv 0$ is an exponentially stable equilibrium point of the average system (3.2), $K \subset D$ is a compact subset of its region of attraction, and $\bar{X}_0 = x \in K$, then, for any $\varsigma \in (0, 1)$, there exist a measurable set $\Omega_\varsigma \subset \Omega$ with $P(\Omega_\varsigma) > 1 - \varsigma$, a class $\mathcal{K}$ function α_ς, and a constant $\varepsilon^*(\varsigma) > 0$ such that if $X^\varepsilon_0 - \bar{X}_0 = O(\alpha_\varsigma(\varepsilon))$, then, for all $0 < \varepsilon < \varepsilon^*(\varsigma)$,*

$$\left| X^\varepsilon_t(\omega) \right| \leq c|x|e^{-\gamma t} + O(\alpha_\varsigma(\varepsilon)) \quad \text{for all } t \in [0, \infty), \tag{3.14}$$

uniformly in $\omega \in \Omega_\varsigma$ for some constants $\gamma, c > 0$.

Remark 3.1 Notice that for any given $\varsigma \in (0, 1)$, α_ς is a class $\mathcal{K}$ function of ε. Then by (3.14), we obtain that, for any $\delta > 0$ and any $\varsigma > 0$, there exists a constant $\varepsilon^*(\varsigma, \delta) > 0$ such that, for all $0 < \varepsilon < \varepsilon^*(\varsigma, \delta)$,

$$P\left\{ \left| X^\varepsilon_t(\omega) \right| \leq c|x|e^{-\gamma t} + \delta, \forall t \in [0, \infty) \right\} > 1 - \varsigma \tag{3.15}$$

for $X^\varepsilon_0 = \bar{X}_0 = x \in K$ and some positive constants γ, c. This can be viewed as a form of exponential practical stability in probability.

Remark 3.2 Since Y_t is a time homogeneous continuous Markov process, if $a(x, y)$ is globally Lipschitz in (x, y), then the solution of Eq. (3.1) exists with probability 1 for any $x \in \mathbb{R}^n$ and it is defined uniquely for all $t \geq 0$ (see Sect. 2 of Chap. 7 of [39]). Here, by Assumption 3.1, $a(x, y)$ is, in general, locally Lipschitz instead of globally

Lipschitz. Notice that the solution of Eq. (3.1) can be defined for every trajectory of the stochastic process $(Y_t, t \geq 0)$. Then by Corollary 3.1, for any sufficiently small positive number ς, there exist a measurable set $\Omega_\varsigma \subset \Omega$ and a positive number $\varepsilon^*(\varsigma)$ such that $P(\Omega_\varsigma) > 1 - \varsigma$ (which can be sufficiently close to 1) and for any $0 < \varepsilon < \varepsilon^*(\varsigma)$ and any $\omega \in \Omega_\varsigma$, the solution $\{X_t^\varepsilon(\omega), t \in [0, \infty)\}$ exists. The uniqueness of $\{X_t^\varepsilon(\omega), t \in [0, \infty)\}$ is ensured by the local Lipschitzness of $a(x, y)$ with respect to x.

Remark 3.3 Assumptions 3.1 and 3.2 guarantee that there exists a deterministic vector function $\bar{a}(x)$ such that

$$\lim_{T \to \infty} \frac{1}{T} \int_t^{t+T} a\left(x, Y_s(\omega)\right) ds = \bar{a}(x) \quad \text{a.s.} \tag{3.16}$$

uniformly in $(t, x) \in [0, \infty) \times D_0$ for any compact subset $D_0 \subset D$. This uniform convergence condition is critical in the proof and a similar condition is required in the deterministic general averaging on infinite time interval for aperiodic functions (see (1.23), (1.24) or [56, Chap. 10]).

In weak convergence methods of stochastic averaging on finite time intervals, some uniform convergence with respect to (t, x) of some integral of $a(x, Y_s)$ is required [57, (3.2)], [39, (9.3), p. 263] and there the boundedness of $a(x, y)$ is assumed. Here we do not need the boundedness of $a(x, y)$, but we do need a stronger convergence (3.16) to obtain a better result—"exponential practical stability" on an infinite time interval.

The separable form in Assumption 3.1 is to guarantee that the limit (3.16) is uniform with respect to x, while the uniform convergence (3.6) in Assumption 3.2 is to guarantee that the limit (3.16) is uniform with respect to t. For the following stochastic processes $(Y_s, s \geq 0)$, we can verify that the uniform convergence (3.6) holds:

1. $dY_s = pY_s \, ds + qY_s \, dW_s$, $p < \frac{q^2}{2}$;
2. $dY_s = -pY_s \, ds + qe^{-s} \, dW_s$, $p, q > 0$;
3. $Y_s = e^{\xi_s} + c$, where c is a constant and ξ_s satisfies $d\xi_s = -ds + dW_s$.

In these three examples, W_s is a 1-dimensional standard Brownian motion defined on some complete probability space and Y_0 is independent of $(W_s, s \geq 0)$. In fact, for these three kinds of stochastic processes, it holds that

$$\lim_{s \to \infty} Y_s = c \quad \text{a.s. for some constant } c, \tag{3.17}$$

which, together with the fact $\lim_{T \to \infty} \frac{1}{T} \int_t^{t+T} b_i(Y_s) \, ds = \lim_{s \to \infty} b_i(Y_s)$ a.s., when the latter limit exists, gives that, for any continuous function b_i,

$$\lim_{T \to \infty} \frac{1}{T} \int_t^{t+T} b_i(Y_s) \, ds = \lim_{s \to \infty} b_i(Y_s) = b_i(c) \quad \text{a.s.} \tag{3.18}$$

uniformly in $t \in [0, \infty)$. If b_i has the form $b_i(y_1 + y_2) = b_{i1}(y_1) + b_{i2}(y_2) + b_{i3}(y_1)b_{i4}(y_2)$ for any $y_1, y_2 \in S_Y$ and b_{ij}, $j = 1, \ldots, 4$, are continuous functions, and

$$Y_s = \sin(s) + g(s) \sin(\xi_s), \tag{3.19}$$

where $(\xi_s, s \geq 0)$ is any continuous stochastic process and $g(s)$ is a function decaying to zero, e.g., e^{-s} or $\frac{1}{1+s}$, then

$$\lim_{T \to \infty} \frac{1}{T} \int_t^{t+T} b_i(Y_s) \, ds$$

$$= \lim_{T \to \infty} \frac{1}{T} \left\{ \int_t^{t+T} \left[b_{i1}(\sin(s)) + b_{i2}(g(s)\sin(\xi_s)) \right. \right.$$

$$\left. \left. + b_{i3}(\sin(s)) b_{i4}(g(s)\sin(\xi_s)) \right] ds \right\}$$

$$= \frac{1}{2\pi} \int_0^{2\pi} b_{i1}(\sin(s)) \, ds + b_{i2}(0)$$

$$+ b_{i4}(0) \cdot \frac{1}{2\pi} \int_0^{2\pi} b_{i3}(\sin(s)) \, ds, \quad \text{a.s.} \tag{3.20}$$

uniformly in $t \in [0, \infty)$.

If the process $(Y_s, s \geq 0)$ is ergodic with invariant measure μ, then (cf., e.g., Theorem 3 on page 9 of [129])

$$\lim_{T \to \infty} \frac{1}{T} \int_0^T b_i(Y_s) \, ds = \bar{b}_i \quad \text{a.s.,} \tag{3.21}$$

where $\bar{b}_i = \int_{S_Y} b_i(y) \mu(dy)$. While one might expect the averaging under condition (3.21) to be applicable on the infinite interval, this is not true. A stronger condition (3.6) on the perturbation process is needed (note the difference between the integration limits; that is the reason why we refer to this kind of perturbation processes as "uniform strong ergodic"). Uniform convergence, as opposed to ergodicity, is essential for the averaging principle on the infinite time interval. The same requirement of uniformity in time is needed for general averaging on the infinite time interval in the deterministic case. Only under the ergodicity (3.21) of the perturbation process, can we obtain a weaker averaging principle on the infinite time interval, which is investigated in the next chapter.

In Sects. 3.4.1 and 3.4.2, we give examples illustrating the theorems of this section.

3.2.2 φ-Mixing Perturbation Process

Let $\mathcal{F}_t^s$ denote the smallest σ-algebra that makes $\{Y_u, t \leq u \leq s\}$ measurable. If there is a function $\phi(s) \to 0$ as $s \to \infty$ such that

$$\sup_{A \in \mathcal{F}_{t+s}^\infty, B \in \mathcal{F}_0^t} \left| P\{A|B\} - P\{B\} \right| \leq \phi(s), \tag{3.22}$$

then $(Y_u, u \geq 0)$ is said to be ϕ-mixing with mixing rate $\phi(\cdot)$ (see [77]).

In this subsection, we assume that the perturbation $(Y_t, t \geq 0)$ is ϕ-mixing and also ergodic with invariant measure μ. The average system of (3.1) is (3.2), where

$$\bar{a}(x) = \int_{S_Y} a(x, y)\mu(dy), \tag{3.23}$$

and S_Y is the living space of the perturbation process $(Y_t, t \geq 0)$.

Assumption 3.3 The process $(Y_t, t \geq 0)$ is continuous, ϕ-mixing with exponential mixing rate $\phi(t)$ and also exponentially ergodic with invariant measure μ.

Remark 3.4 (i) In the weak convergence methods (e.g., [77]), the perturbation process is usually assumed to be ϕ-mixing with mixing rate $\phi(t)$ ($\int_0^\infty \phi^{1/2}(s)\, ds < \infty$). Here we consider infinite time horizon, so exponential ergodicity is needed. (ii) According to [111], ergodic Markov processes on compact state space are examples of ϕ-mixing processes with exponential mixing rates, e.g., the Brownian motion on the unit circle [35] $(Y_t, t \geq 0)$:

$$dY_t = -\frac{1}{2}Y_t\, dt + BY_t\, dW_t, \quad Y_0 = \left[\cos(\vartheta), \sin(\vartheta)\right]^T \quad \forall \vartheta \in \mathbb{R}, \tag{3.24}$$

where $B = \begin{bmatrix} 0 & -1 \\ 1 & 0 \end{bmatrix}$ and W_t is a 1-dimensional standard Brownian motion.

Assumption 3.4 For the average system (3.2), there exist a function $V(x) \in C^2$, positive constants c_i $(i = 1, \ldots, 4)$, δ, and γ such that, for $|x| \leq \delta$,

$$c_1|x|^2 \leq V(x) \leq c_2|x|^2, \tag{3.25}$$

$$\left|\frac{\partial V(x)}{\partial x}\right| \leq c_3|x|, \tag{3.26}$$

$$\left|\frac{\partial^2 V(x)}{\partial x^2}\right| \leq c_4, \tag{3.27}$$

$$\frac{dV(x)}{dt} = \left(\frac{\partial V(x)}{\partial x}\right)^T \bar{a}(x) \leq -\gamma V(x), \tag{3.28}$$

i.e., the average system (3.2) is exponentially stable.

Assumption 3.5 The vector field $a(x, y)$ satisfies

1. $a(x, y)$ and its first-order partial derivatives with respect to x are continuous and $a(0, y) \equiv 0$;
2. For any compact set $D \subset \mathbb{R}^n$, there is a constant $k_D > 0$ such that, for all $x \in D$ and $y \in S_Y$, $\left|\frac{\partial a(x,y)}{\partial x}\right| \leq k_D$.

Theorem 3.4 *Consider the system* (3.1) *satisfying Assumptions* 3.3, 3.4, *and* 3.5. *Then there exists* $\varepsilon^* > 0$ *such that, for all* $0 < \varepsilon \leq \varepsilon^*$, *the solution* $X_t^\varepsilon \equiv 0$ *of the* *original system is asymptotically stable in probability, i.e., for any $r > 0$ and $\varsigma > 0$, there is a constant $\delta_0 > 0$ such that if $|X_0^\varepsilon| = |x| < \delta_0$, then*

$$P\left\{\sup_{t\geq 0}|X_t^\varepsilon| \leq r\right\} \geq 1 - \varsigma, \tag{3.29}$$

and moreover,

$$\lim_{x\to 0} P\left\{\lim_{t\to\infty}|X_t^\varepsilon| = 0\right\} = 1. \tag{3.30}$$

Remark 3.5 This is the first local stability result based on the stochastic averaging approach for locally Lipschitz nonlinear systems, which is an extension from the deterministic general averaging for aperiodic functions [121].

If the local conditions in Theorem 3.4 hold globally, we get global results under the following set of assumptions.

Assumption 3.6 The average system (3.2) is globally exponentially stable, i.e., Assumption 3.4 holds with "for $|x| \leq \delta$" replaced by "for any $x \in \mathbb{R}^n$".

Assumption 3.7 The vector field $a(x, y)$ satisfies

1. $a(x, y)$ and its first-order partial derivatives with respect to x are continuous and $a(0, y) \equiv 0$;
2. There is a constant $k > 0$ such that, for all $x \in \mathbb{R}^n$ and $y \in S_Y$, $|\frac{\partial a(x,y)}{\partial x}| \leq k$.

Assumption 3.8 The vector field $a(x, y)$ satisfies

1. $a(x, y)$ and its first-order partial derivatives with respect to x are continuous and $\sup_{y\in S_Y} |a(0, y)| < \infty$;
2. There is a constant $k > 0$ such that, for all $x \in \mathbb{R}^n$ and $y \in S_Y$, $|\frac{\partial a(x,y)}{\partial x}| \leq k$.

Theorem 3.5 *Consider the system* (3.1) *satisfying Assumptions* 3.3, 3.6, *and* 3.7. *Then there exists $\varepsilon^* > 0$ such that, for $0 < \varepsilon \leq \varepsilon^*$, the solution $X_t^\varepsilon \equiv 0$ of the original system is globally asymptotically stable in probability, i.e., for any $\eta_1 > 0$ and $\eta_2 > 0$, there is a constant $\delta_0 > 0$ such that if $|X_0^\varepsilon| = |x| < \delta_0$, then*

$$P\left\{|X_t^\varepsilon| \leq \eta_2 e^{-\tilde{\gamma}t}, t \geq 0\right\} \geq 1 - \eta_1 \tag{3.31}$$

with a constant $\tilde{\gamma} > 0$, and moreover, for any $x \in \mathbb{R}^n$,

$$P\left\{\lim_{t\to\infty}|X_t^\varepsilon| = 0\right\} = 1. \tag{3.32}$$

If, on the other hand, (3.1) has no equilibrium, we obtain the following result.

Theorem 3.6 *Consider the system* (3.1) *satisfying Assumptions* 3.3, 3.6, *and* 3.8. *Then there exists $\varepsilon^* > 0$ such that, for $0 < \varepsilon \leq \varepsilon^*$, the solution process X_t^ε of the original system is bounded in probability, i.e.,*

$$\lim_{r\to\infty} \sup_{t\geq 0} P\left\{|X_t^\varepsilon| > r\right\} = 0. \tag{3.33}$$

Remark 3.6 Theorems 3.5 and 3.6 are aimed at globally Lipschitz systems and can be viewed as an extension of the deterministic averaging principle [121] to the stochastic case. We present the results for the global case not only for the sake of completeness but also because of the novelty relative to [21]: (i) ergodic Markov process on some compact space is replaced by an exponential ϕ-mixing and exponentially ergodic process; (ii) for the case without equilibrium condition the weak convergence is considered in [21], while here we obtain the result on boundedness in probability.

In Sect. 3.4.3, we present an example that illustrates the theorems of this section.

3.3 Proofs of the Theorems

3.3.1 Proofs for the Case of Uniform Strong Ergodic Perturbation Process

3.3.1.1 Technical Lemma

To prove Theorems 3.1 and 3.2, we first prove one technical lemma. Toward that end, denote

$$F_i(T, \lambda, \omega) = \frac{1}{T} \int_\lambda^{\lambda+T} b_i\big(Y_u(\omega)\big)\, du \tag{3.34}$$

for $T > 0$, $\lambda \geq 0$, $\omega \in \Omega$, $i = 1, \ldots, l$. We can verify that $F_i(T, \lambda, \omega)$ is continuous with respect to (T, λ) for any $i = 1, \ldots, l$.

Lemma 3.1 *Suppose that Assumptions 3.1 and 3.2 hold. Then, for any $\varsigma > 0$, there exists a measurable set $\Omega_\varsigma \subset \Omega$ such that $P(\Omega_\varsigma) > 1 - \varsigma$, and for any $i = 1, \ldots, l$,*

$$\lim_{T \to \infty} \frac{1}{T} \int_\lambda^{\lambda+T} b_i\big(Y_u(\omega)\big)\, du = \bar{b}_i \quad \text{uniformly in } (\omega, \lambda) \in \Omega_\varsigma \times [0, \infty). \tag{3.35}$$

Moreover, there exists a strictly decreasing, continuous, bounded function $\sigma^\varsigma(T)$ such that $\sigma^\varsigma(T) \to 0$ as $T \to \infty$, and for any compact subset $D_0 \subset D$,

$$\left| \frac{1}{T} \int_\lambda^{\lambda+T} a\big(x, Y_u(\omega)\big)\, du - \bar{a}(x) \right|$$
$$\leq k_{D_0} \sigma^\varsigma(T) \quad \forall (\omega, \lambda, x) \in \Omega_\varsigma \times [0, \infty) \times D_0, \tag{3.36}$$

where k_{D_0} is a positive constant.

Proof **Step 1 (Proof of (3.35)).** From (3.6) we know that for any $i = 1, \ldots, l$,

$$\text{for a.e. } \omega \in \Omega, \quad \lim_{T \to \infty} F_i(T, \lambda, \omega) = \bar{b}_i \quad \text{uniformly in } \lambda \geq 0. \tag{3.37}$$

Noticing that

$$\left\{\omega \Big| \lim_{T\to\infty} F_i(T,\lambda,\omega) = \bar{b}_i \text{ uniformly in } \lambda \geq 0\right\}$$

$$= \bigcap_{k=1}^{\infty} \bigcup_{t>0} \bigcap_{T\geq t} \bigcap_{\lambda\geq 0}\left\{\left|F_i(T,\lambda,\omega) - \bar{b}_i\right| < \frac{1}{k}\right\}, \tag{3.38}$$

by (3.37), we get that

$$P\left(\bigcup_{k=1}^{\infty} \bigcap_{t>0} \bigcup_{T\geq t} \bigcup_{\lambda\geq 0}\left\{\left|F_i(T,\lambda,\omega) - \bar{b}_i\right| \geq \frac{1}{k}\right\}\right) = 0. \tag{3.39}$$

Since $F_i(T,\lambda,\omega)$ is continuous with respect to (T,λ), we can easily prove that $\forall k \geq 1, \forall t > 0$, the sets $\bigcup_{\lambda\geq 0}\{|F_i(T,\lambda,\omega) - \bar{b}_i| \geq \frac{1}{k}\}$, $\bigcup_{T\geq t}\bigcup_{\lambda\geq 0}\{|F_i(T,\lambda,\omega) - \bar{b}_i| \geq \frac{1}{k}\}$, and $\bigcap_{t>0}\bigcup_{T\geq t}\bigcup_{\lambda\geq 0}\{|F_i(T,\lambda,\omega) - \bar{b}_i| \geq \frac{1}{k}\}$ are measurable. Then, by (3.39), we obtain that for any $k \geq 1$,

$$P\left(\bigcap_{t>0} \bigcup_{T\geq t} \bigcup_{\lambda\geq 0}\left\{\left|F_i(T,\lambda,\omega) - \bar{b}_i\right| \geq \frac{1}{k}\right\}\right) = 0. \tag{3.40}$$

Since the set $\bigcup_{T\geq t}\bigcup_{\lambda\geq 0}\{|F_i(T,\lambda,\omega) - \bar{b}_i| \geq \frac{1}{k}\}$ is decreasing as t increases, it follows from (3.40) that

$$\lim_{t\to\infty} P\left(\bigcup_{T\geq t} \bigcup_{\lambda\geq 0}\left\{\left|F_i(T,\lambda,\omega) - \bar{b}_i\right| \geq \frac{1}{k}\right\}\right) = 0. \tag{3.41}$$

Thus, for any $\varsigma > 0$ and any $k \geq 1$, there exists $t_k^{(i)} > 0$ such that

$$P\left(\bigcup_{T\geq t_k^{(i)}} \bigcup_{\lambda\geq 0}\left\{\left|F_i(T,\lambda,\omega) - \bar{b}_i\right| \geq \frac{1}{k}\right\}\right) < \frac{\varsigma}{2^k l}. \tag{3.42}$$

Define

$$\Omega_\varsigma = \bigcap_{i=1}^{l} \bigcap_{k=1}^{\infty} \bigcap_{T\geq t_k^{(i)}} \bigcap_{\lambda\geq 0}\left\{\left|F_i(T,\lambda,\omega) - \bar{b}_i\right| < \frac{1}{k}\right\}. \tag{3.43}$$

Then, by (3.42), $P(\Omega_\varsigma) \geq 1 - \varsigma$. Further, by the construction of Ω_ς, we know that for any $i = 1, \ldots, l$,

$$\lim_{T\to\infty} \frac{1}{T} \int_{\lambda}^{\lambda+T} b_i\big(Y_u(\omega)\big) \, du = \bar{b}_i \quad \text{uniformly in } (\omega,\lambda) \in \Omega_\varsigma \times [0,\infty), \tag{3.44}$$

i.e., (3.35) holds.

Step 2 (Proof of (3.36)). By (3.44), for any $k \geq 1$, there exists $t_k(\varsigma) > 0$ (without loss of generality, we can assume that $t_k(\varsigma)$ is increasing with respect to k) such that for any $T \geq t_k(\varsigma)$, any $(\omega,\lambda) \in \Omega_\varsigma \times [0,\infty)$, and any $i = 1, \ldots, l$, we have that

$$\left|\frac{1}{T} \int_{\lambda}^{\lambda+T} b_i\big(Y_u(\omega)\big) \, du - \bar{b}_i\right| < \frac{1}{k}. \tag{3.45}$$

By Assumption 3.1 and (3.6), there exists a constant $M > 1$ such that, for any $i = 1, \ldots, l$,

$$\sup_{y \in S_Y} |b_i(y)| \leq M \quad \text{and} \quad |\bar{b}_i| \leq M. \tag{3.46}$$

Now we define a function $H^\varsigma(T)$ as

$$H^\varsigma(T) = \begin{cases} 2M & \text{if } T \in [0, t_1(\varsigma)); \\ \frac{1}{k} & \text{if } T \in [t_k(\varsigma), t_{k+1}(\varsigma)), k = 1, 2, \ldots. \end{cases} \tag{3.47}$$

Then, by (3.45), for any $(\omega, \lambda) \in \Omega_\varsigma \times [0, \infty)$ and any $i = 1, \ldots, l$, we have

$$\left| \frac{1}{T} \int_\lambda^{\lambda+T} b_i\big(Y_u(\omega)\big) \, du - \bar{b}_i \right| \leq H^\varsigma(T), \tag{3.48}$$

and $H^\varsigma(T) \downarrow 0$ as $T \to \infty$. Noticing that the function $H^\varsigma(T)$ is a piecewise constant (and thus piecewise continuous) function, we construct a strictly decreasing, continuous, bounded function $\sigma^\varsigma(T)$:

$$\sigma^\varsigma(T) = \begin{cases} -\frac{1}{t_1(\varsigma)} T + (2M + 1) & \text{if } T \in [0, t_1(\varsigma)); \\ -\frac{2M-1}{t_2(\varsigma)-t_1(\varsigma)}(T - t_1(\varsigma)) + 2M & \text{if } T \in [t_1(\varsigma), t_2(\varsigma)); \\ -\frac{1/(k-1)-1/k}{t_{k+1}(\varsigma)-t_k(\varsigma)}(T - t_k(\varsigma)) + \frac{1}{k-1} & \text{if } T \in [t_k(\varsigma), t_{k+1}(\varsigma)), \\ & k = 2, 3, \ldots, \end{cases} \tag{3.49}$$

which satisfies $\sigma^\varsigma(T) \geq H^\varsigma(T) \; \forall T \geq 0$, and $\sigma^\varsigma(T) \downarrow 0$ as $T \to \infty$.

For any compact set $D_0 \subset D$, by Assumption 3.1, there exists a positive constant $M_{D_0} > 0$ such that, for any $i = 1, \ldots, l$,

$$|a_i(x)| \leq M_{D_0} \quad \forall x \in D_0. \tag{3.50}$$

Define $k_{D_0} = l M_{D_0}$. Then, by Assumption 3.1, (3.48), (3.50), and the facts that $\bar{a}(x) = \sum_{i=1}^{l} a_i(x)\bar{b}_i$ and $\sigma^\varsigma(T) \geq H^\varsigma(T) \; \forall T \geq 0$, we get that $\forall (\omega, \lambda, x) \in \Omega_\varsigma \times [0, \infty) \times D_0$,

$$\left| \frac{1}{T} \int_\lambda^{\lambda+T} a\big(x, Y_u(\omega)\big) \, du - \bar{a}(x) \right| = \left| \sum_{i=1}^{l} a_i(x) \left(\frac{1}{T} \int_\lambda^{\lambda+T} b_i\big(Y_u(\omega)\big) \, du - \bar{b}_i \right) \right|$$

$$\leq \sum_{i=1}^{l} |a_i(x)| \left| \frac{1}{T} \int_\lambda^{\lambda+T} b_i\big(Y_u(\omega)\big) \, du - \bar{b}_i \right|$$

$$\leq k_{D_0} \sigma^\varsigma(T), \tag{3.51}$$

i.e., (3.36) holds. $\qquad\qquad\square$

3.3.1.2 Proof of Theorem 3.1

The basic idea of the proof comes from [56, Sect. 10.6]. Fix ς and Ω_ς as in Lemma 3.1. For any $\omega \in \Omega_\varsigma$, define

$$\hat{a}(s, x, \omega) = a\big(x, Y_s(\omega)\big). \tag{3.52}$$

Then we simply rewrite the system (3.4) as

$$\frac{dz}{ds} = \varepsilon \hat{a}(s, z, \omega). \tag{3.53}$$

Let

$$h(s, z, \omega) = \hat{a}(s, z, \omega) - \bar{a}(z), \tag{3.54}$$

$$w(s, z, \omega, \eta) = \int_0^s h(\tau, z, \omega) \exp\left[-\eta(s - \tau)\right] d\tau \tag{3.55}$$

for some $\eta > 0$. For any compact set $D_0 \subset D$, by (3.51), we get that, for $z \in D_0$,

$$
\begin{aligned}
\left| w(s + \delta, z, \omega, 0) - w(s, z, \omega, 0) \right| &= \left| \int_0^{s+\delta} h(\tau, z, \omega) d\tau - \int_0^s h(\tau, z, \omega) d\tau \right| \\
&= \left| \int_s^{s+\delta} h(\tau, z, \omega) d\tau \right| \\
&\leq k_{D_0} \delta \sigma^{\varsigma}(\delta).
\end{aligned} \tag{3.56}
$$

This implies, in particular, that

$$\left| w(s, z, \omega, 0) \right| \leq k_{D_0} s \sigma^{\varsigma}(s), \quad \forall (s, z) \in (0, \infty) \times D_0, \tag{3.57}$$

since $w(0, z, \omega, 0) = 0$. Integrating the right-hand side of (3.55) by parts, we obtain

$$
\begin{aligned}
&w(s, z, \omega, \eta) \\
&= w(s, z, \omega, 0) - \eta \int_0^s \exp\left[-\eta(s - \tau)\right] w(\tau, z, \omega, 0) d\tau \\
&= \exp(-\eta s) w(s, z, \omega, 0) \\
&\quad - \eta \int_0^s \exp\left[-\eta(s - \tau)\right]\left[w(\tau, z, \omega, 0) - w(s, z, \omega, 0)\right] d\tau, \tag{3.58}
\end{aligned}
$$

where the second equality is obtained by adding and subtracting

$$\eta \int_0^s \exp\left[-\eta(s - \tau)\right] d\tau \, w(s, z, \omega, 0) \tag{3.59}$$

to the right-hand side. Using (3.56) and (3.57), we obtain that

$$
\begin{aligned}
\left| w(s, z, \omega, \eta) \right| &\leq k_{D_0} s \exp(-\eta s) \sigma^{\varsigma}(s) \\
&\quad + k_{D_0} \eta \int_0^s \exp\left[-\eta(s - \tau)\right](s - \tau) \sigma^{\varsigma}(s - \tau) d\tau. \tag{3.60}
\end{aligned}
$$

For (3.60), we now show that there is a class $\mathcal{K}$ function α_{ς} such that

$$\eta \left| w(s, z, \omega, \eta) \right| \leq k_{D_0} \alpha_{\varsigma}(\eta) \quad \forall (s, z, \omega) \in [0, \infty) \times D_0 \times \Omega_{\varsigma}. \tag{3.61}$$

Let $z \in D_0$. Firstly, for $s \leq \frac{1}{\sqrt{\eta}}$, by (3.60), and the property of the function σ^{ς},

$$\eta\big|w(s,z,\omega,\eta)\big|$$

$$\leq k_{D_0}\left(\eta s e^{-\eta s}\sigma^\varsigma(s)+\eta^2\int_0^s\exp[-\eta(s-\tau)](s-\tau)\sigma^\varsigma(s-\tau)\,d\tau\right)$$

$$=k_{D_0}\left(\eta s e^{-\eta s}\sigma^\varsigma(s)+\eta^2\int_0^s\exp(-\eta u)u\sigma^\varsigma(u)\,du\right)$$

$$\leq k_{D_0}\left(\eta\frac{1}{\sqrt{\eta}}\sigma^\varsigma(0)+\eta^2\int_0^{1/\sqrt{\eta}}e^{-\eta u}\frac{1}{\sqrt{\eta}}\sigma^\varsigma(0)\,du\right)$$

$$\leq k_{D_0}\left(\sqrt{\eta}\sigma^\varsigma(0)+\sqrt{\eta}\left(1-e^{-\sqrt{\eta}}\right)\sigma^\varsigma(0)\right)$$

$$\leq k_{D_0}\left(2\sqrt{\eta}\sigma^\varsigma(0)\right).\tag{3.62}$$

Then, for $s\geq\frac{1}{\sqrt{\eta}}$, by (3.60), (3.62) and the property of the function σ^ς, we obtain

$$\eta\big|w(s,z,\omega,\eta)\big|$$

$$\leq k_{D_0}\left\{\eta s e^{-\eta s}\sigma^\varsigma(s)+\eta^2\int_0^s\exp[-\eta(s-\tau)](s-\tau)\sigma^\varsigma(s-\tau)\,d\tau\right\}$$

$$=k_{D_0}\left\{\eta s e^{-\eta s}\sigma^\varsigma(s)+\eta^2\int_0^s\exp(-\eta u)u\sigma^\varsigma(u)\,du\right\}$$

$$=k_{D_0}\Big\{\eta s e^{-\eta s}\sigma^\varsigma(s)$$

$$+\eta^2\left[\int_0^{1/\sqrt{\eta}}\exp(-\eta u)u\sigma^\varsigma(u)\,du+\int_{1/\sqrt{\eta}}^s\exp(-\eta u)u\sigma^\varsigma(u)\,du\right]\Big\}$$

$$\leq k_{D_0}\left\{\eta s e^{-\eta s}\sigma^\varsigma\left(\frac{1}{\sqrt{\eta}}\right)+\sqrt{\eta}\left(1-e^{-\sqrt{\eta}}\right)\sigma^\varsigma(0)\right.$$

$$\left.+\eta^2\sigma^\varsigma\left(\frac{1}{\sqrt{\eta}}\right)\int_0^s\exp(-\eta u)u\,du\right\}$$

$$\leq k_{D_0}\left\{\sqrt{\eta}\sigma^\varsigma(0)+\eta s e^{-\eta s}\sigma^\varsigma\left(\frac{1}{\sqrt{\eta}}\right)\right.$$

$$\left.+\eta^2\sigma^\varsigma\left(\frac{1}{\sqrt{\eta}}\right)\left[-\frac{1}{\eta}s e^{-\eta s}+\frac{1}{\eta^2}\left(1-e^{-\eta s}\right)\right]\right\}$$

$$\leq k_{D_0}\left(\sqrt{\eta}\sigma^\varsigma(0)+\sigma^\varsigma\left(\frac{1}{\sqrt{\eta}}\right)\right).\tag{3.63}$$

Thus we define

$$\alpha_\varsigma(\eta)=\begin{cases}2\sqrt{\eta}\sigma^\varsigma(0)+\sigma^\varsigma(\frac{1}{\sqrt{\eta}})&\text{if }\eta>0;\\0&\text{if }\eta=0.\end{cases}\tag{3.64}$$

Then $\alpha_\varsigma(\eta)$ is a class $\mathcal{K}$ function of η, and for $\eta\in[0,1]$, $\alpha_\varsigma(\eta)\geq 2\sigma^\varsigma(0)\eta$. By (3.62) and (3.63), we obtain that, for any $\eta\geq 0$, (3.61) holds.

The partial derivatives $\frac{\partial w}{\partial s}$ and $\frac{\partial w}{\partial z}$ are given by

$$\frac{\partial w(s, z, \omega, \eta)}{\partial s} = h(s, z, \omega) - \eta w(s, z, \omega, \eta), \tag{3.65}$$

$$\frac{\partial w(s, z, \omega, \eta)}{\partial z} = \int_0^s \frac{\partial h}{\partial z}(\tau, z, \omega) \exp[-\eta(s - \tau)] d\tau. \tag{3.66}$$

Noticing that

$$\frac{\partial \bar{a}(x)}{\partial x} = \sum_{i=1}^l \frac{\partial a_i(x)}{\partial x} \bar{b}_i = \sum_{i=1}^l \frac{\partial a_i(x)}{\partial x} \lim_{T \to \infty} \int_t^{t+T} b_i(Y_s) \, ds$$

$$= \lim_{T \to \infty} \int_t^{t+T} \frac{\partial a(x, Y_s)}{\partial x} \, ds \quad \text{a.s.,} \tag{3.67}$$

we can build results similar to (3.35) and (3.36) in Lemma 3.1 for $(\frac{\partial a(x,y)}{\partial x}, \frac{\partial \bar{a}(x)}{\partial x})$ instead of $(a(x, y), \bar{a}(x))$. Furthermore, for $\varsigma > 0$, we can take the same measurable set $\Omega_\varsigma \subset \Omega$. Hence, for $\frac{\partial \hat{a}(s,z,\omega)}{\partial z} = \frac{\partial a(z, Y_s(\omega))}{\partial z}$, we can obtain the same property (3.51) as $\hat{a}(s, z, \omega) = a(z, Y_s(\omega))$. Consequently, $\frac{\partial h}{\partial z}(s, z, \omega) = \frac{\partial \hat{a}}{\partial z}(s, z, \omega) - \frac{\partial \bar{a}}{\partial z}(z)$ possesses the same properties as $h(s, z, \omega)$. Thus we can repeat the above derivations to obtain that (3.61) also holds for $\frac{\partial w}{\partial z}$, i.e.,

$$\eta \left| \frac{\partial w}{\partial z}(s, z, \omega, \eta) \right| \le k_{D_0} \alpha_\varsigma(\eta) \quad \forall (s, z, \omega) \in [0, \infty) \times D_0 \times \Omega_\varsigma. \tag{3.68}$$

There is no loss of generality in using the same positive constant k_{D_0} in both (3.61) and (3.68). Since $k_{D_0} = l M_{D_0}$ will differ only in the bound M_{D_0} in (3.50), we can define M_{D_0} by using the larger of the two constants.

Define the change of variable

$$z = \zeta + \varepsilon w(s, \zeta, \omega, \varepsilon), \tag{3.69}$$

where $\varepsilon w(s, \zeta, \omega, \varepsilon)$ is of order $O(\alpha_\varsigma(\varepsilon))$ by (3.61). By (3.68), for sufficiently small ε, the matrix $[I + \varepsilon \frac{\partial w}{\partial \zeta}]$ is nonsingular. Differentiating both sides with respect to s, we obtain

$$\frac{dz}{ds} = \frac{d\zeta}{ds} + \varepsilon \frac{\partial w(s, \zeta, \omega, \varepsilon)}{\partial s} + \varepsilon \frac{\partial w(s, \zeta, \omega, \varepsilon)}{\partial \zeta} \frac{d\zeta}{ds}. \tag{3.70}$$

Substituting for $\frac{dz}{ds}$ from (3.53), by (3.69), (3.65), and (3.54), we find that the new state variable ζ satisfies the equation

$$\left[I + \varepsilon \frac{\partial w}{\partial \zeta} \right] \frac{d\zeta}{ds} = \varepsilon \hat{a}(s, \zeta + \varepsilon w, \omega) - \varepsilon \frac{\partial w(s, \zeta, \omega, \varepsilon)}{\partial s}$$

$$= \varepsilon \hat{a}(s, \zeta + \varepsilon w, \omega) - \varepsilon[\hat{a}(s, \zeta, \omega) - \bar{a}(\zeta)] + \varepsilon^2 w(s, \zeta, \omega, \varepsilon)$$

$$= \varepsilon \bar{a}(\zeta) + p(s, \zeta, \omega, \varepsilon), \tag{3.71}$$

where

$$p(s, \zeta, \omega, \varepsilon) = \varepsilon[\hat{a}(s, \zeta + \varepsilon w, \omega) - \hat{a}(s, \zeta, \omega)] + \varepsilon^2 w(s, \zeta, \omega, \varepsilon). \tag{3.72}$$

Using the mean value theorem, there exists a function f such that $p(s, \zeta, \omega, \varepsilon)$ is expressed as

$$p(s, \zeta, \omega, \varepsilon) = \varepsilon^2 f(s, \zeta, \varepsilon w, \omega) w(s, \zeta, \omega, \varepsilon) + \varepsilon^2 w(s, \zeta, \omega, \varepsilon)$$
$$= \varepsilon^2 \big[f(s, \zeta, \varepsilon w, \omega) + 1 \big] w(s, \zeta, \omega, \varepsilon). \tag{3.73}$$

Notice that

$$\left[I + \varepsilon \frac{\partial w}{\partial \zeta} \right]^{-1} = I + O\big(\alpha_\varsigma(\varepsilon)\big), \tag{3.74}$$

and $\alpha_\varsigma(\varepsilon) \geq 2\sigma^\varsigma(0)\varepsilon$ for $\varepsilon \in [0, 1]$. Then, by (3.71) and (3.73), the state equation for ζ is given by

$$\frac{d\zeta}{ds} = \big[I + O\big(\alpha_\varsigma(\varepsilon)\big) \big] \times \big[\varepsilon \bar{a}(\zeta) + \varepsilon^2 \big(f(s, \zeta, \varepsilon w, \omega) + 1 \big) w(s, \zeta, \omega, \varepsilon) \big]$$
$$\triangleq \varepsilon \bar{a}(\zeta) + \varepsilon \alpha_\varsigma(\varepsilon) q(s, \zeta, \omega, \varepsilon), \tag{3.75}$$

where $q(s, \zeta, \omega, \varepsilon)$ is uniformly bounded on $[0, \infty) \times D_0 \times \Omega_\varsigma$ for sufficiently small ε. The system (3.75) is a perturbation of the average system

$$\frac{d\zeta}{ds} = \varepsilon \bar{a}(\zeta). \tag{3.76}$$

Notice that, for any compact set $D_0 \subset D$, $q(s, \zeta, \omega, \varepsilon)$ is uniformly bounded on $[0, \infty) \times D_0 \times \Omega_\varsigma$ for sufficiently small ε. Then, by the definition of Ω_ς and the averaging principle of deterministic systems (see Theorems 10.5 and 9.1 of [56]), we obtain the result of Theorem 3.1. The proof is completed.

3.3.1.3 Proof of Theorem 3.2

For any $\varsigma > 0$, by Theorem 3.1, there exist a measurable set $\Omega_\varsigma \subset \Omega$ with $P(\Omega_\varsigma) > 1 - \varsigma$, a class $\mathcal{K}$ function α_ς, and a constant $\varepsilon^*(\varsigma) > 0$ such that, for all $0 < \varepsilon < \varepsilon^*(\varsigma)$,

$$\sup_{s \in [0, \infty)} \big| Z_s^\varepsilon(\omega) - \bar{Z}_s^\varepsilon \big| = O\big(\alpha_\varsigma(\varepsilon)\big) \tag{3.77}$$

uniformly in $\omega \in \Omega_\varsigma$. So there exists a positive constant $C_\varsigma > 0$ such that, for any $\omega \in \Omega_\varsigma$ and any $0 < \varepsilon < \varepsilon^*(\varsigma)$,

$$\sup_{s \in [0, \infty)} \big| Z_s^\varepsilon(\omega) - \bar{Z}_s^\varepsilon \big| \leq C_\varsigma \cdot \alpha_\varsigma(\varepsilon). \tag{3.78}$$

Since $\alpha_\varsigma(\varepsilon)$ is continuous and $\alpha_\varsigma(0) = 0$, for any $\delta > 0$, there exists an $\varepsilon'(\varsigma) > 0$ such that, for any $0 < \varepsilon < \varepsilon'(\varsigma)$,

$$C_\varsigma \cdot \alpha_\varsigma(\varepsilon) < \delta. \tag{3.79}$$

Denote $\bar{\varepsilon}(\varsigma) = \min\{\varepsilon^*(\varsigma), \varepsilon'(\varsigma)\}$. Then for any $\omega \in \Omega_\varsigma$ and any $0 < \varepsilon < \bar{\varepsilon}(\varsigma)$, it holds that

$$\sup_{s \in [0, \infty)} \big| Z_s^\varepsilon(\omega) - \bar{Z}_s^\varepsilon \big| < \delta, \tag{3.80}$$

which means that

$$\left\{ \sup_{s \in [0,\infty)} \left| Z_s^\varepsilon(\omega) - \bar{Z}_s^\varepsilon \right| > \delta \right\} \subset (\Omega \setminus \Omega_\varsigma). \tag{3.81}$$

Thus, we obtain that for any $0 < \varepsilon < \bar{\varepsilon}(\varsigma)$,

$$P\left\{ \sup_{s \in [0,\infty)} \left| Z_s^\varepsilon - \bar{Z}_s^\varepsilon \right| > \delta \right\} \leq P(\Omega \setminus \Omega_\varsigma) < \varsigma. \tag{3.82}$$

Hence the limit (3.11) holds. The proof is completed.

3.3.2 Proofs for the Case of ϕ-Mixing Perturbation Process

3.3.2.1 Proof of Theorem 3.4

Throughout this part, we suppose that the initial value $X_0^\varepsilon = x$ satisfies $|x| < \delta$ (δ is stated in Assumption 3.4). Define $D_\delta = \{x' \in \mathbb{R}^n : |x'| \leq \delta\}$. For any $\varepsilon > 0$ and $t \geq 0$, define two stopping times τ_δ^ε and $\tau_\delta^\varepsilon(t)$ by

$$\tau_\delta^\varepsilon = \inf\{s \geq 0 : X_s^\varepsilon \notin D_\delta\} = \inf\{s \geq 0 : |X_s^\varepsilon| > \delta\} \quad \text{and} \quad \tau_\delta^\varepsilon(t) = \tau_\delta^\varepsilon \wedge t. \tag{3.83}$$

Hereafter, we make the convention that $\inf \emptyset = \infty$.

Define the truncated processes $X_t^{\varepsilon,\delta}$ by

$$X_t^{\varepsilon,\delta} = X_{t \wedge \tau_\delta^\varepsilon}^\varepsilon = X_{\tau_\delta^\varepsilon(t)}^\varepsilon, \quad t \geq 0. \tag{3.84}$$

Then for any $t \geq 0$, we have that

$$X_t^{\varepsilon,\delta} = x + \int_0^{\tau_\delta^\varepsilon(t)} a\left(X_s^\varepsilon, Y_{s/\varepsilon}\right) ds. \tag{3.85}$$

For any $t \geq 0$, define the σ-field $\mathcal{F}_t^{\varepsilon,\delta}$ as follows:

$$\mathcal{F}_t^{\varepsilon,\delta} = \sigma\left\{X_s^{\varepsilon,\delta}, Y_{s/\varepsilon} : 0 \leq s \leq t\right\} = \sigma\{Y_{s/\varepsilon} : 0 \leq s \leq t\} \triangleq \mathcal{F}_{t/\varepsilon}^Y. \tag{3.86}$$

Since $\mathcal{F}_t^{\varepsilon,\delta} = \mathcal{F}_{t/\varepsilon}^Y$ is independent of δ, for simplicity, throughout the rest part of this paper we use $\mathcal{F}_t^\varepsilon$ instead of $\mathcal{F}_t^{\varepsilon,\delta}$.

Step 1 (Lyapunov estimates for Theorem 3.4).   For any $x \in \mathbb{R}^n$ with $|x| \leq \delta$, and $t \geq 0$, define $V^\varepsilon(x,t)$ by

$$V^\varepsilon(x,t) = V(x) + V_1^\varepsilon(x,t), \tag{3.87}$$

where

$$
\begin{aligned}
V_1^\varepsilon(x,t) &= \int_{\tau_\delta^\varepsilon(t)}^{\tau_\delta^\varepsilon} \left(\frac{\partial V(x)}{\partial x}\right)^T E\left[a(x, Y_{s/\varepsilon}) - \bar{a}(x) | \mathcal{F}_t^\varepsilon\right] ds \\
&= \varepsilon \int_{\tau_\delta^\varepsilon(t)/\varepsilon}^{\tau_\delta^\varepsilon/\varepsilon} \left(\frac{\partial V(x)}{\partial x}\right)^T E\left[a(x, Y_u) - \bar{a}(x) | \mathcal{F}_t^\varepsilon\right] du
\end{aligned}
$$

$$= \varepsilon \int_{\tau_\delta^\varepsilon(t)/\varepsilon}^{\tau_\delta^\varepsilon/\varepsilon} \left(\frac{\partial V(x)}{\partial x} \right)^T$$

$$\times \left[E\left[a(x, Y_u)|\mathcal{F}_t^\varepsilon\right] - \int_{S_Y} a(x, y)\left[P_u(dy) - P_u(dy) + \mu(dy) \right] \right] du$$

$$= \varepsilon \int_{\tau_\delta^\varepsilon(t)/\varepsilon}^{\tau_\delta^\varepsilon/\varepsilon} \left(\frac{\partial V(x)}{\partial x} \right)^T \left(E\left[a(x, Y_u)|\mathcal{F}_t^\varepsilon\right] - E\left[a(x, Y_u)\right] \right) du$$

$$+ \varepsilon \int_{\tau_\delta^\varepsilon(t)/\varepsilon}^{\tau_\delta^\varepsilon/\varepsilon} \left(\frac{\partial V(x)}{\partial x} \right)^T \left(\int_{S_Y} a(x, y)\left(P_u(dy) - \mu(dy) \right) \right) du$$

$$\triangleq \varepsilon V_{1,1}^\varepsilon(x, t) + \varepsilon V_{1,2}^\varepsilon(x, t), \tag{3.88}$$

and where P_u is the distribution of the random variable Y_u. Next we give some estimates of $\varepsilon V_{1,1}^\varepsilon(x, t)$ and $\varepsilon V_{1,2}^\varepsilon(x, t)$, which imply that $V_1^\varepsilon(x, t)$ is well defined.

By Assumption 3.5, there exists a positive constant k_δ such that, for any $x \in \mathbb{R}^n$ with $|x| \leq \delta$ and $y \in S_Y$,

$$a(0, y) \equiv 0, \quad \left| \frac{\partial a(x, y)}{\partial x} \right| \leq k_\delta \tag{3.89}$$

Then by Taylor's expansion and (3.23), for any $x \in \mathbb{R}^n$ with $|x| \leq \delta$ and $y \in S_Y$,

$$\left| a(x, y) \right| \leq k_\delta |x|, \quad \left| \bar{a}(x) \right| \leq k_\delta |x|. \tag{3.90}$$

Without loss of generality, we assume that the initial condition $Y_0 = y$ is deterministic. By Assumption 3.3, we have

$$\mathrm{var}(P_t - \mu) \leq c_5 e^{-\alpha t} \tag{3.91}$$

for two positive constants c_5 and α, where "var" denotes the total variation norm of a signed measure over the Borel σ-field, and the mixing rate function $\phi(\cdot)$ of the process Y_t satisfies $\phi(s) = c_6 e^{-\beta s}$ for two positive constants c_6 and β.

Thus, by (3.86), (3.26), (3.90), Lemma B.1, and the mixing rate function $\phi(s) = c_6 e^{-\beta s}$ of the process Y_t, we obtain that for $t < \tau_\delta^\varepsilon$,

$$\varepsilon \left| V_{1,1}^\varepsilon(x, t) \right| = \varepsilon \left| \int_{\tau_\delta^\varepsilon(t)/\varepsilon}^{\tau_\delta^\varepsilon/\varepsilon} \left(\frac{\partial V(x)}{\partial x} \right)^T \left(E\left[a(x, Y_u)|\mathcal{F}_t^\varepsilon\right] - E\left[a(x, Y_u)\right] \right) du \right|$$

$$\leq \varepsilon \int_{t/\varepsilon}^{\tau_\delta^\varepsilon/\varepsilon} \left| \frac{\partial V(x)}{\partial x} \right| \cdot \left| E\left[a(x, Y_u)|\mathcal{F}_{t/\varepsilon}^Y\right] - E\left[a(x, Y_u)\right] \right| du$$

$$\leq \varepsilon \int_{t/\varepsilon}^{\tau_\delta^\varepsilon/\varepsilon} c_3 |x| \cdot k_\delta |x| \cdot \phi\left(u - \frac{t}{\varepsilon} \right) du$$

$$\leq \varepsilon c_3 c_6 k_\delta |x|^2 \int_{t/\varepsilon}^{\tau_\delta^\varepsilon/\varepsilon} e^{-\beta(u - t/\varepsilon)} du$$

$$\leq \varepsilon \frac{c_3 c_6 k_\delta}{\beta} |x|^2, \tag{3.92}$$

and for $t \geq \tau_\delta^\varepsilon$,

$$\varepsilon\left|V_{1,1}^{\varepsilon}(x,t)\right| = \varepsilon\left|\int_{\tau_{\delta}^{\varepsilon}/\varepsilon}^{\tau_{\delta}^{\varepsilon}/\varepsilon}\left(\frac{\partial V(x)}{\partial x}\right)^{T}\left(E\left[a(x,Y_u)|\mathcal{F}_t^{\varepsilon}\right] - E\left[a(x,Y_u)\right]\right)du\right|$$

$$= 0. \tag{3.93}$$

Thus for any $t \geq 0$,

$$\varepsilon\left|V_{1,1}^{\varepsilon}(x,t)\right| \leq \varepsilon\frac{c_3 c_6 k_{\delta}}{\beta}|x|^2. \tag{3.94}$$

By Hölder's inequality, (3.26), (3.90), and (3.91), we get that

$$\varepsilon\left|V_{1,2}^{\varepsilon}(x,t)\right| = \varepsilon\left|\int_{\tau_{\delta}^{\varepsilon}(t)/\varepsilon}^{\tau_{\delta}^{\varepsilon}/\varepsilon}\left(\frac{\partial V(x)}{\partial x}\right)^{T}\left(\int_{S_Y}a(x,y)\big(P_u(dy)-\mu(dy)\big)\right)du\right|$$

$$\leq \varepsilon\int_{\tau_{\delta}^{\varepsilon}(t)/\varepsilon}^{\tau_{\delta}^{\varepsilon}/\varepsilon}\left|\int_{S_Y}\left(\frac{\partial V(x)}{\partial x}\right)^{T}a(x,y)\big(P_u(dy)-\mu(dy)\big)\right|du$$

$$\leq \varepsilon\int_{\tau_{\delta}^{\varepsilon}(t)/\varepsilon}^{\tau_{\delta}^{\varepsilon}/\varepsilon}\left(\int_{S_Y}\left|\left(\frac{\partial V(x)}{\partial x}\right)^{T}a(x,y)\right|^{2}\left[P_u(dy)+\mu(dy)\right]\right)^{1/2}$$

$$\cdot\left(\int_{S_Y}|P_u-\mu|(dy)\right)^{1/2}du$$

$$\leq \varepsilon\int_{\tau_{\delta}^{\varepsilon}(t)/\varepsilon}^{\tau_{\delta}^{\varepsilon}/\varepsilon}\left(\int_{S_Y}(k_{\delta}c_3)^2|x|^4\left[P_u(dy)+\mu(dy)\right]\right)^{1/2}$$

$$\cdot\big(\mathrm{var}(P_u-\mu)\big)^{1/2}du$$

$$\leq \varepsilon c_3 k_{\delta}|x|^2\int_{\tau_{\delta}^{\varepsilon}(t)/\varepsilon}^{\tau_{\delta}^{\varepsilon}/\varepsilon}\left(\int_{S_Y}\left[P_u(dy)+\mu(dy)\right]\right)^{1/2}\big(c_5 e^{-\alpha u}\big)^{1/2}du$$

$$= \varepsilon\sqrt{2c_5}c_3 k_{\delta}|x|^2\int_{\tau_{\delta}^{\varepsilon}(t)/\varepsilon}^{\tau_{\delta}^{\varepsilon}/\varepsilon}e^{-(\alpha/2)u}du$$

$$\leq \varepsilon\frac{2\sqrt{2c_5}c_3 k_{\delta}}{\alpha}|x|^2. \tag{3.95}$$

Therefore, by (3.88), (3.94), and (3.95), for any $x \in \mathbb{R}^n$ with $|x| \leq \delta$, and $t \geq 0$,

$$-\varepsilon C_1(\delta)|x|^2 \leq V_1^{\varepsilon}(x,t) \leq \varepsilon C_1(\delta)|x|^2, \tag{3.96}$$

where $C_1(\delta) = \frac{2\sqrt{2c_5}c_3 k_{\delta}}{\alpha} + \frac{c_3 c_6 k_{\delta}}{\beta}$. By (3.25), (3.87), and (3.96), there exists an $\varepsilon_1 > 0$ such that $\frac{\varepsilon_1 C_1}{c_1} < 1$, and for $0 < \varepsilon \leq \varepsilon_1$, $x \in \mathbb{R}^n$ with $|x| \leq \delta$, and $t \geq 0$,

$$k_1(\delta)V(x) \leq V^{\varepsilon}(x,t) \leq k_2(\delta)V(x), \tag{3.97}$$

where $k_1(\delta) = 1 - \frac{\varepsilon_1 C_1(\delta)}{c_1} > 0$, $k_2(\delta) = 1 + \frac{\varepsilon_1 C_1(\delta)}{c_1} > 0$.

Step 2 (Action of the p-infinitesimal operator on Lyapunov function in the case with local conditions). We discuss the action of the p-infinitesimal operator $\hat{\mathcal{A}}_{\delta}^{\varepsilon}$ of the vector process $(X_t^{\varepsilon,\delta}, Y_{t/\varepsilon})$ on the perturbed Lyapunov function $V^{\varepsilon}(x,t)$.

Recall that $\tau_\delta^\varepsilon(t)$ is defined by (3.83). By the continuity of the process X_t^ε, we know that, for any $t \geq 0$, $X_{\tau_\delta^\varepsilon(t)}^\varepsilon \in D_\delta = \{x' \in \mathbb{R}^n : |x'| \leq \delta\}$. Define

$$G(x, y) = \left(\frac{\partial V(x)}{\partial x}\right)^T a(x, y), \qquad \bar{G}(x) = \left(\frac{\partial V(x)}{\partial x}\right)^T \bar{a}(x), \qquad (3.98)$$

$$\tilde{G}(x, y) = G(x, y) - \bar{G}(x). \qquad (3.99)$$

Notice that $X_{\tau_\delta^\varepsilon(t)}^\varepsilon$ is measurable with respect to the σ-field $\mathcal{F}_t^\varepsilon$. Then, by the definition in (3.87),

$$V^\varepsilon\left(X_{\tau_\delta^\varepsilon(t)}^\varepsilon, t\right) = V\left(X_{\tau_\delta^\varepsilon(t)}^\varepsilon\right) + V_1^\varepsilon\left(X_{\tau_\delta^\varepsilon(t)}^\varepsilon, t\right). \qquad (3.100)$$

Now we prove that, for $0 < \varepsilon \leq \varepsilon_1$, $V^\varepsilon(X_{\tau_\delta^\varepsilon(t)}^\varepsilon, t) \in \mathcal{D}(\hat{\mathcal{A}}_\delta^\varepsilon)$, the domain of p-infinitesimal operator $\hat{\mathcal{A}}_\delta^\varepsilon$ (for definitions of p-limit and p-infinitesimal operator, please see Appendix A), and

$$\hat{\mathcal{A}}_\delta^\varepsilon V^\varepsilon\left(X_{\tau_\delta^\varepsilon(t)}^\varepsilon, t\right)$$

$$= I_{\{t < \tau_\delta^\varepsilon\}} \cdot \left\{ \bar{G}\left(X_t^\varepsilon\right) + \int_{\tau_\delta^\varepsilon(t)}^{\tau_\delta^\varepsilon} \left[\frac{\partial E_t^\varepsilon[\tilde{G}(x, Y_{s/\varepsilon})]}{\partial x} \bigg|_{x = X_t^\varepsilon} \right]^T a\left(X_t^\varepsilon, Y_{t/\varepsilon}\right) ds \right\}$$

$$\triangleq g_\delta^\varepsilon(t), \qquad (3.101)$$

where $E_t^\varepsilon[\cdot]$ stands for the conditional expectation $E[\cdot | \mathcal{F}_t^\varepsilon]$, i.e., $E[\cdot | \mathcal{F}_{t/\varepsilon}^Y]$.

Since X_t^ε and Y_t are both continuous processes, we know that $V^\varepsilon(X_{\tau_\delta^\varepsilon(t)}^\varepsilon, t)$ and $g_\delta^\varepsilon(t)$ are progressively measurable with respect to $\{\mathcal{F}_t^\varepsilon\}$. In order to prove (3.101), we need only to prove the following three claims for $0 < \varepsilon \leq \varepsilon_1$:

(i) $V^\varepsilon(X_{\tau_\delta^\varepsilon(t)}^\varepsilon, t) \in \overline{\mathcal{M}}_\delta^\varepsilon$, where $\overline{\mathcal{M}}_\delta^\varepsilon$ is defined with respect to the vector process $(X_t^{\varepsilon,\delta}, Y_{t/\varepsilon})$ similarly as $\overline{\mathcal{M}}^\varepsilon$ defined in Appendix A.

(ii) $g_\delta^\varepsilon(t) \in \overline{\mathcal{M}}_\delta^\varepsilon$.

(iii) $\displaystyle p\text{-}\lim_{\delta' \downarrow 0} \frac{E_t^\varepsilon[V^\varepsilon(X_{\tau_\delta^\varepsilon(t+\delta')}^\varepsilon, t + \delta')] - V^\varepsilon(X_{\tau_\delta^\varepsilon(t)}^\varepsilon, t)}{\delta'} = g_\delta^\varepsilon(t). \qquad (3.102)$

By (3.97) and the definition of $\tau_\delta^\varepsilon(t)$, we get that, for $0 < \varepsilon \leq \varepsilon_1$,

$$\sup_{t \geq 0} E\left[\left|V^\varepsilon\left(X_{\tau_\delta^\varepsilon(t)}^\varepsilon, t\right)\right|\right] \leq \sup_{t \geq 0} E\left[k_2(\delta) V\left(X_{\tau_\delta^\varepsilon(t)}^\varepsilon\right)\right]$$

$$\leq k_2(\delta) \cdot \sup_{x \in D_\delta} V(x) < \infty. \qquad (3.103)$$

Thus (i) holds. For the proofs of (ii) and (iii), see Lemmas B.2 and B.3.

Hence, by (3.101), (3.28), (B.12), and (3.25), for any $t \geq 0$ and $0 < \varepsilon \leq \varepsilon_1$,

$$\hat{\mathcal{A}}_\delta^\varepsilon V^\varepsilon\left(X_{\tau_\delta^\varepsilon(t)}^\varepsilon, t\right) \leq I_{\{t < \tau_\delta^\varepsilon\}}\left(-\gamma V\left(X_{\tau_\delta^\varepsilon(t)}^\varepsilon\right) + \varepsilon\frac{C_2(\delta)}{c_1} V\left(X_{\tau_\delta^\varepsilon(t)}^\varepsilon\right)\right)$$

$$= -\left(\gamma - \varepsilon\frac{C_2(\delta)}{c_1}\right) V\left(X_{\tau_\delta^\varepsilon(t)}^\varepsilon\right) \cdot I_{\{t < \tau_\delta^\varepsilon\}}. \qquad (3.104)$$

Take $\varepsilon_1' > 0$ such that $\gamma - \varepsilon_1' \frac{C_2(\delta)}{c_1} > 0$. Let $\varepsilon_2 = \min\{\varepsilon_1, \varepsilon_1'\}$. Then for $0 < \varepsilon \le \varepsilon_2$ and any $t \ge 0$,

$$\hat{A}_\delta^\varepsilon V^\varepsilon \left(X_{\tau_\delta^\varepsilon(t)}^\varepsilon, t\right) \le 0. \tag{3.105}$$

Step 3 (Proof of stability in probability (3.29)).   Suppose $\varepsilon \in (0, \varepsilon_2]$, $r \in (0, \delta)$, and $X_0^\varepsilon = x$ is such that $|x| \le r$. For $t \ge 0$, define two stopping times τ_r^ε and $\tau_r^\varepsilon(t)$ by

$$\tau_r^\varepsilon = \inf\{s \ge 0 : |X_s^\varepsilon| > r\} \quad \text{and} \quad \tau_r^\varepsilon(t) = \tau_r^\varepsilon \wedge t. \tag{3.106}$$

Then for any $t \ge 0$,

$$\left|X_{\tau_r^\varepsilon(t)}^\varepsilon\right| \le r < \delta, \quad \tau_r^\varepsilon(t) \le \tau_\delta^\varepsilon(t), \tag{3.107}$$

and

$$
\begin{aligned}
\tau_\delta^\varepsilon\left(\tau_r^\varepsilon(t)\right) &= \tau_\delta^\varepsilon \wedge \tau_r^\varepsilon(t) = \tau_\delta^\varepsilon \wedge \left(\tau_r^\varepsilon \wedge t\right) \\
&= \left(\tau_\delta^\varepsilon \wedge t\right) \wedge \left(\tau_r^\varepsilon \wedge t\right) \\
&= \tau_\delta^\varepsilon(t) \wedge \tau_r^\varepsilon(t) = \tau_r^\varepsilon(t).
\end{aligned} \tag{3.108}
$$

Thus by Theorem A.1, the property of conditional expectation, and (3.105),

$$
\begin{aligned}
E\left[V^\varepsilon\left(X_{\tau_r^\varepsilon(t)}^\varepsilon, \tau_r^\varepsilon(t)\right) - V^\varepsilon(x, 0)\right] \\
= E\left[V^\varepsilon\left(X_{\tau_\delta^\varepsilon(\tau_r^\varepsilon(t))}^\varepsilon, \tau_r^\varepsilon(t)\right) - V^\varepsilon(x, 0)\right] \\
= E\left[E\left[V^\varepsilon\left(X_{\tau_\delta^\varepsilon(\tau_r^\varepsilon(t))}^\varepsilon, \tau_r^\varepsilon(t)\right) - V^\varepsilon(x, 0)|\mathcal{F}_0^\varepsilon\right]\right] \\
= E\left[E_0^\varepsilon\left[V^\varepsilon\left(X_{\tau_\delta^\varepsilon(\tau_r^\varepsilon(t))}^\varepsilon, \tau_r^\varepsilon(t)\right)\right] - V^\varepsilon(x, 0)\right] \\
= E\left[E_0^\varepsilon\left[\int_0^{\tau_r^\varepsilon(t)} \hat{A}_\delta^\varepsilon V^\varepsilon\left(X_{\tau_\delta^\varepsilon(u)}^\varepsilon, u\right) du\right]\right] \\
= E\left[\int_0^{\tau_r^\varepsilon(t)} \hat{A}_\delta^\varepsilon V^\varepsilon\left(X_{\tau_\delta^\varepsilon(u)}^\varepsilon, u\right) du\right] \le 0.
\end{aligned} \tag{3.109}
$$

By (3.97) and (3.109),

$$
\begin{aligned}
E\left[k_1(\delta) V\left(X_{\tau_r^\varepsilon(t)}^\varepsilon\right)\right] &\le E\left[V^\varepsilon\left(X_{\tau_r^\varepsilon(t)}^\varepsilon, \tau_r^\varepsilon(t)\right)\right] \\
&\le E\left[V^\varepsilon(x, 0)\right] \le k_2(\delta) V(x).
\end{aligned} \tag{3.110}
$$

Denote

$$V_r = \inf_{r \le |x| \le \delta} V(x). \tag{3.111}$$

Then for any $T > 0$, we have

$$
\begin{aligned}
E\left[V\left(X_{\tau_r^\varepsilon(T)}^\varepsilon\right)\right] &= \int_{\{\tau_r^\varepsilon < T\}} V\left(X_{\tau_r^\varepsilon(T)}^\varepsilon\right) dP + \int_{\{\tau_r^\varepsilon \ge T\}} V\left(X_{\tau_r^\varepsilon(T)}^\varepsilon\right) dP \\
&\ge \int_{\{\tau_r^\varepsilon < T\}} V\left(X_{\tau_r^\varepsilon(T)}^\varepsilon\right) dP \ge \int_{\{\sup_{0 \le t \le T} |X_t^\varepsilon| > r\}} V\left(X_{\tau_r^\varepsilon(T)}^\varepsilon\right) dP \\
&\ge V_r \cdot P\left\{\sup_{0 \le t \le T} |X_t^\varepsilon| > r\right\},
\end{aligned} \tag{3.112}
$$

which, together with (3.110), implies

$$P\left\{\sup_{0\leq t\leq T}\left|X_t^\varepsilon\right|>r\right\}\leq\frac{E[V(X_{\tau_r^\varepsilon(T)}^\varepsilon)]}{V_r}\leq\frac{k_2(\delta)V(x)}{k_1(\delta)V_r}. \qquad (3.113)$$

Letting $T\to\infty$, we get

$$P\left\{\sup_{t\geq 0}\left|X_t^\varepsilon\right|>r\right\}\leq\frac{k_2(\delta)V(x)}{k_1(\delta)V_r}. \qquad (3.114)$$

Hence

$$P\left\{\sup_{t\geq 0}\left|X_t^\varepsilon\right|\leq r\right\}>1-\frac{k_2(\delta)V(x)}{k_1(\delta)V_r}. \qquad (3.115)$$

Since $V(0)=0$ and $V(x)$ is continuous, for any $\varsigma>0$, there exists $\delta_1(r,\varsigma)\in(0,\delta)$ such that $V(x)<\frac{k_1(\delta)V_r}{k_2(\delta)}\varsigma$ for all $|x|<\delta_1(r,\varsigma)$. Thus we obtain that, for any $0<\varepsilon\leq\varepsilon^*$ with $\varepsilon^*=\min\{\varepsilon_1,\varepsilon_2\}=\varepsilon_2$, for any given $r>0$, $\varsigma>0$, there exists $\delta_0=\delta_1(\min(r,\delta/2),\varsigma)\in(0,\delta)$ such that for all $|x|<\delta_0$,

$$P\left\{\sup_{t\geq 0}\left|X_t^\varepsilon\right|\leq r\right\}\geq P\left\{\sup_{t\geq 0}\left|X_t^\varepsilon\right|\leq\min(r,\delta/2)\right\}>1-\varsigma, \qquad (3.116)$$

equivalently, for any $0<\varepsilon\leq\varepsilon^*$, and any given $r>0$,

$$\lim_{x\to 0}P\left\{\sup_{t\geq 0}\left|X_t^\varepsilon\right|>r\right\}=0. \qquad (3.117)$$

Step 4 (Proof of asymptotic convergence property (3.30)). Let $0<\varepsilon<\varepsilon^*$ $(=\varepsilon_2)$. By Theorem A.1, for any $0\leq s\leq t$,

$$E\left[V^\varepsilon\left(X_{\tau_\delta^\varepsilon(t)}^\varepsilon,t\right)|\mathcal{F}_s^\varepsilon\right]$$
$$=V^\varepsilon\left(X_{\tau_\delta^\varepsilon(s)}^\varepsilon,s\right)+\int_s^t E\left[\hat{A}_\delta^\varepsilon V^\varepsilon\left(X_{\tau_\delta^\varepsilon(u)}^\varepsilon,u\right)|\mathcal{F}_s^\varepsilon\right]du \quad\text{a.s.,} \qquad (3.118)$$

where $\mathcal{F}_s^\varepsilon$ is defined by (3.86). By (3.97), we know that, for any $t\geq 0$, $V^\varepsilon(X_{\tau_\delta^\varepsilon(t)}^\varepsilon,t)$ is integrable. By (3.105) and (3.118), we obtain that, for any $0\leq s\leq t$,

$$E\left[V^\varepsilon\left(X_{\tau_\delta^\varepsilon(t)}^\varepsilon,t\right)|\mathcal{F}_s^\varepsilon\right]\leq V^\varepsilon\left(X_{\tau_\delta^\varepsilon(s)}^\varepsilon,s\right) \quad\text{a.s.} \qquad (3.119)$$

Hence by definition $\{V^\varepsilon(X_{\tau_\delta^\varepsilon(t)}^\varepsilon,t):t\geq 0\}$ is a nonnegative supermartingale with respect to $\{\mathcal{F}_t^\varepsilon\}$. By Doob's theorem,

$$\lim_{t\to\infty}V^\varepsilon\left(X_{\tau_\delta^\varepsilon(t)}^\varepsilon,t\right)=\xi \quad\text{a.s.,} \qquad (3.120)$$

and ξ is finite almost surely. Let B_x^ε denote the set of sample paths of $(X_t^\varepsilon:t\geq 0)$ with $X_0^\varepsilon=x$ such that $\tau_\delta^\varepsilon=\infty$. Since $X_t^\varepsilon\equiv 0$ is stable in probability, by (3.117),

$$\lim_{x\to 0}P\left(B_x^\varepsilon\right)=1. \qquad (3.121)$$

Note that $\varepsilon^*=\varepsilon_2=\min\{\varepsilon_1,\varepsilon_1'\}$, and $\varepsilon_1'>0$ satisfies $\gamma-\varepsilon_1'\frac{C_2(\delta)}{c_1}>0$. Then by (3.104), we get that, for any $0<\varepsilon\leq\varepsilon^*$,

$$\hat{A}_\delta^\varepsilon V^\varepsilon\left(X_{\tau_\delta^\varepsilon(t)}^\varepsilon,t\right)\leq-c_\varepsilon V\left(X_{\tau_\delta^\varepsilon(t)}^\varepsilon\right)\cdot I_{\{t<\tau_\delta^\varepsilon\}}, \qquad (3.122)$$

where $c_\varepsilon = \gamma - \varepsilon \frac{C_2(\delta)}{c_1} > 0$. For any $0 < \varsigma < \delta$, let $c_\varepsilon^\varsigma = c_\varepsilon c_1 \varsigma^2$. Notice that, for any $t \geq 0$, $|X_{\tau_\delta^\varepsilon(t)}^\varepsilon| \leq \delta$. Then by (3.25) and (3.122), we obtain that if $0 < \varepsilon \leq \varepsilon^*$ and $|X_{\tau_\delta^\varepsilon(t)}^\varepsilon| \geq \varsigma$, then

$$\hat{A}_\delta^\varepsilon V^\varepsilon \left(X_{\tau_\delta^\varepsilon(t)}^\varepsilon, t \right) \leq -c_\varepsilon^\varsigma \cdot I_{\{t < \tau_\delta^\varepsilon\}}. \tag{3.123}$$

For $0 < \varepsilon \leq \varepsilon^*$, $0 < \varsigma < \delta$, and any $t \geq 0$, define two stopping times $\tau_{\varsigma,\delta}^\varepsilon$ and $\tau_{\varsigma,\delta}^\varepsilon(t)$ by

$$\tau_{\varsigma,\delta}^\varepsilon = \inf \{ t : |X_t^\varepsilon| \notin [\varsigma, \delta] \} = \inf \{ t : |X_t^\varepsilon| < \varsigma \text{ or } |X_t^\varepsilon| > \delta \} \quad \text{and}$$
$$\tau_{\varsigma,\delta}^\varepsilon(t) = \tau_{\varsigma,\delta}^\varepsilon \wedge t. \tag{3.124}$$

Then for any $t \geq 0$, we have that $\tau_{\varsigma,\delta}^\varepsilon(t) \leq \tau_\delta^\varepsilon(t)$. Suppose that $X_0^\varepsilon = x$ and $|x| \in (\varsigma, \delta)$. Then for any $t \in [0, \tau_{\varsigma,\delta}^\varepsilon]$, $|X_t^\varepsilon| \in [\varsigma, \delta]$. If $u \in [0, \tau_{\varsigma,\delta}^\varepsilon(t)]$, then

$$0 \leq \tau_\delta^\varepsilon(u) = \tau_\delta^\varepsilon \wedge u \leq u \leq \tau_{\varsigma,\delta}^\varepsilon(t) \leq \tau_{\varsigma,\delta}^\varepsilon, \tag{3.125}$$

and thus $|X_{\tau_\delta^\varepsilon(u)}^\varepsilon| \in [\varsigma, \delta]$. Hence by Theorem A.1, the property of conditional expectation, and (3.123), we obtain that

$$E\left[V^\varepsilon \left(X_{\tau_\delta^\varepsilon(\tau_{\varsigma,\delta}^\varepsilon(t))}^\varepsilon, \tau_{\varsigma,\delta}^\varepsilon(t) \right) \right] - E\left[V^\varepsilon(x, 0) \right]$$
$$= E\left[V^\varepsilon \left(X_{\tau_\delta^\varepsilon(\tau_{\varsigma,\delta}^\varepsilon(t))}^\varepsilon, \tau_{\varsigma,\delta}^\varepsilon(t) \right) - V^\varepsilon(x, 0) \right]$$
$$= E\left[E\left[V^\varepsilon \left(X_{\tau_\delta^\varepsilon(\tau_{\varsigma,\delta}^\varepsilon(t))}^\varepsilon, \tau_{\varsigma,\delta}^\varepsilon(t) \right) - V^\varepsilon(x, 0) | \mathcal{F}_0^\varepsilon \right] \right]$$
$$= E\left[E_0^\varepsilon \left[V^\varepsilon \left(X_{\tau_\delta^\varepsilon(\tau_{\varsigma,\delta}^\varepsilon(t))}^\varepsilon, \tau_{\varsigma,\delta}^\varepsilon(t) \right) \right] - V^\varepsilon(x, 0) \right]$$
$$= E\left[E_0^\varepsilon \left[\int_0^{\tau_{\varsigma,\delta}^\varepsilon(t)} \hat{A}_\delta^\varepsilon V^\varepsilon \left(X_{\tau_\delta^\varepsilon(u)}^\varepsilon, u \right) du \right] \right]$$
$$= E\left[\int_0^{\tau_{\varsigma,\delta}^\varepsilon(t)} \hat{A}_\delta^\varepsilon V^\varepsilon \left(X_{\tau_\delta^\varepsilon(u)}^\varepsilon, u \right) du \right]$$
$$\leq E\left[\int_0^{\tau_{\varsigma,\delta}^\varepsilon(t)} \left(-c_\varepsilon^\varsigma \cdot I_{\{t < \tau_\delta^\varepsilon\}} \right) du \right]$$
$$= -c_\varepsilon^\varsigma E\left[\tau_{\varsigma,\delta}^\varepsilon(t) \cdot I_{\{t < \tau_\delta^\varepsilon\}} \right], \tag{3.126}$$

where $E_0^\varepsilon[\cdot]$ means the conditional expectation $E[\cdot | \mathcal{F}_0^\varepsilon]$.

Thus by (3.126) and (3.97),

$$E\left[\tau_{\varsigma,\delta}^\varepsilon(t) \cdot I_{\{t < \tau_\delta^\varepsilon\}} \right] \leq \frac{E[V^\varepsilon(x, 0)]}{c_\varepsilon^\varsigma} \leq \frac{k_2(\delta) V(x)}{c_\varepsilon^\varsigma}. \tag{3.127}$$

By the definitions of $\tau_{\varsigma,\delta}^\varepsilon$ and τ_δ^ε, we have that $\tau_{\varsigma,\delta}^\varepsilon \leq \tau_\delta^\varepsilon$. Thus by the property of expectation and (3.127), we have

$$P\{t < \tau_{\varsigma,\delta}^\varepsilon\} = P\{t < \tau_{\varsigma,\delta}^\varepsilon, t < \tau_\delta^\varepsilon\}$$
$$\leq \frac{E[\tau_{\varsigma,\delta}^\varepsilon(t) \cdot I_{\{t < \tau_\delta^\varepsilon\}}]}{t}$$
$$\leq \frac{k_2(\delta) V(x)}{c_\varepsilon^\varsigma t}, \tag{3.128}$$

which means that the solution process X_t^ε beginning in the domain $\varsigma < |x| < \delta$, almost surely reaches the boundary of this domain in finite time. Then by the definition of the set B_x^ε, for all paths contained in the set B_x^ε, except for a set of paths of probability zero, we have $\inf_{t>0} |X_t^\varepsilon| = 0$. Since $a(0, y) \equiv 0$, if $X_s^\varepsilon = 0$ for some $s \geq 0$, then $X_t^\varepsilon = 0$ for all $t \geq s$. Hence we obtain

$$\liminf_{t\to\infty} |X_t^\varepsilon| = 0, \tag{3.129}$$

and then by (3.25) and (3.97), for any $0 < \varepsilon \leq \varepsilon^*$, we have

$$\liminf_{t\to\infty} V^\varepsilon(X_t^\varepsilon, t) = 0. \tag{3.130}$$

But by (3.120) and the definition of the set B_x^ε, the limit

$$\lim_{t\to\infty} V^\varepsilon\left(X_{\tau_\delta^\varepsilon(t)}^\varepsilon, t\right) = \lim_{t\to\infty} V^\varepsilon(X_t^\varepsilon, t) \tag{3.131}$$

exists for almost all paths in B_x^ε. By the above discussion, this limit is equal to zero. Thus by (3.97) and (3.121), we obtain

$$\lim_{x\to 0} P\left\{ \lim_{t\to\infty} |X_t^\varepsilon| = 0 \right\} = 1. \tag{3.132}$$

The proof is completed.

3.3.2.2 Proof of Theorem 3.5

For brevity and to avoid overlap, we refer to parts of the proof of Theorem 3.4 that are adapted in the proof of Theorem 3.5.

Step 1 (Action of the p-infinitesimal operator on Lyapunov function in the case with global conditions). In the proof of Theorem 3.4, take $\delta = M$ for some positive integer M. Then similar to (3.97) and (3.104), we obtain that there exists an $\varepsilon_1 > 0$ such that, for any $0 < \varepsilon < \varepsilon_1$, $x \in \mathbb{R}^n$ with $|x| \leq M$, and $t \geq 0$,

$$k_1 V(x) \leq V^\varepsilon(x, t) \leq k_2 V(x), \tag{3.133}$$

$$\hat{A}_M^\varepsilon V^\varepsilon\left(X_{\tau_M^\varepsilon(t)}^\varepsilon, t\right) \leq -\left(\gamma - \varepsilon \frac{C_2}{c_1}\right) V\left(X_{\tau_M^\varepsilon(t)}^\varepsilon\right) \cdot I_{\{t < \tau_M^\varepsilon\}}, \tag{3.134}$$

where $k_1 = 1 - \frac{\varepsilon_1 C_1}{c_1} > 0$, $k_2 = 1 + \frac{\varepsilon_1 C_1}{c_1}$, $C_1 = \frac{2\sqrt{2c_5}c_3 k}{\alpha} + \frac{c_3 c_6 k}{\beta}$, $C_2 = \frac{c_6(c_3+c_4)k^2}{\beta} + \frac{2\sqrt{2c_5}(c_3+c_4)k^2}{\alpha}$ (independent of M used in the truncation).

Step 2 (Proof of global asymptotical stability in probability). Let $0 < \varepsilon_0' < \min\{\frac{c_1}{C_2}\gamma, \varepsilon_1\}$ and denote

$$\hat{\gamma} = \frac{1}{2k_2}\left(\gamma - \varepsilon_0' \frac{C_2}{c_1}\right). \tag{3.135}$$

Then by (3.133), (3.134), we get that for any $\varepsilon \in (0, \varepsilon_0']$,

$$\hat{\mathcal{A}}_M^{\varepsilon} V^{\varepsilon}\big(X_{\tau_M^{\varepsilon}(t)}^{\varepsilon}, t\big) \leq -2\hat{\gamma} k_2 V\big(X_{\tau_M^{\varepsilon}(t)}^{\varepsilon}\big) \cdot I_{\{t < \tau_M^{\varepsilon}\}}$$

$$\leq -2\hat{\gamma} V^{\varepsilon}\big(X_{\tau_M^{\varepsilon}(t)}^{\varepsilon}, t\big) \cdot I_{\{t < \tau_M^{\varepsilon}\}}. \tag{3.136}$$

By Lemma B.4,

$$\hat{\mathcal{A}}_M^{\varepsilon}\big(V^{\varepsilon}\big(X_{\tau_M^{\varepsilon}(t)}^{\varepsilon}, t\big) \cdot I_{\{t < \tau_M^{\varepsilon}\}}\big) = \hat{\mathcal{A}}_M^{\varepsilon} V^{\varepsilon}\big(X_{\tau_M^{\varepsilon}(t)}^{\varepsilon}, t\big), \tag{3.137}$$

which, together with (3.136), implies that

$$\big(\hat{\mathcal{A}}_M^{\varepsilon} + 2\hat{\gamma}\big)\big(V^{\varepsilon}\big(X_{\tau_M^{\varepsilon}(t)}^{\varepsilon}, t\big) \cdot I_{\{t < \tau_M^{\varepsilon}\}}\big) \leq 0. \tag{3.138}$$

For $t \geq 0$, define

$$M_t^{\varepsilon} = e^{2\hat{\gamma} t} V^{\varepsilon}\big(X_{\tau_M^{\varepsilon}(t)}^{\varepsilon}, t\big) \cdot I_{\{t < \tau_M^{\varepsilon}\}} + e^{2\hat{\gamma} \tau_M^{\varepsilon}} V\big(X_{\tau_M^{\varepsilon}}^{\varepsilon}\big) \cdot I_{\{\tau_M^{\varepsilon} \leq t\}} - V^{\varepsilon}(x, 0)$$

$$- \int_0^t e^{2\hat{\gamma} s}\big(\hat{\mathcal{A}}_M^{\varepsilon} + 2\hat{\gamma}\big)\big(V^{\varepsilon}\big(X_{\tau_M^{\varepsilon}(s)}^{\varepsilon}, s\big) \cdot I_{\{s < \tau_M^{\varepsilon}\}}\big) ds. \tag{3.139}$$

Then by the fact that $\tau_M^{\varepsilon} > 0$ a.s., we know that $M_0^{\varepsilon} = 0$ a.s. By the definition of $V^{\varepsilon}(x, t)$, we can verify that $e^{2\hat{\gamma} t} V^{\varepsilon}(X_{\tau_M^{\varepsilon}(t)}^{\varepsilon}, t) \cdot I_{\{t < \tau_M^{\varepsilon}\}} + e^{2\hat{\gamma} \tau_M^{\varepsilon}} V(X_{\tau_M^{\varepsilon}}^{\varepsilon}) \cdot I_{\{\tau_M^{\varepsilon} \leq t\}}$ is continuous in t, and thus M_t^{ε} is continuous. By (3.133), (3.98), the definition of $\hat{\mathcal{A}}_M^{\varepsilon} V^{\varepsilon}(X_{\tau_M^{\varepsilon}(t)}^{\varepsilon}, t)$ (replace δ by M in (3.101)), (B.13) with δ replaced by M, and the fact that $|X_{\tau_M^{\varepsilon}}^{\varepsilon}| \leq M$, we know that for any $t \geq 0$, M_t^{ε} is integrable. By Lemma B.5, we know that M_t^{ε} is a martingale relative to $\{\mathcal{F}_t^{\varepsilon}\}$, and thus it is a zero-mean, continuous martingale relative to $\{\mathcal{F}_t^{\varepsilon}\}$.

By (3.133), (3.138), and (3.139), we get that

$$0 \leq k_1 e^{2\hat{\gamma} t} V\big(X_{\tau_M^{\varepsilon}(t)}^{\varepsilon}\big) \cdot I_{\{t < \tau_M^{\varepsilon}\}} \leq e^{2\hat{\gamma} t} V^{\varepsilon}\big(X_{\tau_M^{\varepsilon}(t)}^{\varepsilon}, t\big) \cdot I_{\{t < \tau_M^{\varepsilon}\}}$$

$$\leq e^{2\hat{\gamma} t} V^{\varepsilon}\big(X_{\tau_M^{\varepsilon}(t)}^{\varepsilon}, t\big) \cdot I_{\{t < \tau_M^{\varepsilon}\}} + e^{2\hat{\gamma} \tau_M^{\varepsilon}} V\big(X_{\tau_M^{\varepsilon}}^{\varepsilon}\big) \cdot I_{\{\tau_M^{\varepsilon} \leq t\}} \quad (\text{since } V(x) \geq 0)$$

$$= V^{\varepsilon}(x, 0) + M_t^{\varepsilon} + \int_0^t e^{2\hat{\gamma} s}\big(\hat{\mathcal{A}}_M^{\varepsilon} + 2\hat{\gamma}\big)\big(V^{\varepsilon}\big(X_{\tau_M^{\varepsilon}(s)}^{\varepsilon}, s\big) \cdot I_{\{s < \tau_M^{\varepsilon}\}}\big) ds$$

$$\leq V^{\varepsilon}(x, 0) + M_t^{\varepsilon}$$

$$\leq k_2 V(x) + M_t^{\varepsilon}, \tag{3.140}$$

which means $k_2 V(x) + M_t^{\varepsilon}$ is a nonnegative continuous martingale relative to $\{\mathcal{F}_t^{\varepsilon}\}$. By (3.140), and Doob's inequality (cf. Sect. 2.III.9 of [34]), we have that, for any $\eta > 0$ and $T > 0$,

$$P\Big\{\sup_{0 \leq t \leq T} k_1 e^{2\hat{\gamma} t} V\big(X_{\tau_M^{\varepsilon}(t)}^{\varepsilon}\big) \cdot I_{\{t < \tau_M^{\varepsilon}\}} > \eta\Big\} \leq P\Big\{\sup_{0 \leq t \leq T}\big\{k_2 V(x) + M_t^{\varepsilon}\big\} > \eta\Big\}$$

$$\leq \frac{k_2 V(x)}{\eta}. \tag{3.141}$$

Letting $T \uparrow \infty$ in (3.141) yields

$$P\Big\{\sup_{t \geq 0} k_1 e^{2\hat{\gamma} t} V\big(X_{\tau_M^{\varepsilon}(t)}^{\varepsilon}\big) \cdot I_{\{t < \tau_M^{\varepsilon}\}} > \eta\Big\} \leq \frac{k_2 V(x)}{\eta}. \tag{3.142}$$

Notice that under Assumption 3.7, the original system (3.1) is globally Lipschitz. Then we know that the solution process X_t^ε is regular (cf. Sect. 7.2 of [39]), i.e.,

$$\lim_{M \to \infty} \tau_M^\varepsilon = \infty \quad \text{a.s.} \tag{3.143}$$

Notice that k_1, k_2, and $\hat{\gamma}$ are independent of M, and $\tau_M^\varepsilon(t) = t \wedge \tau_M^\varepsilon$. Then by (3.143),

$$\sup_{t \geq 0} k_1 e^{2\hat{\gamma}t} V\left(X_t^\varepsilon\right) \leq \liminf_{M \to \infty} \sup_{t \geq 0} k_1 e^{2\hat{\gamma}t} V\left(X_{\tau_M^\varepsilon(t)}^\varepsilon\right) \cdot I_{\{t < \tau_M^\varepsilon\}} \quad \text{a.s.} \tag{3.144}$$

In fact, let $\Omega^\varepsilon = \{\omega \in \Omega \mid \lim_{M \to \infty} \tau_M^\varepsilon = \infty\}$. Then by (3.143), we have

$$P\left(\Omega^\varepsilon\right) = 1. \tag{3.145}$$

For any $\omega \in \Omega^\varepsilon$, let

$$\lambda(\omega) = \sup_{t \geq 0} k_1 e^{2\hat{\gamma}t} V\left(X_t^\varepsilon\right)(\omega). \tag{3.146}$$

Firstly, we assume that $\lambda(\omega) < \infty$. Then for any $0 < \delta < \lambda(\omega)$, there exists a constant $t_0 = t_0(\delta, \omega) \geq 0$ such that

$$\lambda(\omega) \geq k_1 e^{2\hat{\gamma}t_0} V\left(X_{t_0}^\varepsilon\right)(\omega) > \lambda(\omega) - \delta. \tag{3.147}$$

By the definition of Ω^ε, there exists $M_0 = M_0(t_0, \omega) > 0$ such that for any $M > M_0$, we have that $\tau_M^\varepsilon(\omega) > t_0$, and thus for any $M > M_0$,

$$k_1 e^{2\hat{\gamma}t_0} V\left(X_{\tau_M^\varepsilon(t_0)}^\varepsilon(\omega)\right) \cdot I_{\{t_0 < \tau_M^\varepsilon\}}(\omega) = k_1 e^{2\hat{\gamma}t_0} V\left(X_{t_0}^\varepsilon\right)(\omega). \tag{3.148}$$

By (3.147) and (3.148), we get that for any $M > M_0$,

$$\sup_{t \geq 0} k_1 e^{2\hat{\gamma}t} V\left(X_{\tau_M^\varepsilon(t)}^\varepsilon(\omega)\right) \cdot I_{\{t < \tau_M^\varepsilon\}}(\omega) > \lambda(\omega) - \delta, \tag{3.149}$$

and thus

$$\liminf_{M \to \infty} \sup_{t \geq 0} k_1 e^{2\hat{\gamma}t} V\left(X_{\tau_M^\varepsilon(t)}^\varepsilon(\omega)\right) \cdot I_{\{t < \tau_M^\varepsilon\}}(\omega) \geq \lambda(\omega) - \delta. \tag{3.150}$$

Since δ can be any positive constant, we obtain that

$$\liminf_{M \to \infty} \sup_{t \geq 0} k_1 e^{2\hat{\gamma}t} V\left(X_{\tau_M^\varepsilon(t)}^\varepsilon(\omega)\right) \cdot I_{\{t < \tau_M^\varepsilon\}}(\omega) \geq \lambda(\omega). \tag{3.151}$$

Secondly, if $\lambda(\omega) = \infty$, then for any $M > 0$, there exists a constant $t_1 = t_1(\delta, \omega) \geq 0$ such that

$$k_1 e^{2\hat{\gamma}t_1} V\left(X_{t_1}^\varepsilon\right)(\omega) > M. \tag{3.152}$$

Following the proofs of (3.148)–(3.151), we can obtain that

$$\liminf_{M \to \infty} \sup_{t \geq 0} k_1 e^{2\hat{\gamma}t} V\left(X_{\tau_M^\varepsilon(t)}^\varepsilon(\omega)\right) \cdot I_{\{t < \tau_M^\varepsilon\}}(\omega) \geq M, \tag{3.153}$$

and thus by the arbitrariness of M, it holds that

$$\liminf_{M \to \infty} \sup_{t \geq 0} k_1 e^{2\hat{\gamma}t} V\left(X_{\tau_M^\varepsilon(t)}^\varepsilon(\omega)\right) \cdot I_{\{t < \tau_M^\varepsilon\}}(\omega) = \infty. \tag{3.154}$$

By (3.146), (3.151), and the fact that $P(\Omega^\varepsilon) = 1$, we obtain that (3.144) holds.

Now, by (3.144), Fatou's lemma, and (3.142), we obtain

$$
\begin{aligned}
P\left\{\sup_{t\geq0} k_1 e^{2\hat{\gamma}t} V\left(X_t^\varepsilon\right) > \eta\right\} &= E\left[I_{(\eta,\infty]}\left(\sup_{t\geq0} k_1 e^{2\hat{\gamma}t} V\left(X_t^\varepsilon\right)\right)\right] \\
&\leq E\left[\liminf_{M\to\infty} I_{(\eta,\infty]}\left(\sup_{t\geq0} k_1 e^{2\hat{\gamma}t} V\left(X_{\tau_M^\varepsilon(t)}^\varepsilon\right)\cdot I_{\{t<\tau_M^\varepsilon\}}\right)\right] \\
&\leq \liminf_{M\to\infty} E\left[I_{(\eta,\infty]}\left(\sup_{t\geq0} k_1 e^{2\hat{\gamma}t} V\left(X_{\tau_M^\varepsilon(t)}^\varepsilon\right)\cdot I_{\{t<\tau_M^\varepsilon\}}\right)\right] \\
&= \liminf_{M\to\infty} P\left\{\sup_{t\geq0} k_1 e^{2\hat{\gamma}t} V\left(X_{\tau_M^\varepsilon(t)}^\varepsilon\right)\cdot I_{\{t<\tau_M^\varepsilon\}} > \eta\right\} \\
&\leq \frac{k_2 V(x)}{\eta}.
\end{aligned}
\tag{3.155}
$$

By Assumption 3.6, we have

$$
\left\{c_1\left|X_t^\varepsilon\right|^2 \leq e^{-2\hat{\gamma}t}\frac{\eta}{k_1}, t\geq0\right\} \supseteq \left\{V\left(X_t^\varepsilon\right) \leq e^{-2\hat{\gamma}t}\frac{\eta}{k_1}, t\geq0\right\},
\tag{3.156}
$$

which, together with (3.155), implies

$$
P\left\{\left|X_t^\varepsilon\right| \leq e^{-\hat{\gamma}t}\left(\frac{\eta}{k_1 c_1}\right)^{1/2}, t\geq0\right\} \geq 1 - \frac{k_2 V(x)}{\eta}.
\tag{3.157}
$$

Let $\eta_1 > 0$ and $\eta_2 > 0$ be given. Choose η such that $(\frac{\eta}{k_1 c_1})^{1/2} \leq \eta_2$, and then choose $\delta_0 > 0$ such that if $|x| < \delta_0$, then $\frac{k_2 V(x)}{\eta} \leq \eta_1$. Thus we have

$$
P\left\{\left|X_t^\varepsilon\right| \leq \eta_2 e^{-\hat{\gamma}t}, t\geq0\right\} \geq 1 - \eta_1.
\tag{3.158}
$$

Now, we prove for any $x \in \mathbb{R}^n$, $P\{\lim_{t\to\infty}|X_t^\varepsilon| = 0\} = 1$. Notice that for any $H > 0$,

$$
\left\{\lim_{t\to\infty}\left|X_t^\varepsilon\right| = 0\right\} = \left\{\lim_{t\to\infty} V\left(X_t^\varepsilon\right) = 0\right\} \supseteq \left\{\sup_{t\geq0} k_1 e^{2\hat{\gamma}t} V\left(X_t^\varepsilon\right) \leq H\right\}.
\tag{3.159}
$$

Then by (3.155), we obtain

$$
P\left\{\lim_{t\to\infty}\left|X_t^\varepsilon\right| = 0\right\} \geq 1 - \frac{k_2 V(x)}{H},
\tag{3.160}
$$

and letting $H \uparrow \infty$ yields $P\{\lim_{t\to\infty}|X_t^\varepsilon| = 0\} = 1$. The proof is completed.

3.3.2.3 Proof of Theorem 3.6

The only condition of Theorem 3.6 that is different from the conditions in Theorem 3.5 is $a(0, y) \equiv 0$ which is replaced by $\sup_{y\in S_Y}|a(0, y)| < \infty$. Thus here we use the same approach as in the proof of Theorem 3.5.

Step 1 (Lyapunov estimates for Theorem 3.6). Let $c = (\sup_{y \in S_Y} |a(0, y)|) \vee 1$. Then by Assumption 3.8 (assume $k \geq 1$, otherwise, replace k by $k \vee 1$), we get that, for any $x \in \mathbb{R}^n$ and $y \in S_Y$,

$$|a(x, y)| \leq c + k|x| \leq k(c + |x|). \tag{3.161}$$

By (3.23) and (3.161), we get that, for any $x \in \mathbb{R}^n$,

$$|\bar{a}(x)| \leq k(c + |x|). \tag{3.162}$$

Then following the proofs of Theorem 3.4, we obtain that for $x \in \mathbb{R}^n$ with $|x| \leq M$, and $t \geq 0$,

$$-\varepsilon C_1 |x|(c + |x|) \leq V_1^\varepsilon(x, t) \leq \varepsilon C_1 |x|(c + |x|), \tag{3.163}$$

where $C_1 = \frac{2\sqrt{2c_5}c_3k}{\alpha} + \frac{c_3c_6k}{\beta}$ (the same with the one in the proof of Theorem 3.5).

By Assumption 3.6, the definition of $V^\varepsilon(x, t)$, and (3.163), we get that for any $\varepsilon > 0$, $x \in \mathbb{R}^n$ with $|x| \leq M$, and $t \geq 0$,

$$V(x) - \varepsilon C_1 |x|(c + |x|) \leq V^\varepsilon(x, t) \leq V(x) + \varepsilon C_1 |x|(c + |x|). \tag{3.164}$$

It follows from (3.164) and $c \geq 1$ that if $|x| \leq 1$, then

$$V(x) - 2\varepsilon c C_1 \leq V^\varepsilon(x, t) \leq V(x) + 2\varepsilon c C_1. \tag{3.165}$$

By Assumption 3.6 and $c \geq 1$, we have that if $|x| \geq 1$, then $|x|(c + |x|) \leq 2c|x|^2 \leq \frac{2c}{c_1} V(x)$, and thus by (3.164), if $|x| \geq 1$, then

$$\left(1 - \frac{2\varepsilon c C_1}{c_1}\right) V(x) \leq V^\varepsilon(x, t) \leq \left(1 + \frac{2\varepsilon c C_1}{c_1}\right) V(x). \tag{3.166}$$

Take a positive constant $\varepsilon_1' < \frac{c_1}{2cC_1}$, and define

$$k_1' = 1 - \frac{2cC_1}{c_1}\varepsilon_1', \qquad k_2' = 1 + \frac{2cC_1}{c_1}\varepsilon_1'. \tag{3.167}$$

Then by (3.166), we get that for any $0 < \varepsilon \leq \varepsilon_1'$ and $|x| \geq 1$,

$$k_1' V(x) \leq V^\varepsilon(x, t) \leq k_2' V(x). \tag{3.168}$$

Step 2 (Action of the p-infinitesimal operator on Lyapunov function without equilibrium condition). By (3.161) and Assumptions 3.6, 3.8, we get that, for any $x \in \mathbb{R}^n$, $y \in S_Y$,

$$|Q(x, y)| \leq c_4k(c + |x|) + c_3k|x|, \tag{3.169}$$

where $Q(x, y)$ is given by (B.8). Then by (3.161) and (3.169), following the proof of (B.12), we obtain that

$$\left|\int_{\tau_M^\varepsilon(t)}^{\tau_M^\varepsilon} \left[\frac{\partial E_t^\varepsilon[\tilde{G}(x, Y_{s/\varepsilon})]}{\partial x}\right]^T a(x, Y_{t/\varepsilon}) ds\right|$$
$$\leq \varepsilon \int_{\tau_M^\varepsilon(t)/\varepsilon}^{\tau_M^\varepsilon/\varepsilon} \left|E[Q(x, Y_u)|\mathcal{F}_{t/\varepsilon}^Y] - E[Q(x, Y_u)]\right||a(x, Y_{t/\varepsilon})| du$$

$$+ \varepsilon \int_{\tau_M^\varepsilon(t)/\varepsilon}^{\tau_M^\varepsilon/\varepsilon} \left| \left| \int_{S_Y} Q(x, y)\big(P_u(dy) - \mu(dy)\big) \right| \right| \big| a(x, Y_{t/\varepsilon}) \big| \, du$$

$$\leq \varepsilon \left[\frac{(c_4 k(c + |x|) + c_3 k|x|) \cdot k(c + |x|)}{\beta} \right.$$

$$\left. + \frac{2\sqrt{2c_5}(c_4 k(c + |x|) + c_3 k|x|) \cdot k(c + |x|)}{\alpha} \right]$$

$$\leq \varepsilon k^2 \left[\frac{c_6}{\beta} + \frac{2\sqrt{2c_5}}{\alpha} \right]$$

$$\cdot \big[c_4 c^2 + (c_3 + 2c_4)c|x| + (c_3 + c_4)|x|^2 \big]. \tag{3.170}$$

By Assumption 3.6 and $c \geq 1$, we have that if $|x| \geq 1$, then

$$c_4 c^2 + (c_3 + 2c_4)c|x| + (c_3 + c_4)|x|^2$$

$$\leq \big[c_4 c^2 + (c_3 + 2c_4)c + (c_3 + c_4) \big]|x|^2$$

$$= \frac{c_4 c^2 + (c_3 + 2c_4)c + (c_3 + c_4)}{c_1} c_1 |x|^2$$

$$\leq \frac{c_4 c^2 + (c_3 + 2c_4)c + (c_3 + c_4)}{c_1} V(x). \tag{3.171}$$

Denote

$$C_2' = k^2 \left[\frac{c_6}{\beta} + \frac{2\sqrt{2c_5}}{\alpha} \right] \frac{c_4 c^2 + (c_3 + 2c_4)c + (c_3 + c_4)}{c_1}. \tag{3.172}$$

Then by (3.170), we obtain that if $|x| \geq 1$, then

$$\left| \int_{\tau_M^\varepsilon(t)}^{\tau_M^\varepsilon} \left[\frac{\partial E_t^\varepsilon[\tilde{G}(x, Y_{s/\varepsilon})]}{\partial x} \right]^T a(x, Y_{t/\varepsilon}) \, ds \right| \leq \varepsilon C_2' V(x); \tag{3.173}$$

if $|x| < 1$, then

$$\left| \int_{\tau_M^\varepsilon(t)}^{\tau_M^\varepsilon} \left[\frac{\partial E_t^\varepsilon[\tilde{G}(x, Y_{s/\varepsilon})]}{\partial x} \right]^T a(x, Y_{t/\varepsilon}) \, ds \right| \leq \varepsilon c_1 C_2'. \tag{3.174}$$

By the definition of $\hat{\mathcal{A}}_M^\varepsilon V^\varepsilon(X_{\tau_M^\varepsilon(t)}^\varepsilon, t)$, Assumption 3.6, (3.173), and (3.174), for any $t \geq 0$,

$$\hat{\mathcal{A}}_M^\varepsilon V^\varepsilon \big(X_{\tau_M^\varepsilon(t)}^\varepsilon, t \big)$$

$$\leq \begin{cases} (-\gamma V(X_{\tau_M^\varepsilon(t)}^\varepsilon) + \varepsilon c_1 C_2') \cdot I_{\{t < \tau_M^\varepsilon\}} & \text{if } |X_{\tau_M^\varepsilon(t)}^\varepsilon| < 1; \\ -(\gamma - \varepsilon C_2') V(X_{\tau_M^\varepsilon(t)}^\varepsilon) \cdot I_{\{t < \tau_M^\varepsilon\}} & \text{if } |X_{\tau_M^\varepsilon(t)}^\varepsilon| \geq 1. \end{cases} \tag{3.175}$$

Step 3 (Proof of boundedness in probability).   Let $0 < \varepsilon^* < \min\{\frac{\gamma}{C_2'}, \varepsilon_1'\}$ and denote $\hat{\gamma} = \frac{\gamma - \varepsilon^* C_2'}{k_2'}$. Then by (3.175) and (3.168), we get that for any $\varepsilon \in (0, \varepsilon^*]$, if

$|X^\varepsilon_{\tau^\varepsilon_M(t)}| \geq 1$, then

$$\hat{\mathcal{A}}^\varepsilon_M V^\varepsilon\left(X^\varepsilon_{\tau^\varepsilon_M(t)}, t\right) \leq -\hat{\gamma} k'_2 V\left(X^\varepsilon_{\tau^\varepsilon_M(t)}\right) \cdot I_{\{t < \tau^\varepsilon_M\}}$$
$$\leq -\hat{\gamma} V^\varepsilon\left(X^\varepsilon_{\tau^\varepsilon_M(t)}, t\right) \cdot I_{\{t < \tau^\varepsilon_M\}}. \tag{3.176}$$

Since $\hat{\mathcal{A}}^\varepsilon_M(V^\varepsilon(X^\varepsilon_{\tau^\varepsilon_M(t)}, t) \cdot I_{\{t < \tau^\varepsilon_M\}}) = \hat{\mathcal{A}}^\varepsilon_M V^\varepsilon(X^\varepsilon_{\tau^\varepsilon_M(t)}, t)$ (see Lemma B.4) and due to (3.176), we get that if $|X^\varepsilon_{\tau^\varepsilon_M(t)}| \geq 1$, then

$$\left(\hat{\mathcal{A}}^\varepsilon_M + \hat{\gamma}\right)\left(V^\varepsilon\left(X^\varepsilon_{\tau^\varepsilon_M(t)}, t\right) \cdot I_{\{t < \tau^\varepsilon_M\}}\right) \leq 0. \tag{3.177}$$

For $t \geq 0$, define

$$M^\varepsilon_t = e^{\hat{\gamma} t} V^\varepsilon\left(X^\varepsilon_{\tau^\varepsilon_M(t)}, t\right) \cdot I_{\{t < \tau^\varepsilon_M\}} + e^{\hat{\gamma} \tau^\varepsilon_M} V\left(X^\varepsilon_{\tau^\varepsilon_M}\right) \cdot I_{\{\tau^\varepsilon_M \leq t\}} - V^\varepsilon(x, 0)$$
$$- \int_0^t e^{\hat{\gamma} s}\left(\hat{\mathcal{A}}^\varepsilon_M + \hat{\gamma}\right)\left(V^\varepsilon\left(X^\varepsilon_{\tau^\varepsilon_M(s)}, s\right) \cdot I_{\{s < \tau^\varepsilon_M\}}\right) ds. \tag{3.178}$$

As in the proof of Theorem 3.5, we can prove that M^ε_t is a zero-mean, continuous martingale relative to $\{\mathcal{F}^\varepsilon_t\}$. Thus since $\hat{\mathcal{A}}^\varepsilon_M(V^\varepsilon(X^\varepsilon_{\tau^\varepsilon_M(t)}, t) \cdot I_{\{t < \tau^\varepsilon_M\}}) = \hat{\mathcal{A}}^\varepsilon_M V^\varepsilon(X^\varepsilon_{\tau^\varepsilon_M(t)}, t)$, due to (3.165), (3.175), (3.177), and the fact that $\gamma > \hat{\gamma}$, we have, for any $0 < \varepsilon \leq \varepsilon^*$,

$$E\left[e^{\hat{\gamma} t} V^\varepsilon\left(X^\varepsilon_{\tau^\varepsilon_M(t)}, t\right) I_{\{t < \tau^\varepsilon_M\}}\right]$$
$$\leq E\left[e^{\hat{\gamma} t} V^\varepsilon\left(X^\varepsilon_{\tau^\varepsilon_M(t)}, t\right) \cdot I_{\{t < \tau^\varepsilon_M\}} + e^{\hat{\gamma} \tau^\varepsilon_M} V\left(X^\varepsilon_{\tau^\varepsilon_M}\right) I_{\{\tau^\varepsilon_M \leq t\}}\right]$$
$$= V^\varepsilon(x, 0) + \int_0^t E\left[e^{\hat{\gamma} s}\left(\hat{\mathcal{A}}^\varepsilon_M + \hat{\gamma}\right)\left(V^\varepsilon\left(X^\varepsilon_{\tau^\varepsilon_M(s)}, s\right) \cdot I_{\{s < \tau^\varepsilon_M\}}\right)\right] ds$$
$$= V^\varepsilon(x, 0) + \int_0^t E\left[e^{\hat{\gamma} s}\left(\hat{\mathcal{A}}^\varepsilon_M + \hat{\gamma}\right)\left(V^\varepsilon\left(X^\varepsilon_{\tau^\varepsilon_M(s)}, s\right) \cdot I_{\{s < \tau^\varepsilon_M\}}\right) I_{\{|X^\varepsilon_{\tau^\varepsilon_M(s)}| < 1\}}\right] ds$$
$$+ \int_0^t E\left[e^{\hat{\gamma} s}\left(\hat{\mathcal{A}}^\varepsilon_M + \hat{\gamma}\right)\left(V^\varepsilon\left(X^\varepsilon_{\tau^\varepsilon_M(s)}, s\right) \cdot I_{\{s < \tau^\varepsilon_M\}}\right) I_{\{|X^\varepsilon_{\tau^\varepsilon_M(s)}| \geq 1\}}\right] ds$$
$$\leq V^\varepsilon(x, 0) + \int_0^t E\left[e^{\hat{\gamma} s}\left(\hat{\mathcal{A}}^\varepsilon_M + \hat{\gamma}\right)\left(V^\varepsilon\left(X^\varepsilon_{\tau^\varepsilon_M(s)}, s\right) \cdot I_{\{s < \tau^\varepsilon_M\}}\right) I_{\{|X^\varepsilon_{\tau^\varepsilon_M(s)}| < 1\}}\right] ds$$
$$\leq V^\varepsilon(x, 0) + \int_0^t E\left[e^{\hat{\gamma} s}\left(-\gamma V\left(X^\varepsilon_{\tau^\varepsilon_M(s)}\right) + \varepsilon c_1 C'_2\right.\right.$$
$$\left.\left. + \hat{\gamma} V^\varepsilon\left(X^\varepsilon_{\tau^\varepsilon_M(s)}, s\right)\right) I_{\{s < \tau^\varepsilon_M\}} I_{\{|X^\varepsilon_{\tau^\varepsilon_M(s)}| < 1\}}\right] ds$$
$$\leq V^\varepsilon(x, 0) + \int_0^t E\left[e^{\hat{\gamma} s}\left(-\gamma V\left(X^\varepsilon_{\tau^\varepsilon_M(s)}\right) + \varepsilon c_1 C'_2\right.\right.$$
$$\left.\left. + \hat{\gamma}\left(V\left(X^\varepsilon_{\tau^\varepsilon_M(s)}\right) + 2\varepsilon c C_1\right)\right) I_{\{s < \tau^\varepsilon_M\}} I_{\{|X^\varepsilon_{\tau^\varepsilon_M(s)}| < 1\}}\right] ds$$
$$\leq V^\varepsilon(x, 0) + \int_0^t E\left[e^{\hat{\gamma} s}\left(\varepsilon c_1 C'_2 + 2\hat{\gamma} \varepsilon c C_1\right)\right] ds$$

$$= V^\varepsilon(x,0) + \frac{\varepsilon c_1 C_2' + 2\hat{\gamma}\varepsilon c C_1}{\hat{\gamma}}\left(e^{\hat{\gamma}t} - 1\right)$$

$$\le V^\varepsilon(x,0) + \frac{\varepsilon c_1 C_2' + 2\hat{\gamma}\varepsilon c C_1}{\hat{\gamma}} e^{\hat{\gamma}t}, \tag{3.179}$$

where in the first equality we used Fubini's theorem and the integrability condition

$$\int_0^t E\big[\big|e^{\hat{\gamma}s}(\hat{A}_M^\varepsilon + \hat{\gamma})V^\varepsilon(X_{\tau_M^\varepsilon(s)}^\varepsilon, s)\big|\big]\, ds < \infty, \tag{3.180}$$

which can be verified by (3.101) (with δ changed to M), (3.98), (3.99), (B.8), (3.165), and (3.168). Thus we have that

$$E\big[V^\varepsilon\big(X_{\tau_M^\varepsilon(t)}^\varepsilon, t\big)\cdot I_{\{t<\tau_M^\varepsilon\}}\big] \le e^{-\hat{\gamma}t}V^\varepsilon(x,0) + \frac{\varepsilon c_1 C_2' + 2\hat{\gamma}\varepsilon c C_1}{\hat{\gamma}}. \tag{3.181}$$

By (3.168), Assumption 3.6, and the property of expectation, we get that for any $r > 1$,

$$P\big\{\big|X_{\tau_M^\varepsilon(t)}^\varepsilon\big| > r, t < \tau_M^\varepsilon\big\}$$

$$= P\big\{\big|X_{\tau_M^\varepsilon(t)}^\varepsilon\big| > r, k_1' V\big(X_{\tau_M^\varepsilon(t)}^\varepsilon\big) \le V^\varepsilon\big(X_{\tau_M^\varepsilon(t)}^\varepsilon, t\big) \le k_2' V\big(X_{\tau_M^\varepsilon(t)}^\varepsilon\big), t < \tau_M^\varepsilon\big\}$$

$$= P\big\{\big|X_{\tau_M^\varepsilon(t)}^\varepsilon\big| > r, V\big(X_{\tau_M^\varepsilon(t)}^\varepsilon\big) > c_1 r^2,$$

$$\qquad k_1' V\big(X_{\tau_M^\varepsilon(t)}^\varepsilon\big) \le V^\varepsilon\big(X_{\tau_M^\varepsilon(t)}^\varepsilon, t\big) \le k_2' V\big(X_{\tau_M^\varepsilon(t)}^\varepsilon\big), t < \tau_M^\varepsilon\big\}$$

$$\le P\big\{\big|X_{\tau_M^\varepsilon(t)}^\varepsilon\big| > 1, V\big(X_{\tau_M^\varepsilon(t)}^\varepsilon\big) > c_1 r^2,$$

$$\qquad k_1' V\big(X_{\tau_M^\varepsilon(t)}^\varepsilon\big) \le V^\varepsilon\big(X_{\tau_M^\varepsilon(t)}^\varepsilon, t\big) \le k_2' V\big(X_{\tau_M^\varepsilon(t)}^\varepsilon\big), t < \tau_M^\varepsilon\big\}$$

$$\le P\big\{\big|X_{\tau_M^\varepsilon(t)}^\varepsilon\big| > 1, V^\varepsilon\big(X_{\tau_M^\varepsilon(t)}^\varepsilon, t\big) > c_1 k_1' r^2, t < \tau_M^\varepsilon\big\}$$

$$\le \frac{1}{c_1 k_1' r^2} E\big[V^\varepsilon\big(X_{\tau_M^\varepsilon(t)}^\varepsilon, t\big)\cdot I_{\{t<\tau_M^\varepsilon\}}\cdot I_{\{|X_{\tau_M^\varepsilon(t)}^\varepsilon|>1\}}\big]. \tag{3.182}$$

Thus by (3.181), (3.165), and (3.168), we obtain, for any $0 < \varepsilon \le \varepsilon^*$ and any $t \ge 0$,

$$E\big[V^\varepsilon\big(X_{\tau_M^\varepsilon(t)}^\varepsilon, t\big)\cdot I_{\{t<\tau_M^\varepsilon\}}\cdot I_{\{|X_{\tau_M^\varepsilon(t)}^\varepsilon|>1\}}\big]$$

$$= E\big[V^\varepsilon\big(X_{\tau_M^\varepsilon(t)}^\varepsilon, t\big)\cdot I_{\{t<\tau_M^\varepsilon\}}\big] - E\big[V^\varepsilon\big(X_{\tau_M^\varepsilon(t)}^\varepsilon, t\big)\cdot I_{\{t<\tau_M^\varepsilon\}}\cdot I_{\{|X_{\tau_M^\varepsilon(t)}^\varepsilon|\le 1\}}\big]$$

$$\le e^{-\hat{\gamma}t}V^\varepsilon(x,0) + \frac{\varepsilon c_1 C_2' + 2\hat{\gamma}\varepsilon c C_1}{\hat{\gamma}}$$

$$\qquad - E\big[\big(V\big(X_{\tau_M^\varepsilon(t)}^\varepsilon\big) - 2\varepsilon c C_1\big)\cdot I_{\{t<\tau_M^\varepsilon\}}\cdot I_{\{|X_{\tau_M^\varepsilon(t)}^\varepsilon|\le 1\}}\big]$$

$$\le e^{-\hat{\gamma}t}V^\varepsilon(x,0) + \frac{\varepsilon c_1 C_2' + 2\hat{\gamma}\varepsilon c C_1}{\hat{\gamma}} + 2\varepsilon c C_1$$

$$\le \max\big\{V(x) + 2\varepsilon^* c C_1, k_2' V(x)\big\} + \frac{\varepsilon^* c_1 C_2'}{\hat{\gamma}} + 4\varepsilon^* c C_1 \triangleq C, \tag{3.183}$$

where C is a positive constant dependent on x, ε^*, c, c_1, C_1, C_2', k_2', and $\hat{\gamma}$. Thus by (3.182) and (3.183), we get that, for any $0 < \varepsilon \leq \varepsilon^*$, any $r > 1$, and any $t \geq 0$,

$$P\{|X^\varepsilon_{\tau^\varepsilon_M(t)}| > r, t < \tau^\varepsilon_M\} \leq \frac{C}{c_1 k_1' r^2}. \tag{3.184}$$

By the fact that $\lim_{M \to \infty} \tau^\varepsilon_M = \infty$ a.s. (see (3.143)), the dominated convergence theorem, and (3.184), we get that, for any $0 < \varepsilon \leq \varepsilon^*$ and any $r > 1$,

$$\sup_{t \geq 0} P\{|X^\varepsilon_t| > r\} = \sup_{t \geq 0} E\left[I_{(r,\infty]}(|X^\varepsilon_t|)\right]$$

$$= \sup_{t \geq 0} E\left[\lim_{M \to \infty} I_{(r,\infty]}(|X^\varepsilon_{\tau^\varepsilon_M(t)}|) \cdot I_{\{t < \tau^\varepsilon_M\}}\right]$$

$$= \sup_{t \geq 0}\left(\lim_{M \to \infty} E\left[I_{(r,\infty]}(|X^\varepsilon_{\tau^\varepsilon_M(t)}|) \cdot I_{\{t < \tau^\varepsilon_M\}}\right]\right)$$

$$= \sup_{t \geq 0}\left(\lim_{M \to \infty} P\{|X^\varepsilon_{\tau^\varepsilon_M(t)}| > r, t < \tau^\varepsilon_M\}\right)$$

$$\leq \frac{C}{c_1 k_1' r^2}, \tag{3.185}$$

which implies that

$$\lim_{r \to \infty} \sup_{t \geq 0} P\{|X^\varepsilon_t| > r\} = 0, \tag{3.186}$$

i.e., the solution process X^ε_t is bounded in probability. The proof is completed.

3.4 Examples

3.4.1 Perturbation Process Is Asymptotically Periodic

Consider the following system

$$\frac{dx^\varepsilon_t}{dt} = \xi^2_{t/\varepsilon}(x^\varepsilon_t + 1) - \frac{1}{2}(x^\varepsilon_t + 1)^2, \tag{3.187}$$

where the perturbation process is

$$dY_t = -pY_t\,dt + q\,dW_t, \qquad \xi_t = \sin t + e^{-at}\sin Y_t \tag{3.188}$$

with $p, q, a > 0$, W_t is a 1-dimensional standard Brownian motion defined on some complete probability space. Noticing that for any $t \geq 0$

$$\lim_{T \to \infty} \frac{1}{T} \int_t^{t+T} \xi^2_s\,ds = \lim_{T \to \infty} \frac{1}{T} \int_t^{t+T} \left(\sin^2 s + 2\sin s\,e^{-as}\sin Y_s + e^{-2as}\sin^2 Y_s\right) ds$$

$$= \lim_{T \to \infty} \frac{1}{T} \int_t^{t+T} \sin^2 s\,ds$$

Fig. 3.1 States of the
original and average systems
for system (3.187)–(3.188) to
illustrate Theorems 3.2
and 3.3

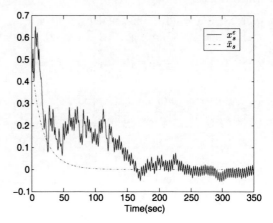

$$= \frac{1}{2\pi} \int_0^{2\pi} \sin^2 s \, ds$$

$$= \frac{1}{2} \quad \text{a.s.,} \tag{3.189}$$

we obtain the average system of (3.187) as

$$\frac{d\bar{x}_t}{dt} = -\frac{1}{2}(\bar{x}_t + \bar{x}_t^2), \quad \bar{x}_0 = x, \tag{3.190}$$

which is locally exponentially stable at $\bar{x}_t = 0$. Figure 3.1 shows the simulation
results with $\bar{x}_0 = x_0^\varepsilon = 0.5$, $p = 1$, $q = 2$, $a = 0.01$, $\varepsilon = 0.09$, from which we can
see that the solution of the original system (3.187) converges (in probability) to the
solution of the average system $\frac{d\bar{x}_t}{dt} = -(\bar{x}_t + \bar{x}_t^2)/2$ (see (3.11) in Theorem 3.2)
and the solution of system (3.187) is exponentially practically stable in probability
(Theorem 3.3).

3.4.2 Perturbation Process Is Almost Surely Exponentially Stable

Consider the following system

$$\frac{dx_t^\varepsilon}{dt} = -\sin^2(\xi_{t/\varepsilon}) + \left(\sin(\xi_{t/\varepsilon}) - \frac{1}{2}\right)(x_t^\varepsilon)^2 - x_t^\varepsilon,$$

$$d\xi_t = p\xi_t \, dt + q\xi_t \, dW_t, \tag{3.191}$$

where $p < q^2/2$. We know that $\xi_t = \xi_0 e^{(p-q^2/2)t + qW_t}$. By the law of iterated logarithm for Brownian motion (see Theorem 2.9.23 of [55]), we know that $\xi_t \to 0$ a.s.
as $t \to \infty$. Noticing that

$$\lim_{T \to \infty} \frac{1}{T} \int_t^{t+T} f(s) \, ds = \lim_{s \to \infty} f(s) \tag{3.192}$$

Fig. 3.2 States of the
original and average systems
for system (3.191). *Top*: for
$\varepsilon = 0.01$, which is small (the
average approximation is
tight). *Bottom*: for $\varepsilon = 0.64$,
which is large (the average
approximation is qualitatively
correct, but it is not very
accurate since the condition
on the smallness of ε in
Corollary 3.2 and
Theorem 3.3 is not met)

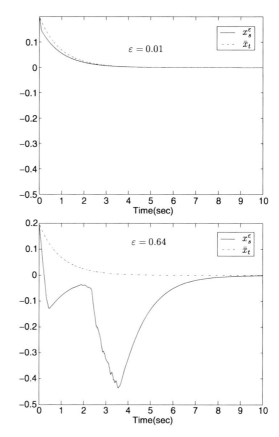

for continuous function f when the latter limit exists, we have that for any $t \geq 0$

$$\lim_{T \to \infty} \frac{1}{T} \int_t^{t+T} \sin^2(\xi_s)\,ds = 0 \quad \text{a.s.,} \tag{3.193}$$

$$\lim_{T \to \infty} \frac{1}{T} \int_t^{t+T} \left(\left(\sin(\xi_s) - \frac{1}{2} \right) x^2 - x \right) ds = -\frac{1}{2}x^2 - x \quad \text{a.s.} \tag{3.194}$$

Thus we obtain the average system of (3.191) as

$$\frac{d\bar{x}_t}{dt} = -\bar{x}_t - \frac{1}{2}\bar{x}_t^2, \quad \bar{x}_0 = x, \tag{3.195}$$

which is locally exponentially stable at $\bar{x}_t = 0$. Figure 3.2 shows the simulation
results with $\bar{x}_0 = x_0^\varepsilon = 0.2$, $\xi_0 = 1$, $p = 0.4$, $q = 1$, from which we can see that the
solution of the original system (3.191) converges (in probability) to the solution of
the average system $\frac{d\bar{x}_t}{dt} = -\bar{x}_t - \bar{x}_t^2/2$ (see (3.13) in Corollary 3.2) and the solution
of system (3.191) is exponentially practically stable in probability (Theorem 3.3).

3.4.3 Perturbation Process Is Brownian Motion on the Unit Circle

While in Sects. 3.4.1 and 3.4.2 we illustrated the theorems in Sect. 3.2.1 for uniform strong ergodic perturbation processes, in this section we illustrate the theorems in Sect. 3.2.2 for ϕ-mixing perturbation process. Consider the system

$$\frac{dx_t^\varepsilon}{dt} = -[0 \quad 1]\left[Y_1^2(t/\varepsilon) \quad Y_2^2(t/\varepsilon)\right]^T x_t^\varepsilon$$

$$+ \left([0 \quad 1]\left[Y_1(t/\varepsilon) \quad Y_2(t/\varepsilon)\right]^T - \frac{1}{2}\right)(x_t^\varepsilon)^2, \qquad (3.196)$$

where the perturbation process $Y(t) = [Y_1(t), Y_2(t)]^T$ is a Brownian motion on the unit circle,

$$dY_t = -\frac{1}{2}Y_t\,dt + BY_t\,dW_t, \quad Y_0 = \left[\cos(\vartheta), \sin(\vartheta)\right]^T \quad \text{for all } \vartheta \in \mathbb{R}, \quad (3.197)$$

with $B = \left[\begin{smallmatrix} 0 & -1 \\ 1 & 0 \end{smallmatrix}\right]$. In fact, we have the simple expression [109, Example 5.4, p. 63]

$$Y(t) = \left[\cos(\vartheta + W_t), \sin(\vartheta + W_t)\right]^T = e^{i(\vartheta + W_t)}. \qquad (3.198)$$

We know that the stochastic process $(Y(t), t \geq 0)$ is ϕ-mixing with exponential mixing rate and exponentially ergodic with invariant distribution $\mu(dS) = \frac{l(S)}{2\pi}$ for any set $S \subset T$, where $T = \{(x, y) \in \mathbb{R}^2 | x^2 + y^2 = 1\}$, and $l(S)$ denotes the length (Lebesgue measure) of S. Corresponding to system (3.196), we have the function

$$a(x, y_1, y_2) = -y_2^2 x + \left(y_2 - \frac{1}{2}\right)x^2. \qquad (3.199)$$

Noticing that

$$\int_T -y_2^2 \mu(dy_1, dy_2) = -\int_0^{2\pi} \sin^2(\theta)\frac{1}{2\pi}\,d\theta = -\frac{1}{2} \qquad (3.200)$$

and

$$\int_T \left(y_2 - \frac{1}{2}\right)\mu(dy_1, dy_2) = \int_0^{2\pi} \left(\sin\theta - \frac{1}{2}\right)\frac{1}{2\pi}\,d\theta = -\frac{1}{2}, \qquad (3.201)$$

we obtain the average system of (3.196) as

$$\frac{d\bar{x}_t}{dt} = -\frac{1}{2}(\bar{x}_t + \bar{x}_t^2), \quad \bar{x}_0 = x, \qquad (3.202)$$

which is locally exponentially stable at $\bar{x}_t = 0$. Figure 3.3 shows the simulation results with $\bar{x}_0 = x_0^\varepsilon = 0.1$, $\varepsilon = 0.64$, $Y_0 = [1, 0]^T$, from which we can see that the solution $x_t^\varepsilon \equiv 0$ of the system (3.196) is asymptotically stable (in probability) (see (3.29) and (3.30) in Theorem 3.4).

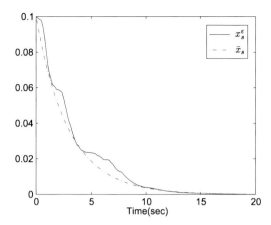

Fig. 3.3 States of the original and average systems for system (3.196), (3.198) to illustrate Theorem 3.4

3.5 Notes and References

In this chapter, which is based on results that we introduced in [88], we developed several basic theorems of stochastic infinite-time averaging for a class of nonlinear systems with uniform strong ergodic stochastic perturbations and ϕ-mixing perturbations. For the former class, under the condition of exponential stability of average equilibrium, the original system is exponentially practically stable in probability. For the latter class, under the condition of exponential stability of average equilibrium, which is also an equilibrium of the original system, the original system is asymptotically stable in probability. This is the first set of results on infinite-time stochastic averaging for locally (rather than globally) Lipschitz systems and represents an extension of the deterministic general averaging for systems with aperiodic vector fields.

Chapter 4
Stochastic Averaging for Practical Stability

In this chapter, we present new stochastic averaging theorems that relax the key limiting conditions in the existing stochastic averaging theory. We first introduce the notion of weak stability under random perturbation for general nonlinear systems. This stability notion is a stability robustness property for a deterministic system, relative to perturbations involving a stochastic process, and in the presence of a small parameter. Then we formulate and study some stability-like properties for the original system by investigating the weak stability under the random perturbation of the equilibrium of the average system. We present the detailed proofs for the general theorems.

4.1 General Stochastic Averaging

4.1.1 Problem Formulation

Consider the following system

$$\frac{dX_t^\varepsilon}{dt} = a\left(X_t^\varepsilon, Y_{t/\varepsilon}\right), \quad X_0^\varepsilon = x, \tag{4.1}$$

where $X_t^\varepsilon \in \mathbb{R}^n$, $Y_t \in \mathbb{R}^m$ is a time homogeneous continuous Markov process defined on a complete probability space $(\Omega, \mathcal{F}, P)$, where Ω is the sample space, $\mathcal{F}$ is a σ-field, and P is a probability measure. The initial condition $X_0^\varepsilon = x$ is deterministic. ε is a small parameter in $(0, \varepsilon_0)$ with fixed $\varepsilon_0 > 0$. Let $S_Y \subset \mathbb{R}^m$ be the living space of the perturbation process $(Y_t, t \geq 0)$ and note that S_Y may be a proper (e.g., compact) subset of $\mathbb{R}^m$.

The following assumptions are made.

Assumption 4.1 The vector field $a(x, y)$ is a continuous function of (x, y), and for any $x \in \mathbb{R}^n$, it is a bounded function of y. Furthermore, $a(x, y)$ satisfies the local

S.-J. Liu, M. Krstic, *Stochastic Averaging and Stochastic Extremum Seeking*,
Communications and Control Engineering,
DOI 10.1007/978-1-4471-4087-0_4, © Springer-Verlag London 2012

Lipschitz condition in $x \in \mathbb{R}^n$ uniformly in $y \in S_Y$, i.e., for any compact subset $D \subset \mathbb{R}^n$, there is a constant k_D such that for all $x_1, x_2 \in D$ and all $y \in S_Y$,

$$|a(x_1, y) - a(x_2, y)| \leq k_D |x_1 - x_2|. \tag{4.2}$$

Assumption 4.2 The perturbation process $(Y_t, t \geq 0)$ is ergodic with invariant distribution μ.

Assumption 4.2 is in contrast to most of the stochastic averaging theory, where, in addition to this assumption, the perturbation process is required to satisfy some form of a strong mixing property. The meaning of ergodicity, in simple terms, is that the time average of a function of the process along the trajectories exists almost surely and equals the space average:

$$\lim_{T \to \infty} \frac{1}{T} \int_0^T f(Y_s) \, ds = \int_{S_Y} f(y) \mu(dy) \quad \text{a.s.} \tag{4.3}$$

for any integrable function $f(\cdot)$. The following are two examples of ergodic stochastic processes (one is a 1-dimensional process and the other is a 2-dimensional process):

1. The Ornstein–Uhlenbeck (OU) process $(Y_t, t \geq 0)$:

$$dY_t = -pY_t \, dt + q \, dW_t, \tag{4.4}$$

   where W_t is a 1-dimensional standard Brownian motion on some probability space $(\Omega, \mathcal{F}, P)$. It is known [112] that the OU process is ergodic with invariant distribution $\mu(dx) = \frac{\sqrt{p}}{\sqrt{\pi}q} e^{-px^2/q^2} \, dx$.

2. Brownian motion on the unit circle $(Y_t, t \geq 0)$:

$$Y_t = e^{jW_t} = \left[\cos(W_t), \sin(W_t) \right]^T, \tag{4.5}$$

   where j is the imaginary unit and W_t is a 1-dimensional Brownian motion which is not necessarily standard in the form $W_0 = 0$. By Ito's formula, its coordinates Y_{1t} and Y_{2t} satisfy

$$\begin{cases} dY_{1t} = -\frac{1}{2} \cos(W_t) \, dt - \sin(W_t) \, dW_t, \\ dY_{2t} = -\frac{1}{2} \sin(W_t) \, dt + \cos(W_t) \, dW_t. \end{cases} \tag{4.6}$$

Thus the process $Y_t = [Y_{1t}, Y_{2t}]^T$ is the solution of the following stochastic differential equations with initial condition $Y_{10} = \cos(W_0)$ and $Y_{20} = \sin(W_0)$:

$$\begin{cases} dY_{1t} = -\frac{1}{2}Y_{1t} \, dt - Y_{2t} \, dW_t, \\ dY_{2t} = -\frac{1}{2}Y_{2t} \, dt + Y_{1t} \, dW_t, \end{cases} \tag{4.7}$$

or, in matrix notation,

$$dY_t = -\frac{1}{2}Y_t \, dt + BY_t \, dW_t, \tag{4.8}$$

where $B = \begin{bmatrix} 0 & -1 \\ 1 & 0 \end{bmatrix}$. On the other hand, the solution of (4.8) with initial value $Y_0 = [\cos(\vartheta), \sin(\vartheta)]^T$ $(\vartheta \in \mathbb{R})$ is

$$Y_t = e^{(-(1/2)I - (1/2)B^2)t + BW_t} Y_0$$

$$= e^{BW_t} Y_0 = \sum_{k=0}^{\infty} \frac{B^k W^k(t)}{k!} Y_0 \quad (B^2 = -I)$$

$$= \big(I \cos(W_t) + B \sin(W_t) \big) Y_0$$

$$= \big[\cos(\vartheta + W_t), \sin(\vartheta + W_t) \big]^T$$

$$= e^{j(\vartheta + W_t)}. \tag{4.9}$$

Therefore, Brownian motion on the unit circle $Y_t = [\cos(W_t), \sin(W_t)]^T$ is equivalent to the solution of stochastic differential equation

$$dY_t = -\frac{1}{2} Y_t \, dt + B Y_t \, d\check{W}_t, \tag{4.10}$$

with initial condition $Y_0 = [\cos(W_0), \sin(W_0)]^T$, where $\check{W}_t$ is a 1-dimensional standard Brownian motion with $\check{W}_0 = 0$. It is known [13] that Brownian motion on the unit circle $(Y_t, t \geq 0)$ is exponentially ergodic and its invariant distribution μ is the uniform measure on $T = \{(x, y) \in \mathbb{R}^2 | x^2 + y^2 = 1\}$, i.e., $\mu(S) = \frac{l(S)}{2\pi}$ for any set $S \subset T$, and $l(S)$ denotes the length (Lebesgue measure) of S.

In the extremum seeking applications in this book, we use the ergodic processes (4.4) and (4.10) as the excitation signals to develop stochastic extremum seeking algorithms.

Assumption 4.3 For any $x \in \mathbb{R}^n$ and the perturbation process $(Y_t, t \geq 0)$, system (4.1) has a unique (almost surely) continuous solution on $[0, \infty)$.

Since Y_t is a time homogeneous continuous Markov process, if $a(x, y)$ is globally Lipschitz in (x, y), then the solution of system (4.1) exists with probability 1 for any $x \in \mathbb{R}^n$ and it is defined uniquely for all $t \geq 0$ (see Sect. 2 of Chap. 7 of [39]). Here, we firstly do not emphasize how to guarantee or prove the existence of the solution of system (4.1) but just assume that system (4.1) has a unique (almost surely) continuous solution on $[0, \infty)$. In fact, by Assumption 4.1, we know that for any trajectory of the perturbing process $(Y_t, t \geq 0)$ and for any $\varepsilon > 0$, system (4.1) has a unique solution up to a possible explosion time. Assumption 4.3 implies that there is no finite explosion time for system (4.1), so that (4.1) has a continuous solution defined on the whole time interval $[0, +\infty)$.

Under Assumption 4.2, we obtain the average system of system (4.1) as follows:

$$\frac{d\bar{X}_t}{dt} = \bar{a}(\bar{X}_t), \quad \bar{X}_0 = x, \tag{4.11}$$

where

$$\bar{a}(x) = \int_{S_Y} a(x, y) \mu(dy). \tag{4.12}$$

By Assumption 4.1, $a(x, y)$ is bounded with respect to y, thus $y \to a(x, y)$ is μ-integrable. So $\bar{a}$ is well defined. For the average system (4.11), we make the following assumption.

Assumption 4.4 The average system (4.11) has a solution on $[0, \infty)$.

For the original system (4.1) and the average system (4.11), we introduce the following definitions.

Definition 4.1 A solution X_t^ε of system (4.1) is said to satisfy the property of

1. *Weak boundedness* if there exists a constant $M > 0$ such that

$$\lim_{\varepsilon \to 0} \inf\{t \geq 0 : |X_t^\varepsilon| > M\} = +\infty \quad \text{a.s.;} \tag{4.13}$$

2. *Weak attractivity* if there exists a point $x^* \in \mathbb{R}^n$ such that for any $\delta > 0$, there exists a constant $T_\delta > 0$ such that

$$\lim_{\varepsilon \to 0} \inf\{t \geq T_\delta : |X_t^\varepsilon - x^*| > \delta\} = +\infty, \quad \text{a.s.} \tag{4.14}$$

By convention, $\inf \emptyset = +\infty$.

Since it is not assumed that system (4.1) has an equilibrium, we cannot necessarily study the stability of an equilibrium solution of system (4.1). However, the average system (4.11) may have stable equilibria. We consider system (4.1) as a perturbation of the average system (4.11) and analyze suitably defined stability properties by studying equilibrium stability of (4.11). To this end, we rewrite system (4.1) as

$$\frac{dX_t^\varepsilon}{dt} = \bar{a}(X_t^\varepsilon) + R(X_t^\varepsilon, Y_{t/\varepsilon}), \quad X_0^\varepsilon = x, \tag{4.15}$$

where $R(X_t^\varepsilon, Y_{t/\varepsilon}) = a(X_t^\varepsilon, Y_{t/\varepsilon}) - \bar{a}(X_t^\varepsilon)$, and consider system (4.15) as a random perturbation of the average system (4.11). We assume that $\bar{a}(0) = 0$, and $\bar{X}_t \equiv 0$ is a stable (resp., asymptotically stable, exponentially stable) solution of system (4.11).

Definition 4.2 The solution $\bar{X}_t \equiv 0$ of system (4.11) is called

1. *Weakly stable* under random perturbation $R(\cdot, Y_{t/\varepsilon})$ if, for any $\delta > 0$, there exists a constant $r_\delta > 0$ such that, for any initial condition $x \in \{\check{x} \in \mathbb{R}^n : |\check{x}| < r_\delta\}$, the solution of system (4.1) satisfies

$$\lim_{\varepsilon \to 0} \inf\{t \geq 0 : |X_t^\varepsilon| > \delta\} = +\infty \quad \text{a.s.;} \tag{4.16}$$

2. *Weakly asymptotically stable* under random perturbation $R(\cdot, Y_{t/\varepsilon})$, if it is weakly stable under random perturbation $R(\cdot, Y_{t/\varepsilon})$ and there exists $r > 0$ such that, for any initial condition $x \in \{\check{x} \in \mathbb{R}^n : |\check{x}| < r\}$, the solution X_t^ε of system (4.1) is weakly attracted to the point 0;

3. *Weakly exponentially stable* under random perturbation $R(\cdot, Y_{t/\varepsilon})$, if there exist constants $r > 0$, $c > 0$, and $\gamma > 0$ such that, for any initial condition $x \in \{\check{x} \in \mathbb{R}^n : |\check{x}| < r\}$ and any $\delta > 0$, the solution of system (4.1) satisfies

$$\lim_{\varepsilon \to 0} \inf\{t \geq 0 : |X_t^\varepsilon| > c|x|e^{-\gamma t} + \delta\} = +\infty \quad \text{a.s.} \tag{4.17}$$

In Definitions 4.1 and 4.2, we use the term "weakly" because the properties in question involve $\lim_{\varepsilon \to 0}$ and are defined through the first exit time from a set. In [58], stability concepts that are similarly defined under random perturbations are introduced for a nonlinear system perturbed by a stochastic process. In this chapter, the system perturbation also comes from a small parameter ε.

4.1.2 Statements of General Results on Stochastic Averaging

Lemma 4.1 *Consider system* (4.1) *under Assumptions* 4.1, 4.2, 4.3, *and* 4.4. *Then for any* $T > 0$,

$$\lim_{\varepsilon \to 0} \sup_{0 \leq t \leq T} |X_t^\varepsilon - \bar{X}_t| = 0 \quad \text{a.s.} \tag{4.18}$$

This result extends the stochastic averaging principle for globally Lipschitz systems [86] to locally Lipschitz systems. The result (4.18) means that $\sup_{0 \leq t \leq T} |X_t^\varepsilon - \bar{X}_t|$ converges to 0 almost surely as $\varepsilon \to 0$, and thus it converges to 0 in probability as $\varepsilon \to 0$, i.e., for any $\delta > 0$,

$$\lim_{\varepsilon \to 0} P\left\{ \sup_{0 \leq t \leq T} |X_t^\varepsilon - \bar{X}_t| > \delta \right\} = 0, \tag{4.19}$$

which is a finite-time stochastic averaging result in [39] for globally Lipschitz systems. Here we obtain a stronger result (4.18) for locally Lipschitz systems by using ergodic perturbation process but assuming the existence and uniqueness of the solution.

Let ρ be the metric on the space $\mathbb{C}([0, \infty), \mathbb{R}^n)$ of all the continuous vector functions $f, g \in \mathbb{C}([0, \infty), \mathbb{R}^n)$, defined as

$$\rho(f, g) = \sum_{k=1}^{\infty} \frac{1}{2^k} \left(1 \wedge \left(\sup_{0 \leq t \leq k} |f(t) - g(t)| \right) \right). \tag{4.20}$$

Suppose that the conditions of Lemma 4.1 hold, and denote $X^\varepsilon(\omega) = (X_t^\varepsilon(\omega), t \geq 0)$, $\bar{X} = (\bar{X}_t, t \geq 0)$. Then, by (4.18), we have

$$\lim_{\varepsilon \to 0} \rho(X^\varepsilon(\omega), \bar{X}) = 0 \quad \text{a.s.,} \tag{4.21}$$

i.e., X^ε converges almost surely to $\bar{X}$ as $\varepsilon \to 0$. By [77], X^ε also converges weakly to $\bar{X}$ as $\varepsilon \to 0$.

Next, we extend the finite-time approximation result in Lemma 4.1 to arbitrarily long time intervals.

Theorem 4.1 *Consider system* (4.1) *under Assumptions* 4.1, 4.2, 4.3, *and* 4.4. *Then*

(i) *For any* $\delta > 0$,

$$\lim_{\varepsilon \to 0} \inf\{t \geq 0 : |X_t^\varepsilon - \bar{X}_t| > \delta\} = +\infty \quad a.s.; \tag{4.22}$$

(ii) *There exists a function* $T(\varepsilon) : (0, \varepsilon_0) \to \mathbb{N}$ *such that, for any* $\delta > 0$,

$$\lim_{\varepsilon \to 0} P\left\{ \sup_{0 \leq t \leq T(\varepsilon)} |X_t^\varepsilon - \bar{X}_t| > \delta \right\} = 0, \tag{4.23}$$

where

$$\lim_{\varepsilon \to 0} T(\varepsilon) = +\infty. \tag{4.24}$$

This is an "approximation theorem" of stochastic averaging for locally Lipschitz systems: as ε tends to zero, the solutions to the original and average systems will remain δ-close for arbitrarily long time in the sense of both almost surely (4.22) and in probability (4.23). Based on this result, we investigate the solution property of the original system (4.1) under the stability of the average system (4.11).

Theorem 4.2 *Consider system* (4.1) *under Assumptions* 4.1, 4.2, 4.3, *and* 4.4. *Then*

(i) *(Boundedness) If the solution of the average system* (4.11) *with the initial condition* $\bar{X}_0 = x$ *is bounded, then the solution of system* (4.1) *with* $X_0^\varepsilon = x$ *is weakly bounded, more precisely, for any* $c > 0$,

$$\lim_{\varepsilon \to 0} \inf\{t \geq 0 : |X_t^\varepsilon| > M + c\} = +\infty \quad a.s., \tag{4.25}$$

where $M = \sup_{t \geq 0} |\bar{X}_t| < +\infty$.

(ii) *(Attractivity) If the solution of the average system* (4.11) *with the initial condition* $\bar{X}_0 = x$ *converges to* $x^* \in \mathbb{R}^n$, *i.e.,* $\lim_{t \to \infty} \bar{X}_t = x^*$, *then for the system* (4.1) *with* $X_0^\varepsilon = x$, *whose solution is* X_t^ε, *the point* x^* *is weakly attractive, i.e., for any* $\delta > 0$, *there exists a constant* $T_\delta > 0$ *such that the solution of system* (4.1) *satisfies*

$$\lim_{\varepsilon \to 0} \inf\{t \geq T_\delta : |X_t^\varepsilon - x^*| > \delta\} = +\infty \quad a.s. \tag{4.26}$$

(iii) *(Stability) If the equilibrium* $\bar{X}_t \equiv 0$ *of the average system* (4.11) *is stable, then it is weakly stable under random perturbation* $R(\cdot, Y_{t/\varepsilon})$, *i.e., for any* $\delta > 0$, *there exists a constant* $r_\delta > 0$ *such that for any initial condition* $x \in \{\check{x} \in \mathbb{R}^n : |\check{x}| < r_\delta\}$, *the solution of system* (4.1) *satisfies*

$$\lim_{\varepsilon \to 0} \inf\{t \geq 0 : |X_t^\varepsilon| > \delta\} = +\infty \quad a.s. \tag{4.27}$$

(iv) *(Asymptotic stability) If the equilibrium* $\bar{X}_t \equiv 0$ *of the average system* (4.11) *is asymptotically stable, then it is weakly asymptotically stable under random perturbation* $R(\cdot, Y_{t/\varepsilon})$, *i.e., for any* $\delta > 0$, *there exists a constant* $r_\delta > 0$, *such that, for any initial condition* $x \in \{\check{x} \in \mathbb{R}^n : |\check{x}| < r_\delta\}$, *the solution of system* (4.1) *satisfies*

$$\lim_{\varepsilon \to 0} \inf\{t \geq 0 : |X_t^\varepsilon| > \delta\} = +\infty \quad a.s., \tag{4.28}$$

and moreover, for any $0 < c < \delta$, *there exists a constant* $T_\delta^c > 0$ *such that*

$$\lim_{\varepsilon \to 0} \inf\{t \geq T_\delta^c : |X_t^\varepsilon| > c\} = +\infty \quad a.s. \tag{4.29}$$

(v) (*Exponential stability*) *If the equilibrium* $\bar{X}_t \equiv 0$ *of the average system* (4.11) *is exponentially stable, then it is weakly exponentially stable under random perturbation* $R(\cdot, Y_{t/\varepsilon})$, *i.e., there exist constants* $r > 0$, $c > 0$, *and* $\gamma > 0$ *such that, for any initial condition* $x \in \{\check{x} \in \mathbb{R}^n : |\check{x}| < r\}$ *and any* $\delta > 0$, *the solution of system* (4.1) *satisfies*

$$\lim_{\varepsilon \to 0} \inf\{t \geq 0 : |X_t^\varepsilon| > c|x|e^{-\gamma t} + \delta\} = +\infty \quad a.s. \tag{4.30}$$

Moreover, there exists a function $T(\varepsilon) : (0, \varepsilon_0) \to \mathbb{N}$ *such that under the conditions of* (i)–(v), *the respective results* (4.25)–(4.30) *can be replaced by*

(i) *The boundedness result*

$$\lim_{\varepsilon \to 0} P\left\{ \sup_{0 \leq t \leq T(\varepsilon)} |X_t^\varepsilon| > M + c \right\} = 0; \tag{4.31}$$

(ii) *The attractivity result*

$$\lim_{\varepsilon \to 0} P\left\{ \sup_{T_\delta \leq t \leq T(\varepsilon)} |X_t^\varepsilon - x^*| > \delta \right\} = 0; \tag{4.32}$$

(iii) *The stability result*

$$\lim_{\varepsilon \to 0} P\left\{ \sup_{0 \leq t \leq T(\varepsilon)} |X_t^\varepsilon| > \delta \right\} = 0; \tag{4.33}$$

(iv) *The asymptotic stability result*

$$(4.33) \quad and \quad \lim_{\varepsilon \to 0} P\left\{ \sup_{T_\delta^c \leq t \leq T(\varepsilon)} |X_t^\varepsilon| > c \right\} = 0; \tag{4.34}$$

(v) *The exponential stability result*

$$\lim_{\varepsilon \to 0} P\left\{ \sup_{0 \leq t \leq T(\varepsilon)} \{|X_t^\varepsilon| - c|x|e^{-\gamma t}\} > \delta \right\} = 0. \tag{4.35}$$

Furthermore, (4.35) *is equivalent to*

$$\lim_{\varepsilon \to 0} P\{|X_t^\varepsilon| \leq c|x|e^{-\gamma t} + \delta, \forall t \in [0, T(\varepsilon)]\} = 1.$$

According to the approximation result (4.22), we obtain the almost sure stabilities: (4.25)–(4.30) in Theorem 4.2, while by the approximation result (4.23), we obtain the stabilities in probabilities: (4.31)–(4.35) in Theorem 4.2. It should be pointed out that the two approximation results, (4.22) and (4.23), together with the corresponding two kinds of stability results in Theorem 4.2, are independent, but to make the content more compact, we combine them in one theorem.

The stability results in Theorem 4.2 are weaker than the stability in probability results in Chap. 3, where stronger conditions, not satisfied in stochastic extremum seeking applications, are imposed. Compared with other results on stochastic averaging on the infinite time interval [21, 65, 68], we remove or weaken the following restrictions: global Lipschitzness of the nonlinear vector field, equilibrium condition, global exponential stability of the average system, and compactness of the state space of the perturbation process, but impose the assumption of the existence and uniqueness of the solution of the original system.

4.2 Proofs of the General Theorems on Stochastic Averaging

4.2.1 Proof of Lemma 4.1

Fix $T > 0$, and denote

$$M' = \sup_{0 \le t \le T} |\bar{X}_t|. \tag{4.36}$$

Since $(\bar{X}_t, t \ge 0)$ is continuous and $[0, T]$ is a compact set, we have that $M' < +\infty$. Denote $M = M' + 1$. For any $\varepsilon \in (0, \varepsilon_0)$, define a stopping time τ_ε by

$$\tau_\varepsilon = \inf\{t \ge 0 : |X_t^\varepsilon| > M\}. \tag{4.37}$$

By the definition of M (noting that $|x| = |\bar{X}_0| \le M'$) and the continuity of the sample path of $(X_t^\varepsilon, t \ge 0)$, we know that $0 < \tau_\varepsilon \le +\infty$, and if $\tau_\varepsilon < +\infty$, then

$$|X_{\tau_\varepsilon}^\varepsilon| = M. \tag{4.38}$$

From (4.1) and (4.11), we have that, for any $t \ge 0$,

$$\begin{aligned}
X_t^\varepsilon - \bar{X}_t &= \int_0^t \left[a(X_s^\varepsilon, Y_{s/\varepsilon}) - \bar{a}(\bar{X}_s) \right] ds \\
&= \int_0^t \left[a(X_s^\varepsilon, Y_{s/\varepsilon}) - a(\bar{X}_s, Y_{s/\varepsilon}) \right] ds \\
&\quad + \int_0^t \left[a(\bar{X}_s, Y_{s/\varepsilon}) - \bar{a}(\bar{X}_s) \right] ds.
\end{aligned} \tag{4.39}$$

By Assumption 4.1, we obtain that, for any $s \le \tau_\varepsilon \wedge T$,

$$\left| a(X_s^\varepsilon, Y_{s/\varepsilon}) - a(\bar{X}_s, Y_{s/\varepsilon}) \right| \le k_M |X_s^\varepsilon - \bar{X}_s|, \tag{4.40}$$

where k_M is the Lipschitz constant of $a(x, y)$ with respect to the compact subset $\{x \in \mathbb{R}^n : |x| \le M\}$ of $\mathbb{R}^n$.

Thus by (4.39) and (4.40), we have that if $t \le \tau_\varepsilon \wedge T$, then

$$|X_t^\varepsilon - \bar{X}_t| \le k_M \int_0^t |X_s^\varepsilon - \bar{X}_s| ds + \left| \int_0^t \left[a(\bar{X}_s, Y_{s/\varepsilon}) - \bar{a}(\bar{X}_s) \right] ds \right|. \tag{4.41}$$

Define

$$\Delta_t^\varepsilon = \left| X_t^\varepsilon - \bar{X}_t \right|, \tag{4.42}$$

$$\alpha(\varepsilon) = \sup_{0 \le t \le T} \left| \int_0^t \left[a(\bar{X}_s, Y_{s/\varepsilon}) - \bar{a}(\bar{X}_s) \right] ds \right|. \tag{4.43}$$

Then by (4.41) and Gronwall's inequality, we have

$$\sup_{0 \le t \le \tau_\varepsilon \wedge T} \Delta_t^\varepsilon \le \alpha(\varepsilon) e^{k_M(\tau_\varepsilon \wedge T)} \le \alpha(\varepsilon) e^{k_M T}. \tag{4.44}$$

Since $(\bar{X}_t, t \ge 0)$ is a deterministic continuous function, by Assumption 4.1 and Birkhoff ergodic theorem (see, e.g., Liptser and Shiryaev [85]), we have that

$$\lim_{\varepsilon \to 0} \alpha(\varepsilon) = 0 \quad \text{a.s.} \tag{4.45}$$

For the reader's convenience, we give the detailed proof of (4.45) in Sect. 4.2.6.
    It follows from (4.42), (4.44), and (4.45) that

$$\limsup_{\varepsilon \to 0} \sup_{0 \le t \le \tau_\varepsilon \wedge T} \left| X_t^\varepsilon - \bar{X}_t \right| = 0 \quad \text{a.s.} \tag{4.46}$$

Thus by (4.36) and (4.46), we have

$$\begin{aligned}
\limsup_{\varepsilon \to 0} \sup_{0 \le t \le \tau_\varepsilon \wedge T} \left| X_t^\varepsilon \right| &\le \limsup_{\varepsilon \to 0} \left[\sup_{0 \le t \le \tau_\varepsilon \wedge T} \left| X_t^\varepsilon - \bar{X}_t \right| + \sup_{0 \le t \le \tau_\varepsilon \wedge T} \left| \bar{X}_t \right| \right] \\
&\le \limsup_{\varepsilon \to 0} \sup_{0 \le t \le \tau_\varepsilon \wedge T} \left| X_t^\varepsilon - \bar{X}_t \right| + M' \\
&= M' < M \quad \text{a.s.}
\end{aligned} \tag{4.47}$$

By (4.38) and (4.47), we obtain that, for almost every $\omega \in \Omega$, there exists an $\varepsilon_0(\omega) > 0$ such that, for any $0 < \varepsilon < \varepsilon_0(\omega)$,

$$\tau_\varepsilon(\omega) > T. \tag{4.48}$$

Thus by (4.46) and (4.48), we obtain that

$$\limsup_{\varepsilon \to 0} \sup_{0 \le t \le T} \left| X_t^\varepsilon - \bar{X}_t \right| = 0 \quad \text{a.s.} \tag{4.49}$$

Hence (4.18) holds. The proof is completed.

4.2.2 Proof of Approximation Result (4.22) of Theorem 4.1

Define

$$\Omega' = \left\{ \omega : \limsup_{\varepsilon \to 0} \sup_{0 \le t \le T} \left| X_t^\varepsilon(\omega) - \bar{X}_t \right| = 0, \forall T \in \mathbb{N} \right\}. \tag{4.50}$$

Then by Lemma 4.1, we have

$$P\left(\Omega' \right) = 1. \tag{4.51}$$

Let $\delta > 0$. For $\varepsilon \in (0, \varepsilon_0)$, define a stopping time τ_ε^δ by

$$\tau_\varepsilon^\delta = \inf\{t \geq 0 : |X_t^\varepsilon - \bar{X}_t| > \delta\}. \tag{4.52}$$

By the fact that $X_0^\varepsilon - \bar{X}_0 = 0$, and the continuity of the sample paths of $(X_t^\varepsilon, t \geq 0)$ and $(\bar{X}_t, t \geq 0)$, we know that $0 < \tau_\varepsilon^\delta \leq +\infty$, and if $\tau_\varepsilon^\delta < +\infty$, then

$$\left| X_{\tau_\varepsilon^\delta}^\varepsilon - \bar{X}_{\tau_\varepsilon^\delta} \right| = \delta. \tag{4.53}$$

For any $\omega \in \Omega'$, by (4.50) and (4.53), we get that for any $T \in \mathbb{N}$, there exists $\varepsilon_0(\omega, \delta, T) > 0$ such that, for any $0 < \varepsilon < \varepsilon_0(\omega, \delta, T)$,

$$\tau_\varepsilon^\delta(\omega) > T, \tag{4.54}$$

which implies that

$$\lim_{\varepsilon \to 0} \tau_\varepsilon^\delta(\omega) = +\infty. \tag{4.55}$$

Thus it follows from (4.51) and (4.55) that

$$\lim_{\varepsilon \to 0} \tau_\varepsilon^\delta = +\infty \quad \text{a.s.} \tag{4.56}$$

The proof is completed.

4.2.3 Preliminary Lemmas for the Proof of Approximation Result (4.23) of Theorem 4.1

Lemma 4.2 *Consider system (4.1) under Assumptions 4.1, 4.2, 4.3, and 4.4. Then for any $\delta > 0$, $0 < \check{\delta} < 1$, there exists a decreasing sequence $\{\varepsilon_T\}_{T \in \mathbb{N}}$ of positive real numbers satisfying $\varepsilon_T \downarrow 0$ as $T \to \infty$ such that*

$$P\left(\bigcap_{T=1}^{\infty} \bigcap_{\varepsilon \in (0, \varepsilon_T]} \left\{ \sup_{0 \leq t \leq T} |X_t^\varepsilon - \bar{X}_t| \leq \delta \right\} \right) > 1 - \check{\delta}, \tag{4.57}$$

or equivalently,

$$P\left\{ \sup_{T \in \mathbb{N}} \sup_{0 < \varepsilon \leq \varepsilon_T} \sup_{0 \leq t \leq T} |X_t^\varepsilon - \bar{X}_t| > \delta \right\} < \check{\delta}. \tag{4.58}$$

Proof Let τ_ε^δ be defined by (4.52). Since

$$\left\{ \lim_{\varepsilon \to 0} \tau_\varepsilon^\delta = +\infty \right\} = \bigcap_{T=1}^{+\infty} \bigcup_{\check{\varepsilon} \in (0, \varepsilon_0)} \bigcap_{\varepsilon \in (0, \check{\varepsilon}]} \{ \tau_\varepsilon^\delta \geq T \}, \tag{4.59}$$

by Theorem 4.1, we have

$$P\left(\bigcup_{T=1}^{+\infty} \bigcap_{\check{\varepsilon} \in (0, \varepsilon_0)} \bigcup_{\varepsilon \in (0, \check{\varepsilon}]} \{ \tau_\varepsilon^\delta < T \} \right) = 0. \tag{4.60}$$

We show that the set $\bigcup_{\varepsilon \in (0,\check{\varepsilon}]} \{\tau_\varepsilon^\delta < T\}$ is measurable. Let Q denote the set of all rational numbers. Then by the definition of τ_ε^δ, and the continuity of X_t^ε and $\bar{X}_t$ with respect to ε and t, we have

$$\bigcup_{\varepsilon \in (0,\check{\varepsilon}]} \{\tau_\varepsilon^\delta < T\} = \{\exists \varepsilon \in (0,\check{\varepsilon}] \text{ s.t. } \tau_\varepsilon^\delta < T\}$$

$$= \{\exists \varepsilon \in (0,\check{\varepsilon}], \exists t \in [0, T) \text{ s.t. } |X_t^\varepsilon - \bar{X}_t| > \delta\}$$

$$= \{\exists \varepsilon \in (0,\check{\varepsilon}] \cap Q, \exists t \in [0, T) \cap Q \text{ s.t. } |X_t^\varepsilon - \bar{X}_t| > \delta\}$$

$$= \bigcup_{(0,\check{\varepsilon}] \cap Q} \bigcup_{[0,T) \cap Q} \{|X_t^\varepsilon - \bar{X}_t| > \delta\}, \tag{4.61}$$

which is measurable. Since the set $\bigcup_{\varepsilon \in (0,\check{\varepsilon}]} \{\tau_\varepsilon^\delta < T\}$ is increasing relative to $\check{\varepsilon}$, we have

$$\bigcap_{\check{\varepsilon} \in (0,\varepsilon_0)} \bigcup_{\varepsilon \in (0,\check{\varepsilon}]} \{\tau_\varepsilon^\delta < T\} = \bigcap_{\check{\varepsilon} \in (0,\varepsilon_0) \cap Q} \bigcup_{\varepsilon \in (0,\check{\varepsilon}]} \{\tau_\varepsilon^\delta < T\}, \tag{4.62}$$

and hence the set $\bigcap_{\check{\varepsilon} \in (0,\varepsilon_0)} \bigcup_{\varepsilon \in (0,\check{\varepsilon}]} \{\tau_\varepsilon^\delta < T\}$ is also measurable. Thus by (4.60), we obtain that, for any $T \in \mathbb{N}$,

$$P\left(\bigcap_{\check{\varepsilon} \in (0,\varepsilon_0)} \bigcup_{\varepsilon \in (0,\check{\varepsilon}]} \{\tau_\varepsilon^\delta < T\} \right) = 0, \tag{4.63}$$

which implies that, for any $T \in \mathbb{N}$,

$$\lim_{\check{\varepsilon} \to 0} P\left(\bigcup_{\varepsilon \in (0,\check{\varepsilon}]} \{\tau_\varepsilon^\delta < T\} \right) = 0, \tag{4.64}$$

and thus there exists $\varepsilon_T \in (0, \varepsilon_0)$ (without loss of generality, we assume that ε_T decreases to 0 as $T \to \infty$) such that

$$P\left(\bigcup_{\varepsilon \in (0,\varepsilon_T]} \{\tau_\varepsilon^\delta < T\} \right) < \frac{\check{\delta}}{2^T}. \tag{4.65}$$

Define

$$N = \bigcup_{T=1}^{+\infty} \bigcup_{\varepsilon \in (0,\varepsilon_T]} \{\tau_\varepsilon^\delta < T\}. \tag{4.66}$$

Then by (4.65), we have

$$P(N) < \check{\delta}, \tag{4.67}$$

and thus $P(N^c) > 1 - \check{\delta}$, where

$$N^c = \bigcap_{T=1}^{+\infty} \bigcap_{\varepsilon \in (0,\varepsilon_T]} \{\tau_\varepsilon^\delta \geq T\}. \tag{4.68}$$

By the definition of τ_ε^δ, we have

$$\left\{ \sup_{0 \leq t \leq T} \left| X_t^\varepsilon - \bar{X}_t \right| \leq \delta \right\} \supseteq \left\{ \tau_\varepsilon^\delta \geq T \right\}. \tag{4.69}$$

Hence (4.57) holds. The proof is completed.   □

Lemma 4.3 *Consider system* (4.1) *under Assumptions* 4.1, 4.2, *and* 4.3. *Then for any* $\delta > 0$, *there exists a function* $T_\delta(\varepsilon) : (0, \varepsilon_0) \to \mathbb{N}$ *such that*

$$\lim_{\varepsilon \to 0} P \left\{ \sup_{0 \leq t \leq T_\delta(\varepsilon)} \left| X_t^\varepsilon - \bar{X}_t \right| > \delta \right\} = 0 \tag{4.70}$$

and

$$\lim_{\varepsilon \to 0} T_\delta(\varepsilon) = +\infty. \tag{4.71}$$

Proof For $\delta > 0$, $0 < \check{\delta} < 1$, we use $\varepsilon_T(\delta, \check{\delta})$ instead of ε_T in Lemma 4.2. Now fix $\delta > 0$. For any $k = 2, 3, \ldots$, by Lemma 4.2 we obtain a decreasing sequence $\{\varepsilon_T(\delta, \frac{1}{k})\}_{T \in \mathbb{N}}$ of positive real numbers, $\varepsilon_T(\delta, \frac{1}{k}) \downarrow 0$ as $T \to \infty$ such that

$$P \left\{ \sup_{T \in \mathbb{N}} \sup_{0 < \varepsilon \leq \varepsilon_T(\delta, 1/k)} \sup_{0 \leq t \leq T} \left| X_t^\varepsilon - \bar{X}_t \right| > \delta \right\} < \frac{1}{k}. \tag{4.72}$$

By the proof of Lemma 4.2, we assume that for any $T \in \mathbb{N}$, $\varepsilon_T(\delta, \frac{1}{k})$ is a nonincreasing function of k, and thus for any $k = 2, 3, \ldots$,

$$0 < \varepsilon_{k+1}\left(\delta, \frac{1}{k+1}\right) < \varepsilon_k\left(\delta, \frac{1}{k+1}\right)$$

$$\leq \varepsilon_k\left(\delta, \frac{1}{k}\right) \leq \varepsilon_k\left(\delta, \frac{1}{2}\right). \tag{4.73}$$

It follows from (4.73) and $\lim_{k \to +\infty} \varepsilon_k(\delta, \frac{1}{2}) = 0$ that

$$\varepsilon_k\left(\delta, \frac{1}{k}\right) \downarrow 0 \quad \text{as } k \to +\infty. \tag{4.74}$$

Now we define the desired function $T_\delta(\varepsilon)$ as follows:

$$T_\delta(\varepsilon) := \begin{cases} 1 & \text{if } \varepsilon \in (\varepsilon_2(\delta, \frac{1}{2}), \varepsilon_0), \\ k & \text{if } \varepsilon \in (\varepsilon_{k+1}(\delta, \frac{1}{k+1}), \varepsilon_k(\delta, \frac{1}{k})], \ k = 2, 3, \ldots. \end{cases} \tag{4.75}$$

Then for any $k = 2, 3, \ldots$, by (4.72) and (4.75), we get that

$$\sup_{\varepsilon_{k+1}(\delta, 1/(k+1)) < \varepsilon \leq \varepsilon_k(\delta, 1/k)} P \left\{ \sup_{0 \leq t \leq T_\delta(\varepsilon)} \left| X_t^\varepsilon - \bar{X}_t \right| > \delta \right\} \leq \frac{1}{k}, \tag{4.76}$$

and for $j = k + 1, k + 2, \ldots$, we have

$$\sup_{\varepsilon_{j+1}(\delta, 1/(j+1)) < \varepsilon \leq \varepsilon_j(\delta, 1/j)} P \left\{ \sup_{0 \leq t \leq T_\delta(\varepsilon)} \left| X_t^\varepsilon - \bar{X}_t \right| > \delta \right\} \leq \frac{1}{j} < \frac{1}{k}. \tag{4.77}$$

By (4.74), (4.76), and (4.77), we get that for any $k = 2, 3, \ldots,$

$$\sup_{0 < \varepsilon \leq \varepsilon_k(\delta, 1/k)} P\left\{ \sup_{0 \leq t \leq T_\delta(\varepsilon)} \left| X_t^\varepsilon - \bar{X}_t \right| > \delta \right\} \leq \frac{1}{k}, \tag{4.78}$$

which implies (4.70). By (4.74) and (4.75), we obtain (4.71). The proof is completed. □

4.2.4 Proof of Approximation Result (4.23) of Theorem 4.1

For $k = 1, 2, \ldots,$ by Lemma 4.3, there exists a function $T_{1/k}(\varepsilon) : (0, \varepsilon_0) \to \mathbb{N}$ such that

$$\lim_{\varepsilon \to 0} P\left\{ \sup_{0 \leq t \leq T_{1/k}(\varepsilon)} \left| X_t^\varepsilon - \bar{X}_t \right| > \frac{1}{k} \right\} = 0, \tag{4.79}$$

and

$$\lim_{\varepsilon \to 0} T_{1/k}(\varepsilon) = +\infty. \tag{4.80}$$

Without loss of generality, we assume that for any $k \in \mathbb{N}$, we have

$$T_{1/(k+1)}(\varepsilon) \leq T_{1/k}(\varepsilon) \quad \forall \varepsilon \in (0, \varepsilon_0). \tag{4.81}$$

In fact, we can replace the function $T_{1/(k+1)}(\varepsilon)$ by $T_{1/(k+1)}(\varepsilon) \wedge T_{1/k}(\varepsilon)$. Let $\varepsilon_1 = 1$. For $k = 2, 3, \ldots,$ define

$$\varepsilon_k := \sup\left\{ \varepsilon \in (0, \varepsilon_{k-1}) : T_{1/k}(\varepsilon) = k \right\}. \tag{4.82}$$

Now we define the desired function $T(\varepsilon) : (0, \varepsilon_0) \to \mathbb{N}$ as follows:

$$T(\varepsilon) = \begin{cases} T_1(\varepsilon) & \text{if } \varepsilon \in (\varepsilon_2 \wedge \frac{1}{2}, \varepsilon_0), \\ T_{1/k}(\varepsilon) & \text{if } \varepsilon \in (\varepsilon_{k+1} \wedge \frac{1}{k+1}, \varepsilon_k \wedge \frac{1}{k}], k = 2, 3, \ldots. \end{cases} \tag{4.83}$$

Since $\lim_{k \to \infty} \varepsilon_k \wedge \frac{1}{k} = 0$, the function $T(\varepsilon)$ is defined on $(0, \varepsilon_0)$. By (4.82) and the definition of $T_{1/k}(\varepsilon)$ ($k \in \mathbb{N}$) stated in the proof of Lemma 4.3 ($T_{1/k}(\varepsilon)$ is increasing when ε decreases to 0), we have that for any $0 < \varepsilon \leq \varepsilon_k \wedge \frac{1}{k}$,

$$T(\varepsilon) \geq k, \tag{4.84}$$

and thus (4.24) holds.

Next, we prove (4.23). For any $\delta > 0$, take $\check{k} \in \mathbb{N}$ such that $\frac{1}{\check{k}} \leq \delta$. Then for $j = \check{k}, \check{k}+1, \check{k}+2, \ldots,$ by (4.81) and (4.83), we get that

$$\sup_{\varepsilon \in (\varepsilon_{j+1} \wedge 1/(j+1), \varepsilon_j \wedge 1/j]} P\left\{ \sup_{0 \leq t \leq T(\varepsilon)} \left| X_t^\varepsilon - \bar{X}_t \right| > \delta \right\}$$

$$= \sup_{\varepsilon \in (\varepsilon_{j+1} \wedge 1/(j+1), \varepsilon_j \wedge 1/j]} P\left\{ \sup_{0 \leq t \leq T_{1/j}(\varepsilon)} \left| X_t^\varepsilon - \bar{X}_t \right| > \delta \right\}$$

$$\leq \sup_{\varepsilon \in (\varepsilon_{j+1} \wedge 1/(j+1), \varepsilon_j \wedge 1/j]} P\left\{ \sup_{0 \leq t \leq T_{1/\check{k}}(\varepsilon)} \left| X_t^\varepsilon - \bar{X}_t \right| > \delta \right\}$$

$$\leq \sup_{\varepsilon \in (\varepsilon_{j+1} \wedge 1/(j+1), \varepsilon_j \wedge 1/j]} P\left\{ \sup_{0 \leq t \leq T_{1/\check{k}}(\varepsilon)} \left| X_t^\varepsilon - \bar{X}_t \right| > \frac{1}{k} \right\}$$

$$\leq \sup_{\varepsilon \in (0, \varepsilon_j \wedge 1/j]} P\left\{ \sup_{0 \leq t \leq T_{1/\check{k}}(\varepsilon)} \left| X_t^\varepsilon - \bar{X}_t \right| > \frac{1}{k} \right\}, \tag{4.85}$$

and thus for any $l = j + 1, j + 2, \ldots,$

$$\sup_{\varepsilon \in (\varepsilon_{l+1} \wedge 1/(l+1), \varepsilon_l \wedge 1/l]} P\left\{ \sup_{0 \leq t \leq T(\varepsilon)} \left| X_t^\varepsilon - \bar{X}_t \right| > \delta \right\}$$

$$\leq \sup_{\varepsilon \in (0, \varepsilon_l \wedge 1/l]} P\left\{ \sup_{0 \leq t \leq T_{1/\check{k}}(\varepsilon)} \left| X_t^\varepsilon - \bar{X}_t \right| > \frac{1}{k} \right\}$$

$$\leq \sup_{\varepsilon \in (0, \varepsilon_j \wedge 1/j]} P\left\{ \sup_{0 \leq t \leq T_{1/\check{k}}(\varepsilon)} \left| X_t^\varepsilon - \bar{X}_t \right| > \frac{1}{k} \right\}, \tag{4.86}$$

where in the second inequality of (4.86) we use the fact that $\varepsilon_l \wedge \frac{1}{l} \leq \varepsilon_j \wedge \frac{1}{j}$ for any $l = j + 1, j + 2, \ldots$. Hence by (4.85), (4.86), and the fact that $\lim_{k \to \infty} \varepsilon_k \wedge \frac{1}{k} = 0$, we obtain that, for $j = \check{k}, \check{k} + 1, \check{k} + 2, \ldots,$

$$\sup_{\varepsilon \in (0, \varepsilon_j \wedge 1/j]} P\left\{ \sup_{0 \leq t \leq T(\varepsilon)} \left| X_t^\varepsilon - \bar{X}_t \right| > \delta \right\}$$

$$\leq \sup_{\varepsilon \in (0, \varepsilon_j \wedge 1/j]} P\left\{ \sup_{0 \leq t \leq T_{1/\check{k}}(\varepsilon)} \left| X_t^\varepsilon - \bar{X}_t \right| > \frac{1}{k} \right\}. \tag{4.87}$$

By the fact that $\lim_{k \to \infty} \varepsilon_k \wedge \frac{1}{k} = 0$, (4.79), and (4.87), we obtain that, for any $\delta > 0$,

$$\lim_{\varepsilon \to 0} P\left\{ \sup_{0 \leq t \leq T(\varepsilon)} \left| X_t^\varepsilon - \bar{X}_t \right| > \delta \right\} = 0. \tag{4.88}$$

The proof is completed.

4.2.5 Proof of Theorem 4.2

(i) We prove boundedness. Notice that $M = \sup_{t \geq 0} |\bar{X}_t|$ and

$$\left\{ \left| X_t^\varepsilon \right| > M + c \right\} \subseteq \left\{ \left| X_t^\varepsilon - \bar{X}_t \right| > c \right\}. \tag{4.89}$$

Then by the continuity of the sample paths of $(X_t^\varepsilon, t \geq 0)$ (we don't mention this fact in the following proofs again), we have

$$\inf\left\{ t \geq 0 : \left| X_t^\varepsilon \right| > M + c \right\} \geq \inf\left\{ t \geq 0 : \left| X_t^\varepsilon - \bar{X}_t \right| > c \right\}. \tag{4.90}$$

Thus by Theorem 4.1, (4.25) holds.

(ii) We prove attractivity. Since $\lim_{t \to \infty} \bar{X}_t = x^*$, we have

$$\lim_{t \to \infty} \left| \bar{X}_t - x^* \right| = 0, \tag{4.91}$$

and thus for any $\delta > 0$, there exists a constant $T_\delta > 0$ such that

$$\sup_{t \geq T_\delta} \left| \bar{X}_t - x^* \right| < \frac{\delta}{2}, \tag{4.92}$$

by which we obtain that, for any $t \geq T_\delta$,

$$\left\{ \left| X_t^\varepsilon - x^* \right| > \delta \right\} = \left\{ \left| (X_t^\varepsilon - \bar{X}_t) + (\bar{X}_t - x^*) \right| > \delta \right\} \subseteq \left\{ \left| X_t^\varepsilon - \bar{X}_t \right| > \frac{\delta}{2} \right\}, \tag{4.93}$$

and thus

$$\inf\left\{ t \geq T_\delta : \left| X_t^\varepsilon - x^* \right| > \delta \right\} \geq \inf\left\{ t \geq T_\delta : \left| X_t^\varepsilon - \bar{X}_t \right| > \frac{\delta}{2} \right\}$$

$$\geq \inf\left\{ t \geq 0 : \left| X_t^\varepsilon - \bar{X}_t \right| > \frac{\delta}{2} \right\}, \tag{4.94}$$

which, together with Theorem 4.1, implies (4.26).

(iii) We prove stability. If $\bar{X}_t \equiv 0 \in \mathbb{R}^n$ is a stable equilibrium of the average system (4.11), then for any $\delta > 0$, there exists a constant $r_\delta > 0$ such that

$$\left| \bar{X}_0 \right| < r_\delta \quad \Rightarrow \quad \sup_{t \geq 0} \left| \bar{X}_t \right| < \frac{\delta}{2}, \tag{4.95}$$

which, together with Theorem 4.1, implies that, for $|x| < r_\delta$,

$$\lim_{\varepsilon \to 0} \inf\left\{ t \geq 0 : \left| X_t^\varepsilon \right| > \delta \right\}$$

$$= \lim_{\varepsilon \to 0} \inf\left\{ t \geq 0 : \left| (X_t^\varepsilon - \bar{X}_t) + \bar{X}_t \right| > \delta \right\}$$

$$\geq \lim_{\varepsilon \to 0} \inf\left\{ t \geq 0 : \left| X_t^\varepsilon - \bar{X}_t \right| > \frac{\delta}{2} \right\} = +\infty \quad \text{a.s.} \tag{4.96}$$

Hence (4.27) holds.

(iv) For asymptotic stability, the proof follows directly from (ii) and (iii) above.

(v) We prove exponential stability. Since the equilibrium $\bar{X}_t = 0$ of the average system is exponentially stable, there exist constants $r > 0$, $c > 0$, and $\gamma > 0$ such that, for any $|x| < r$,

$$\left| \bar{X}_t \right| < c |x| e^{-\gamma t} \quad \forall t > 0. \tag{4.97}$$

Thus for any $\delta > 0$, we have

$$\left\{ \left| X_t^\varepsilon \right| > c |x| e^{-\gamma t} + \delta \right\} \subseteq \left\{ \left| X_t^\varepsilon - \bar{X}_t \right| > \delta \right\}, \tag{4.98}$$

which, together with Theorem 4.1, implies that

$$\lim_{\varepsilon \to 0} \inf\left\{ t \geq 0 : \left| X_t^\varepsilon \right| > c |x| e^{-\gamma t} + \delta \right\}$$

$$\geq \lim_{\varepsilon \to 0} \inf\left\{ t \geq 0 : \left| X_t^\varepsilon - \bar{X}_t \right| > \delta \right\} = +\infty \quad \text{a.s.} \tag{4.99}$$

Hence (4.30) holds.

Let the function $T(\varepsilon)$ be defined in Theorem 4.1. Thus $\lim_{\varepsilon \to 0} T(\varepsilon) = +\infty$. For the stability results (4.31)–(4.35) with respect to the approximation result (4.23), we only prove (4.35). The proofs for (4.31)–(4.34) are similar.

Since the equilibrium $\bar{X}_t = 0$ of the average system is exponentially stable, there exist constants $r > 0$, $c > 0$, and $\gamma > 0$ such that, for any $|x| < r$,

$$|\bar{X}_t| < c|x|e^{-\gamma t} \quad \forall t > 0. \tag{4.100}$$

Thus for any $\delta > 0$, we have that, for any $|x| < r$,

$$\left\{ \sup_{0 \le t \le T(\varepsilon)} \left\{ |X_t^{\varepsilon}| - c|x|e^{-\gamma t} \right\} > \delta \right\}$$

$$= \bigcup_{0 \le t \le T(\varepsilon)} \left\{ |X_t^{\varepsilon}| - c|x|e^{-\gamma t} > \delta \right\}$$

$$\subseteq \bigcup_{0 \le t \le T(\varepsilon)} \left\{ |X_t^{\varepsilon} - \bar{X}_t| > \delta \right\}$$

$$= \left\{ \sup_{0 \le t \le T(\varepsilon)} |X_t^{\varepsilon} - \bar{X}_t| > \delta \right\}, \tag{4.101}$$

which, together with result (4.23) of Theorem 4.1, gives that

$$\limsup_{\varepsilon \to 0} P \left\{ \sup_{0 \le t \le T(\varepsilon)} \left\{ |X_t^{\varepsilon}| - c|x|e^{-\gamma t} \right\} > \delta \right\}$$

$$\le \lim_{\varepsilon \to 0} P \left\{ \sup_{0 \le t \le T(\varepsilon)} |X_t^{\varepsilon} - \bar{X}_t| > \delta \right\} = 0. \tag{4.102}$$

Hence (4.35) holds. The whole proof is completed.

4.2.6 Proof of (4.45)

We give a detailed proof of (4.45), i.e.,

$$\lim_{\varepsilon \to 0} \sup_{0 \le t \le T} \left| \int_0^t \left(a(\bar{X}_s, Y_{s/\varepsilon}) - \bar{a}(\bar{X}_s) \right) ds \right| = 0 \quad \text{a.s.} \tag{4.103}$$

Proof We follow the proof of Theorem 5 of Chap. 3 of [130] for the globally Lipschitz case. Notice that

$$M' = \sup_{0 \le t \le T} |\bar{X}_t|, \quad M = M' + 1, \tag{4.104}$$

and k_M is the Lipschitz constant of $a(x, y)$ with respect to the compact subset $D_M := \{x \in \mathbb{R}^n : |x| \le M\}$ of $\mathbb{R}^n$, i.e., for any $x, \check{x} \in D_M$ and any $y \in S_Y$ (see Assumption 4.1),

$$\left| a(x, y) - a(\check{x}, y) \right| \le k_M |x - \check{x}|. \tag{4.105}$$

Then by (4.12) and (4.105), we have that, for any $x, \check{x} \in D_M$,

$$\left|\bar{a}(x) - \bar{a}(\check{x})\right| \le k_M |x - \check{x}|. \tag{4.106}$$

For any $n \in \mathbb{N}$, define a function $\bar{X}_s^n$, $s \ge 0$, by

$$\bar{X}_s^n = \sum_{k=0}^{\infty} \bar{X}_{k/n} I_{\{k/n \le s < (k+1)/n\}}. \tag{4.107}$$

Then for any $n \in \mathbb{N}$, we have

$$\sup_{0 \le s \le T} \left|\bar{X}_s^n\right| \le \sup_{0 \le s \le T} |\bar{X}_s| \le M' < M. \tag{4.108}$$

By (4.105)–(4.108), we obtain that

$$\sup_{0 \le t \le T} \left| \int_0^t \left(a(\bar{X}_s, Y_{s/\varepsilon}) - \bar{a}(\bar{X}_s) \right) ds \right|$$

$$= \sup_{0 \le t \le T} \left| \int_0^t \left[\left(a(\bar{X}_s, Y_{s/\varepsilon}) - a\left(\bar{X}_s^n, Y_{s/\varepsilon}\right) \right) \right. \right.$$

$$\left. \left. + \left(a\left(\bar{X}_s^n, Y_{s/\varepsilon}\right) - \bar{a}\left(\bar{X}_s^n\right) \right) + \left(\bar{a}\left(\bar{X}_s^n\right) - \bar{a}(\bar{X}_s) \right) \right] ds \right|$$

$$\le \sup_{0 \le t \le T} \int_0^t \left| a(\bar{X}_s, Y_{s/\varepsilon}) - a\left(\bar{X}_s^n, Y_{s/\varepsilon}\right) \right| ds$$

$$+ \sup_{0 \le t \le T} \left| \int_0^t \left(a\left(\bar{X}_s^n, Y_{s/\varepsilon}\right) - \bar{a}\left(\bar{X}_s^n\right) \right) ds \right|$$

$$+ \sup_{0 \le t \le T} \int_0^t \left| \bar{a}\left(\bar{X}_s^n\right) - \bar{a}(\bar{X}_s) \right| ds$$

$$\le 2k_M T \sup_{0 \le s \le T} \left| \bar{X}_s - \bar{X}_s^n \right|$$

$$+ \sup_{0 \le t \le T} \left| \int_0^t \left(a\left(\bar{X}_s^n, Y_{s/\varepsilon}\right) - \bar{a}\left(\bar{X}_s^n\right) \right) ds \right|. \tag{4.109}$$

Next, we focus on the second term on the right-hand side of (4.109). We have

$$\sup_{0 \le t \le T} \left| \int_0^t \left(a\left(\bar{X}_s^n, Y_{s/\varepsilon}\right) - \bar{a}\left(\bar{X}_s^n\right) \right) ds \right|$$

$$= \sup_{0 \le t \le T} \left| \int_0^t \left(a\left(\bar{X}_s^n, Y_{s/\varepsilon}\right) - \bar{a}\left(\bar{X}_s^n\right) \right) \sum_{k=0}^{\infty} I_{\{k/n \le s < (k+1)/n\}} ds \right|$$

$$= \sup_{0 \le t \le T} \left| \int_0^t \sum_{k=0}^{\infty} \left(a(\bar{X}_{k/n}, Y_{s/\varepsilon}) - \bar{a}(\bar{X}_{k/n}) \right) I_{\{k/n \le s < (k+1)/n\}} ds \right|$$

$$= \sup_{0 \le t \le T} \left| \sum_{k=0}^{n(\lfloor t \rfloor + 1)} \int_0^t \left(a(\bar{X}_{k/n}, Y_{s/\varepsilon}) - \bar{a}(\bar{X}_{k/n}) \right) I_{\{k/n \le s < (k+1)/n\}} ds \right|$$

$$= \sup_{0 \le t \le T} \left| \sum_{k=0}^{n([t]+1)} \int_{k/n \wedge t}^{(k+1)/n \wedge t} \left(a(\bar{X}_{k/n}, Y_{s/\varepsilon}) - \bar{a}(\bar{X}_{k/n}) \right) ds \right|$$

$$\le \sup_{0 \le t \le T} \sum_{k=0}^{n([t]+1)} \left| \int_{k/n \wedge t}^{(k+1)/n \wedge t} \left(a(\bar{X}_{k/n}, Y_{s/\varepsilon}) - \bar{a}(\bar{X}_{k/n}) \right) ds \right|, \qquad (4.110)$$

where $[t]$ is the largest integer not greater than t. For fixed n and k with $k \le n([T]+1)$, we have

$$\sup_{0 \le t \le T} \left| \int_{k/n \wedge t}^{(k+1)/n \wedge t} \left(a(\bar{X}_{k/n}, Y_{s/\varepsilon}) - \bar{a}(\bar{X}_{k/n}) \right) ds \right|$$

$$\le \sup_{0 \le t \le T} \left(\left| \int_0^{(k+1)/n \wedge t} \left(a(\bar{X}_{k/n}, Y_{s/\varepsilon}) - \bar{a}(\bar{X}_{k/n}) \right) ds \right| \right.$$

$$\left. + \left| \int_0^{k/n \wedge t} \left(a(\bar{X}_{k/n}, Y_{s/\varepsilon}) - \bar{a}(\bar{X}_{k/n}) \right) ds \right| \right)$$

$$\le 2 \sup_{0 \le t \le (k+1)/n} \left| \int_0^t \left(a(\bar{X}_{k/n}, Y_{s/\varepsilon}) - \bar{a}(\bar{X}_{k/n}) \right) ds \right|$$

$$= 2 \sup_{0 \le t \le (k+1)/n} \varepsilon \left| \int_0^{t/\varepsilon} \left(a(\bar{X}_{k/n}, Y_s) - \bar{a}(\bar{X}_{k/n}) \right) ds \right|. \qquad (4.111)$$

Then by the Birkhoff ergodic theorem and [86, Problem 5.3.2], we obtain that

$$\lim_{\varepsilon \to 0} \sup_{0 \le t \le (k+1)/n} \varepsilon \left| \int_0^{t/\varepsilon} \left(a(\bar{X}_{k/n}, Y_s) - \bar{a}(\bar{X}_{k/n}) \right) ds \right| = 0 \quad \text{a.s.,} \qquad (4.112)$$

which, together with (4.110) and (4.111), gives that, for any $n \in \mathbb{N}$,

$$\lim_{\varepsilon \to 0} \sup_{0 \le t \le T} \left| \int_0^t \left(a(\bar{X}_s^n, Y_{s/\varepsilon}) - \bar{a}(\bar{X}_s^n) \right) ds \right| = 0 \quad \text{a.s.} \qquad (4.113)$$

Thus by (4.109), (4.113), and

$$\lim_{n \to \infty} \sup_{0 \le s \le T} \left| \bar{X}_s - \bar{X}_s^n \right| = 0, \qquad (4.114)$$

we obtain

$$\lim_{\varepsilon \to 0} \sup_{0 \le t \le T} \left| \int_0^t \left(a(\bar{X}_s, Y_{s/\varepsilon}) - \bar{a}(\bar{X}_s) \right) ds \right| = 0 \quad \text{a.s.} \qquad (4.115)$$

The proof is completed.   □

4.3 Discussions of the Existence of Solution

In Sect. 4.1.2, to obtain the general stochastic averaging theorems, the existence of the solution of the original system is assumed. But in fact, owing to the close relationship of the original system and its average system, this condition can be removed when the solution of the average system has some good property.

Now, we consider a weaker condition on the original system (4.1):

Assumption 4.5 The vector field $a(x, y)$ is a continuous function of (x, y), and for any $x \in D$, it is a bounded function of y. Furthermore, $a(x, y)$ satisfies the local Lipschitz condition in $x \in D$ uniformly in $y \in S_Y$, i.e., for any compact subset $D_0 \subset D$, there is a constant k_{D_0} such that, for all $x', x'' \in D_0$ and all $y \in S_Y$, $|a(x', y) - a(x'', y)| \le k_{D_0}|x' - x''|$.

Before presenting the main results, we give two lemmas. To this end, for any point $x' \in D$, we define by $d(x', \partial D)$ the distance between x' and the boundary ∂D of the domain D, i.e.,

$$d(x', \partial D) = \inf\{|x' - y| : y \in \partial D\}. \tag{4.116}$$

By convention $d(x', \emptyset) = \infty$. Since D is a domain, for any $x' \in D$, we have that $d(x', \partial D) > 0$. If A is a subset of D, we define by $d(A, \partial D)$ the distance between A and ∂D as follows:

$$d(A, \partial D) = \inf_{x' \in A} d(x', \partial D) = \inf\{|x' - y| : x' \in A, y \in \partial y\}. \tag{4.117}$$

Throughout this section, we assume that $x \in D$, where x is the initial value of system (4.1). System (4.1) is a stochastic ordinary differential equation (stochastic ODE), and its solution can be defined for each sample path of the perturbation process $(Y_{t/\varepsilon} : t \ge 0)$. If system (4.1) satisfies Assumptions 4.5, then for any compact subset $D_0 \subset D$ and the constant k_{D_0} stated in Assumptions 4.5, it holds that for any $\omega \in \Omega$, any $t \ge 0$, any $\varepsilon \in (0, \varepsilon_0)$, and all $x', x'' \in D_0$,

$$\left|a(x', Y_{t/\varepsilon}(\omega)) - a(x'', Y_{t/\varepsilon}(\omega))\right| \le k_{D_0}|x' - x''|. \tag{4.118}$$

Thus by the theorem on local existence and uniqueness of solutions of nonlinear systems (see, e.g., Theorem 3.1 of [56]), for any $\varepsilon \in (0, \varepsilon_0)$ and any $\omega \in \Omega$, system (4.1) has a unique solution $X_t^\varepsilon(\omega)$ with the life time $l_\varepsilon(\omega) > 0$, where $l_\varepsilon(\omega) = \inf\{t \ge 0 : X_t^\varepsilon(\omega) \in \partial D\}$. For $t > l_\varepsilon(\omega)$, we define $X_t^\varepsilon(\omega) = X_{l_\varepsilon(\omega)}^\varepsilon(\omega)$, i.e., as soon as the solution reaches the boundary of the domain D, we fix it and maintain it at that constant value thereafter.

Lemma 4.4 *Consider system* (4.1) *under Assumptions* 4.5 *and* 4.2. *If* $d(\{\bar{X}_t, t \ge 0\}, \partial D) > 0$, *then for any* $T > 0$, *we have that*

$$\lim_{\varepsilon \to 0} \sup_{0 \le t \le T} |X_t^\varepsilon - \bar{X}_t| = 0 \quad a.s. \tag{4.119}$$

Proof Fix $T > 0$ and define

$$A_T = \{|\bar{X}_t| : 0 \le t \le T\}. \tag{4.120}$$

Then by the assumption that $d(\{\bar{X}_t, t \ge 0\}, \partial D) > 0$, we have that

$$\delta_T := d(A_T, \partial D) > 0. \tag{4.121}$$

For any $\varepsilon \in (0, \varepsilon_0)$, define a stopping time τ_ε by

$$\tau_\varepsilon = \inf\left\{t \geq 0 : \left|X_t^\varepsilon - \bar{X}_t\right| > \frac{\delta_T}{2}\right\}. \tag{4.122}$$

Notice that $X_0^\varepsilon = \bar{X}_0 = x$. Then by the continuity of the sample paths of $(X_t^\varepsilon, t \geq 0)$ and $(\bar{X}_t, t \geq 0)$, we know that $0 < \tau_\varepsilon \leq l_\varepsilon$, and if $\tau_\varepsilon < +\infty$, then

$$\left|X_{\tau_\varepsilon}^\varepsilon - \bar{X}_{\tau_\varepsilon}\right| = \frac{\delta_T}{2}. \tag{4.123}$$

Thus $d(X_{\tau_\varepsilon}^\varepsilon, \partial D) \geq \frac{\delta_T}{2} > 0$, and so in this case $\tau_\varepsilon < l_\varepsilon$.

From (4.1) and (4.11), we have that, for any $0 \leq t < l_\varepsilon$,

$$X_t^\varepsilon - \bar{X}_t = \int_0^t \left[a\left(X_s^\varepsilon, Y_{s/\varepsilon}\right) - a(\bar{X}_s, Y_{s/\varepsilon})\right] ds$$
$$+ \int_0^t \left[a(\bar{X}_s, Y_{s/\varepsilon}) - \bar{a}(\bar{X}_s)\right] ds. \tag{4.124}$$

Since $\bar{X}_t$ is continuous, A_T is a compact subset of D. Further, by the assumption that $d(\{\bar{X}_t, t \geq 0\}, \partial D) > 0$, we know that the set $D_T := \{x' \in D : d(x', A_T) \leq \frac{\delta_T}{2}\}$ is a compact subset of D. Then by Assumption 4.5, we obtain that, for any $0 \leq s \leq \tau_\varepsilon \wedge T$,

$$\left|a\left(X_s^\varepsilon, Y_{s/\varepsilon}\right) - a(\bar{X}_s, Y_{s/\varepsilon})\right| \leq k_T\left|X_s^\varepsilon - \bar{X}_s\right|, \tag{4.125}$$

where k_T is the Lipschitz constant of $a(x, y)$ with respect to the compact subset D_T of D. Thus by (4.124) and (4.125), we have that if $0 \leq t \leq \tau_\varepsilon \wedge T$, then

$$\left|X_t^\varepsilon - \bar{X}_t\right| \leq k_T \int_0^t \left|X_s^\varepsilon - \bar{X}_s\right| ds + \left|\int_0^t \left[a(\bar{X}_s, Y_{s/\varepsilon}) - \bar{a}(\bar{X}_s)\right] ds\right|. \tag{4.126}$$

Define

$$\Delta_t^\varepsilon = \left|X_t^\varepsilon - \bar{X}_t\right|,$$
$$\alpha(\varepsilon) = \sup_{0 \leq t \leq T}\left|\int_0^t \left[a(\bar{X}_s, Y_{s/\varepsilon}) - \bar{a}(\bar{X}_s)\right] ds\right|. \tag{4.127}$$

Then by (4.126) and Gronwall's inequality, we have

$$\sup_{0 \leq t \leq \tau_\varepsilon \wedge T} \Delta_t^\varepsilon \leq \alpha(\varepsilon)e^{k_T(\tau_\varepsilon \wedge T)} \leq \alpha(\varepsilon)e^{k_T T}. \tag{4.128}$$

Since $(\bar{X}_t, t \geq 0)$ is a deterministic continuous function, by Assumption 4.2 and the Birkhoff ergodic theorem (see, e.g., Chap. 1 of [130]), we have that

$$\lim_{\varepsilon \to 0} \alpha(\varepsilon) = 0 \quad \text{a.s.} \tag{4.129}$$

It follows from (4.127), (4.128), and (4.129) that

$$\limsup_{\varepsilon \to 0} \sup_{0 \leq t \leq \tau_\varepsilon \wedge T}\left|X_t^\varepsilon - \bar{X}_t\right| = 0, \quad \text{a.s.} \tag{4.130}$$

By (4.123) and (4.130), we obtain that, for a.e. $\omega \in \Omega$, there exists $\varepsilon_0(\omega) > 0$ such that, for any $0 < \varepsilon < \varepsilon_0(\omega)$,

$$\tau_\varepsilon(\omega) > T. \tag{4.131}$$

Thus by (4.130) and (4.131), we obtain

$$\limsup_{\varepsilon \to 0} \sup_{0 \le t \le T} |X_t^\varepsilon - \bar{X}_t| = 0 \quad \text{a.s.} \tag{4.132}$$

Hence (4.119) holds. The proof is completed. □

Lemma 4.5 *Consider system* (4.1) *under Assumptions* 4.5 *and* 4.2. *If* $d(\{\bar{X}_t, t \ge 0\}, \partial D) > 0$, *then for any* $\delta > 0$, *we have*

$$\liminf_{\varepsilon \to 0} \{t \ge 0 : |X_t^\varepsilon - \bar{X}_t| > \delta\} = +\infty \quad \text{a.s.} \tag{4.133}$$

Proof Define

$$\Omega' = \left\{ \omega : \limsup_{\varepsilon \to 0} \sup_{0 \le t \le T} |X_t^\varepsilon(\omega) - \bar{X}_t| = 0, \forall T \in \mathbb{N} \right\}. \tag{4.134}$$

Then by Lemma 4.1, we have

$$P(\Omega') = 1. \tag{4.135}$$

Let $\delta > 0$. Without loss of generality, we can assume that $\delta < \frac{1}{2} d(\{\bar{X}_t, t \ge 0\}, \partial D)$ since if $0 < a < b$, we have

$$\inf\{t \ge 0 : |X_t^\varepsilon - \bar{X}_t| > b\} \ge \inf\{t \ge : |X_t^\varepsilon - \bar{X}_t| > a\}. \tag{4.136}$$

For $\varepsilon \in (0, \varepsilon_0)$, define a stopping time τ_ε^δ by

$$\tau_\varepsilon^\delta = \inf\{t \ge 0 : |X_t^\varepsilon - \bar{X}_t| > \delta\}. \tag{4.137}$$

By the fact that $X_0^\varepsilon - \bar{X}_0 = 0$, and the continuity of the sample paths of $(X_t^\varepsilon, t \ge 0)$ and $(\bar{X}_t, t \ge 0)$, we know that $0 < \tau_\varepsilon^\delta \le +\infty$, and if $\tau_\varepsilon^\delta < +\infty$, then

$$|X_{\tau_\varepsilon^\delta}^\varepsilon - \bar{X}_{\tau_\varepsilon^\delta}| = \delta. \tag{4.138}$$

For any $\omega \in \Omega'$, by (4.134), (4.138) and $\delta < \frac{1}{2} d(\{\bar{X}_t, t \ge 0\}, \partial D)$, we get that for any $T \in \mathbb{N}$, there exists $\varepsilon_0(\omega, \delta, T) > 0$ such that for any $0 < \varepsilon < \varepsilon_0(\omega, \delta, T)$, $\tau_\varepsilon^\delta(\omega) > T$, which implies that

$$\lim_{\varepsilon \to 0} \tau_\varepsilon^\delta(\omega) = +\infty. \tag{4.139}$$

Thus it follows from (4.135) and (4.139) that $\lim_{\varepsilon \to 0} \tau_\varepsilon^\delta = +\infty$ a.s. This completes the proof. □

Now, by Lemmas 4.4 and 4.5, following the corresponding proofs in Sect. 4.2, we obtain the following two theorems.

Theorem 4.3 *Consider system* (4.1) *under Assumptions* 4.5 *and* 4.2. *If the equilibrium* $\bar{X}_t \equiv \bar{x} \in D$ *of the average system* (4.11) *is exponentially stable, then there exist constants* $r > 0$, $c > 0$, *and* $\gamma > 0$ *such that, for any initial condition* $x \in \{x' \in D : |x' - \bar{x}| < r\}$ *and any* $\delta > 0$, *the solution of system* (4.1) *satisfies*

$$\lim_{\varepsilon \to 0} \inf \{t \geq 0 : |X_t^\varepsilon - \bar{x}| > c|x|e^{-\gamma t} + \delta\} = +\infty \quad a.s. \tag{4.140}$$

Theorem 4.4 *Consider system* (4.1) *under Assumptions* 4.5 *and* 4.2. *If the equilibrium* $\bar{X}_t \equiv \bar{x} \in D$ *of the average system* (4.11) *is exponentially stable, then there exist constants* $r > 0$, $c > 0$, $\gamma > 0$, *and a function* $T(\varepsilon) : (0, \varepsilon_0) \to \mathbb{N}$ *such that, for any initial condition* $x \in \{x' \in D : |x' - \bar{x}| < r\}$ *and any* $\delta > 0$,

$$\lim_{\varepsilon \to 0} P \left\{ \sup_{0 \leq t \leq T(\varepsilon)} \left\{ |X_t^\varepsilon - \bar{x}| - c|x|e^{-\gamma t} \right\} > \delta \right\} = 0, \tag{4.141}$$

or equivalently,

$$\lim_{\varepsilon \to 0} P \left\{ |X_t^\varepsilon - \bar{x}| \leq c|x|e^{-\gamma t} + \delta, \forall t \in [0, T(\varepsilon)] \right\} = 1 \tag{4.142}$$

with $\lim_{\varepsilon \to 0} T(\varepsilon) = +\infty$.

4.4 Notes and References

In this chapter, which is based on our results in [89], we establish stability properties for stochastically perturbed differential equations using the averaging approach. We remove the following restrictions in previous results: (a) the average system or approximating diffusion system is globally exponentially stable; (b) the nonlinear vector field of the original system has bounded derivative or is dominated by some forms of Lyapunov function of the average system; (c) the nonlinear vector field of the original system vanishes at the origin for any value of perturbation process (equilibrium condition); and (d) the state space of the perturbation process is a compact space. The theorems developed in this chapter allow us to design stochastic extremum seeking results and to study their stability properties.

Chapter 5
Single-parameter Stochastic Extremum Seeking

The goal of extremum seeking is to find the optimizing input to an unknown operating map that has at least a local extremum. In addition to static operating maps, dynamic input–output maps are also allowable, provided the dynamics are sufficiently fast, or provided extremum seeking is tuned to operate slowly enough relative to the time constants of the dynamics.

Extremum seeking has traditionally been developed as a deterministic approach, employing sinusoidal perturbations for estimating the map's unknown gradient. For a brief historical account of deterministic extremum seeking, the reader is referred to [71] and to the Preface and Chap. 1 of [6].

Extremum seeking is easier to understand for single-input problems than for multivariable problems. For this reason, we start our presentation of stochastic extremum seeking in this chapter by considering single-input problems.

The simplest version of deterministic extremum seeking employs an additive sinusoidal perturbation at the input of an unknown map and generates an estimate of the unknown derivative of the map by multiplying the measured output of the map with the same sinusoid that is applied additively at the input. Though it is not obvious that this set of operations generates an estimate of the unknown slope of the map, an elementary analysis, under the assumption that the amplitude of the sinusoid is small, shows that, on the average (over the period of the sinusoidal perturbation), the estimate of the map's slope, generated in the manner described above, closely approximates the actual slope of the unknown map. By feeding the estimate of the map's slope into an integrator, the output of the integrator serves as the estimate of the optimizing input into the map, and the integrator's output converges, on the average, to the actual optimizing input of the unknown map.

This book proposes algorithms for stochastic extremum seeking in which the principal change is the replacement of the sinusoidal perturbation by a random noise input. The specific noise input that we employ is the white noise passed through a low pass filter with a high cutoff frequency. Such a signal is often called the Ornstein–Uhlenbeck (OU) process. Sometimes it is simply referred to as "colored noise." The key properties that the OU process has in common with the sinusoid

S.-J. Liu, M. Krstic, *Stochastic Averaging and Stochastic Extremum Seeking*,
Communications and Control Engineering,
DOI 10.1007/978-1-4471-4087-0_5, © Springer-Verlag London 2012

is that its mean, defined in an appropriate sense, is zero, whereas the mean of its square is positive.

For technical reasons (the OU process being unbounded and the unknown map having a nonlinear dependence on the perturbation signal), we actually cannot simply apply the OU signal as a perturbation, but we must pass this signal through a bounded nonlinearity which has a zero value and a positive slope at zero. For example, a saturation function or a sine nonlinearity can be applied to the OU process before it is injected as an additive perturbation in the extremum seeking algorithm. The sine nonlinearity is particularly convenient in the analysis because it facilitates the calculation of certain averaging integrals. The periodicity of the sine function, as a function dependent on the OU signal as its argument, is of no particular significance in the extremum seeking algorithm except that it yields explicit formulae in the averaging calculations, which in turn yield explicit convergence rates for the extremum seeking algorithms.

Hence, the difference between deterministic and stochastic extremum seeking is not conceptually substantial. The algorithm structures are the same and the perturbation signals in both cases are zero in the mean, whereas their squares are positive in the mean. The main difference is in the operation of the two algorithms, where the deterministic algorithm has a predictable, nearly periodic evolution of the input and output, whereas the stochastic algorithm generates inputs and outputs that, to an untrained eye, appear completely random. In certain applications, this randomness offers an advantage.

As in deterministic extremum seeking, certain filters and other modifications can be introduced in the stochastic extremum seeking algorithms, as well illustrate in this and other chapters of this book. Stochastic extremum has several features in common with the methods of stochastic approximation, which are covered in detail in the books [79] and [132] and the references therein. Both methods deal with optimization of unknown maps and employ stochastic perturbations. The key difference is that stochastic extremum seeking, as formulated and analyzed in this book, permits the incorporation of the search and optimization algorithms in continuous-time dynamic processes, as illustrated in this book through mobile robotic vehicles. Our stochastic extremum seeking algorithms operate simultaneously with the dynamic systems to which they are applied and the overall convergence rates are determined for the coupled systems consisting of the search algorithm and the plant being controlled, using the averaging and stability theorems that we develop in Chap. 4.

This chapter has two major sections. In Sect. 5.1, we develop stochastic extremum seeking for the case of a static single-parameter (single-input) map. Since this is the reader's first encounter with extremum seeking, we develop slowly all the details of the stability analysis for our algorithm, based on the stability results in Chap. 4, and illustrate the algorithms with a numerical example. In Sect. 5.2, we extend the analysis to the case of a system that contains dynamics and whose operating map is the equilibrium map of those dynamics.

5.1 Extremum Seeking for a Static Map

Consider the quadratic function

$$\varphi(\theta) = \varphi^* + \frac{\varphi''}{2}(\theta - \theta^*)^2, \tag{5.1}$$

where θ^*, φ^*, and φ'' are unknown. Any $\mathbb{C}^2$ function $\varphi(\theta)$ with an extremum at $\theta = \theta^*$ and with $\varphi'' \neq 0$ can be locally approximated by (5.1). Without loss of generality, we assume that $\varphi'' > 0$. In this section, we design an algorithm to make $\theta - \theta^*$ as small as possible, so that the output $y = \varphi(\theta)$ is driven to its minimum φ^*.

Denote $\hat{\theta}(t)$ as the estimate of the unknown optimal input θ^*. Let

$$\tilde{\theta}(t) = \theta^* - \hat{\theta}(t) \tag{5.2}$$

denote the estimation error. Instead of the deterministic periodic perturbation [6], here we use a stochastic perturbation to develop a gradient estimate. Let

$$\theta(t) = \hat{\theta}(t) + a \sin(\eta(t)), \tag{5.3}$$

where $a > 0$ and $(\eta(t), t \geq 0)$ is a stochastic process satisfying

$$\eta = \frac{\sqrt{\varepsilon q}}{\varepsilon s + 1}[\dot{W}], \quad \text{or} \quad \varepsilon \, d\eta = -\eta \, dt + \sqrt{\varepsilon q} \, dW, \tag{5.4}$$

where $q > 0$, $W(t)$, $t \geq 0$, is a 1-dimensional standard Brownian motion defined on some complete probability space $(\Omega, \mathcal{F}, P)$ and $\frac{\sqrt{\varepsilon q}}{\varepsilon s + 1}[\dot{W}]$ denotes a time domain signal obtained as the output of the transfer function $\frac{\sqrt{\varepsilon q}}{\varepsilon s + 1}$ when the input is $\dot{W}(t)$. Thus, by (5.2) and (5.3), we have

$$\theta - \theta^* = a \sin(\eta) - \tilde{\theta}. \tag{5.5}$$

Substituting (5.5) into (5.1), we have the output

$$y = \varphi^* + \frac{\varphi''}{2}(a \sin(\eta) - \tilde{\theta})^2. \tag{5.6}$$

Now, similar to the deterministic case [6], we design the parameter update law as follows

$$\frac{d\hat{\theta}}{dt} = -k \sin(\eta)(y - \zeta), \tag{5.7}$$

$$\frac{d\zeta}{dt} = -h\zeta + hy, \tag{5.8}$$

$$\varepsilon \, d\eta = -\eta \, dt + \sqrt{\varepsilon q} \, dW, \tag{5.9}$$

where $k > 0$, $h > 0$ are scalar design parameters.

From (5.9), we have

$$\eta(t) = \eta(0) - \int_0^t \frac{1}{\varepsilon} \eta(s) \, ds + \int_0^t \frac{q}{\sqrt{\varepsilon}} \, dW(s). \tag{5.10}$$

Thus it holds that

$$\eta(\varepsilon t) = \eta(0) - \int_0^{\varepsilon t} \frac{1}{\varepsilon} \eta(s)\,ds + \int_0^{\varepsilon t} \frac{q}{\sqrt{\varepsilon}}\,dW(s)$$

$$= \eta(0) - \int_0^t \eta(\varepsilon u)\,du + \int_0^t \frac{q}{\sqrt{\varepsilon}}\,dW(\varepsilon u). \tag{5.11}$$

Define $\chi(t) = \eta(\varepsilon t)$ and $B(t) = \frac{1}{\sqrt{\varepsilon}} W(\varepsilon t)$. Then we have

$$d\chi(t) = -\chi(t)\,dt + q\,dB(t), \tag{5.12}$$

where $B(t)$ is a 1-dimensional standard Brownian motion.

Define the output error variable

$$e = \frac{h}{s+h}[y] - \varphi^*. \tag{5.13}$$

Then we have the following error dynamics

$$\frac{d\tilde{\theta}^\varepsilon}{dt} = -\dot{\hat{\theta}} = k \sin\!\big(\chi(t/\varepsilon)\big)\left(\frac{\varphi''}{2}\big(a\sin\!\big(\chi(t/\varepsilon)\big) - \tilde{\theta}^\varepsilon\big)^2 - e^\varepsilon\right), \tag{5.14}$$

$$\frac{de^\varepsilon}{dt} = h\left(\frac{\varphi''}{2}\big(a\sin\!\big(\chi(t/\varepsilon)\big) - \tilde{\theta}^\varepsilon\big)^2 - e^\varepsilon\right). \tag{5.15}$$

Now we calculate the average system. From Sect. 4.1.1, we known that the stochastic process $(\chi(t), t \geq 0)$ (OU process) is ergodic and has invariant distribution

$$\mu(dx) = \frac{1}{\sqrt{\pi}q} e^{-x^2/q^2}\,dx. \tag{5.16}$$

Notice that e^{-x^2/q^2} is an even function and

$$\int_{-\infty}^{+\infty} \cos(2xt)e^{-bt^2}\,dt = \sqrt{\frac{\pi}{b}} e^{-x^2/b}, \tag{5.17}$$

where x, b are parameters. Thus we have

$$\int_{\mathbb{R}} \sin^{2k+1}(x)\mu(dx) = \int_{-\infty}^{+\infty} \sin^{2k+1}(x)\frac{1}{\sqrt{\pi}q} e^{-x^2/q^2}\,dx$$

$$= 0, \quad k = 0, 1, \ldots, \tag{5.18}$$

$$\int_{\mathbb{R}} \sin^2(x)\mu(dx) = \int_{-\infty}^{+\infty} \sin^2(x)\frac{1}{\sqrt{\pi}q} e^{-x^2/q^2}\,dx$$

$$= \frac{1}{2}\big(1 - e^{-q^2}\big). \tag{5.19}$$

Therefore, by (4.12), we obtain that the average system of (5.14)–(5.15) is

$$\frac{d\tilde{\theta}^{\text{ave}}}{dt} = -\frac{k\varphi'' a}{2}\big(1 - e^{-q^2}\big)\tilde{\theta}^{\text{ave}}, \tag{5.20}$$

$$\frac{de^{\text{ave}}}{dt} = h\left(\frac{\varphi'' a^2}{4}\big(1 - e^{-q^2}\big) + \frac{\varphi''}{2}\tilde{\theta}^{\text{ave}^2} - e^{\text{ave}}\right). \tag{5.21}$$

Fig. 5.1 Stochastic extremum seeking scheme for a static map

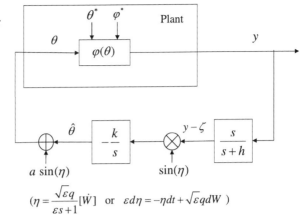

$$\left(\eta = \frac{\sqrt{\varepsilon q}}{\varepsilon s + 1}[\dot{W}] \quad \text{or} \quad \varepsilon d\eta = -\eta dt + \sqrt{\varepsilon q} dW \right)$$

By simple calculation, we get the following equilibrium of the above average system

$$\tilde{\theta}^{a,e} = 0, \qquad e^{a,e} = \frac{a^2 \varphi''}{4}\left(1 - e^{-q^2}\right) \tag{5.22}$$

with the corresponding Jacobian matrix

$$\begin{bmatrix} -\frac{k\varphi'' a}{2}(1 - e^{-q^2}) & 0 \\ 0 & -h \end{bmatrix}. \tag{5.23}$$

Noticing that $\varphi'' > 0$, $k > 0$, $a > 0$, and $h > 0$, we know that the above Jacobian is Hurwitz, i.e., the equilibrium $(0, \frac{a^2 \varphi''}{4}(1 - e^{-q^2}))$ of the average system is exponentially stable.

According to Theorems 4.3 and 4.4 for the stochastic extremum seeking algorithm in Fig. 5.1, we have the following result.

Theorem 1 *Consider the static map* (5.1) *under the parameter update law* (5.7)–(5.9). *Then there exist constants* $r > 0$, $c > 0$, $\gamma > 0$, *and a function* $T(\varepsilon) : (0, \varepsilon_0) \to \mathbb{N}$ *such that for any initial condition* $|\Lambda^\varepsilon(0)| < r$ *and any* $\delta > 0$,

$$\lim_{\varepsilon \to 0} \inf\{t \geq 0 : |\Lambda^\varepsilon(t)| > c|\Lambda^\varepsilon(0)|e^{-\gamma t} + \delta\} = \infty \quad a.s. \tag{5.24}$$

and

$$\lim_{\varepsilon \to 0} P\{|\Lambda^\varepsilon(t)| \leq c|\Lambda^\varepsilon(0)|e^{-\gamma t} + \delta, \forall t \in [0, T(\varepsilon)]\} = 1 \quad \text{with}$$

$$\lim_{\varepsilon \to 0} T(\varepsilon) = \infty, \tag{5.25}$$

where $\Lambda^\varepsilon(t) \triangleq (\tilde{\theta}^\varepsilon(t), e^\varepsilon(t)) - (0, \frac{a^2 \varphi''}{4}(1 - e^{-q^2}))$.

These two results imply that the norm of the error vector $\Lambda^\varepsilon(t)$ exponentially converges, both almost surely and in probability, to below an arbitrarily small residual value δ, over an arbitrarily long time interval, which tends to infinity as ε goes to zero. In particular, the $\tilde{\theta}^{\varepsilon}(t)$-component of the error vector converges to below δ. To quantify the output convergence to the extremum, for any $\varepsilon > 0$, define a stopping time

$$\tau_{\varepsilon}^{\delta} = \inf\{t \geq 0 : |\Lambda^{\varepsilon}(t)| > c|\Lambda^{\varepsilon}(0)|e^{-\gamma t} + \delta\}. \tag{5.26}$$

Then by (5.24) and the definition of $\Lambda^{\varepsilon}(t)$, we know that $\lim_{\varepsilon \to 0} \tau_{\varepsilon}^{\delta} = \infty$ a.s., and

$$|\tilde{\theta}^{\varepsilon}(t)| \leq c|\Lambda^{\varepsilon}(0)|e^{-\gamma t} + \delta, \quad \forall t \leq \tau_{\varepsilon}^{\delta}. \tag{5.27}$$

Since $y(t) = \varphi(\theta^* + \tilde{\theta}^{\varepsilon}(t) + a\sin(\eta(t)))$ and $\varphi'(\theta^*) = 0$, we have

$$y(t) - \varphi(\theta^*) = \frac{\varphi''(\theta^*)}{2}(\tilde{\theta}^{\varepsilon}(t) + a\sin(\eta(t)))^2$$
$$+ O((\tilde{\theta}^{\varepsilon}(t) + a\sin(\eta(t)))^3). \tag{5.28}$$

Thus by (5.27), it holds that

$$|y(t) - \varphi(\theta^*)| \leq O(a^2) + O(\delta^2) + C|\Lambda^{\varepsilon}(0)|^2 e^{-2\gamma t}, \quad \forall t \leq \tau_{\varepsilon}^{\delta}, \tag{5.29}$$

for some positive constant C. Similarly, by (5.25),

$$\lim_{\varepsilon \to 0} P\{|y(t) - \varphi(\theta^*)| \leq O(a^2) + O(\delta^2) + C|\Lambda^{\varepsilon}(0)|^2 e^{-2\gamma t},$$
$$\forall t \in [0, T(\varepsilon)]\} = 1, \tag{5.30}$$

where $T(\varepsilon)$ is a deterministic function with $\lim_{\varepsilon \to 0} T(\varepsilon) = \infty$.

Inequalities (5.29) and (5.30) characterize the asymptotic performance of extremum seeking in Fig. 5.1 and explain why it is not only important that the perturbation parameter ε be small but also that the perturbation gain a be small.

In the gradient-based estimator (5.7), stochastic excitation is chosen in the form of $\sin(\eta(t))$. The use of the sinusoidal nonlinearity should not be confused with the use of sinusoidal perturbation signals in deterministic extremum seeking [6]. In the present stochastic design, the sinusoidal nonlinearity is simply used as a bounded function whose role is to guarantee that the vector field of the error system (5.14)–(5.15) is a bounded function of the perturbation process. We can choose other bounded odd functions to replace sinusoidal functions, such as, $g(x) = xe^{-x^2}$. Corresponding to (5.19) in calculating the average system, the following integral is computed: $\int_{-\infty}^{+\infty} x^2 e^{-2x^2} \frac{1}{\sqrt{\pi}q} e^{-x^2/q^2} \, dx = \frac{1}{2q(2+1/q^2)^{3/2}}$.

Figure 5.2 displays the simulation results with $\varphi^* = 1$, $\varphi'' = 2$, $\theta^* = 0$ in the static map (5.1) and $a = 0.1$, $h = k = q = 1$, $\varepsilon = 0.25$ in the parameter update law (5.7)–(5.9) and initial condition $\tilde{\theta}^{\varepsilon}(0) = 1$, $e^{\varepsilon}(0) = 0.99$, $\hat{\theta}(0) = -1$, $\zeta(0) = 1.99$. The simulation result is robust to design parameters, and similar results are obtained for values on this order of magnitude.

The only requirements on the perturbation process in our averaging theorems are ergodicity and the bounded dependence of the vector field on the perturbation. The OU process satisfies these requirements. Brownian motion on the unit circle can also be used as the excitation signal. In the extremum seeking algorithm, we replace the bounded signal $\sin(\eta(t)) = \sin(\chi(t/\varepsilon))$ with the signal $H^T \check{\eta}(t/\varepsilon)$, where $\check{\eta}(t) =$

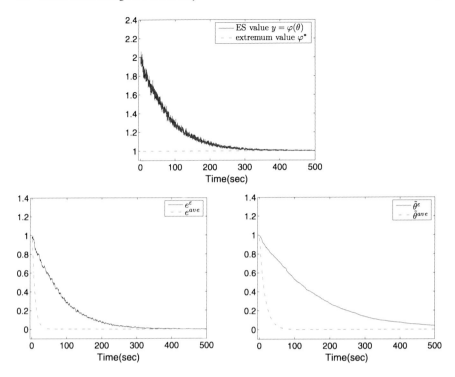

Fig. 5.2 Stochastic extremum seeking with an OU process perturbation. *Top*: output and extremum values. *Bottom*: solutions of the error system and average system

$[\cos(W(t)), \sin(W(t))]^T$ is Brownian motion on the unit circle and $H = [h_1, h_2]^T$ is a constant vector. By a similar analysis, we obtain results as in (5.24) and (5.25), where $\Lambda^{\varepsilon}(t) \triangleq (\tilde{\theta}^{\varepsilon}(t), e^{\varepsilon}(t)) - (0, \frac{a^2 \varphi''}{4}(h_1^2 + h_2^2))$.

For Brownian motion on the unit circle as the stochastic perturbation, Fig. 5.3 shows the simulation results with $\varphi^* = 1$, $\varphi'' = 2$, $\theta^* = 0$ in the static map (5.1), $a = 0.1$, $h = k = h_1 = h_2 = 1$, $\varepsilon = 0.02$ in the parameter update law (5.7)–(5.9) and initial condition $\tilde{\theta}^{\varepsilon}(0) = 1$, $e^{\varepsilon}(0) = 0.99$, $\hat{\theta}(0) = -1$, $\zeta(0) = 1.99$. The simulation is made under the time scale $s = t/\varepsilon$.

By comparing Figs. 5.2 and 5.3, we observe that faster convergence is obtained with the Brownian motion on the unit circle as compared to the convergence rate of the average system, whereas with the OU process the actual convergence is poorer than predicted with the average system (this observation is generic and independent of the fact that different parameters were used for the two perturbation processes). The difference between the effects of the two perturbation processes may be due to the "exponentially decaying form" of the invariant distribution of the OU process, in contrast to the uniform distribution of Brownian motion on the unit circle.

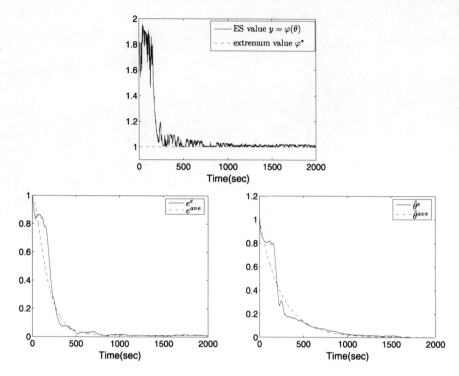

Fig. 5.3 Stochastic extremum seeking with perturbation based on the Brownian motion on the unit circle. *Top*: output and extremum values. *Bottom*: solutions of the error system and average system

5.2 Stochastic Extremum Seeking Feedback for General Nonlinear Dynamic Systems

Consider a general SISO nonlinear model

$$\dot{x} = f(x, u), \tag{5.31}$$

$$y = h(x), \tag{5.32}$$

where $x \in \mathbb{R}^n$ is the state, $u \in \mathbb{R}$ is the input, $y \in \mathbb{R}$ is the output, and $f : \mathbb{R}^n \times \mathbb{R} \to \mathbb{R}^n$ and $h : \mathbb{R}^n \to \mathbb{R}$ are smooth. Suppose that we know a smooth control law

$$u = \alpha(x, \theta) \tag{5.33}$$

parameterized by a scalar parameter θ. Then the closed-loop system

$$\dot{x} = f\big(x, \alpha(x, \theta)\big) \tag{5.34}$$

has equilibria parameterized by θ. As the deterministic case [6], we make the following assumptions about the closed-loop system.

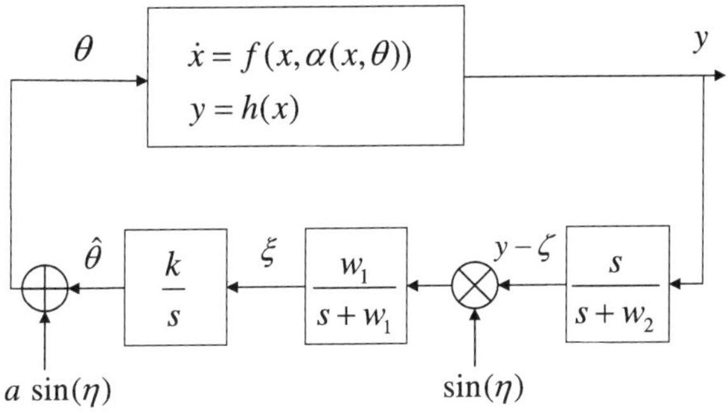

$$(\eta = \frac{\sqrt{\varepsilon q}}{\varepsilon s+1}[\dot{W}] \quad \text{or} \quad \varepsilon d\eta = -\eta dt + \sqrt{\varepsilon q} dW)$$

Fig. 5.4 Stochastic extremum seeking scheme for nonlinear dynamics

Assumption 5.1 There exists a smooth function $l : \mathbb{R} \to \mathbb{R}^n$ such that

$$f\big(x, \alpha(x, \theta)\big) = 0 \quad \text{if and only if} \quad x = l(\theta). \tag{5.35}$$

Assumption 5.2 For each $\theta \in \mathbb{R}$, the equilibrium $x = l(\theta)$ of system (5.34) is exponentially stable with decay and overshoot constant uniform in θ.

Assumption 5.3 There exists $\theta^* \in \mathbb{R}$ such that

$$(h \circ l)'\big(\theta^*\big) = 0, \tag{5.36}$$

$$(h \circ l)''\big(\theta^*\big) < 0. \tag{5.37}$$

Thus, we assume that the output equilibrium map $y = h(l(\theta))$ has a local maximum at $\theta = \theta^*$.

Our objective is to develop a feedback mechanism which makes the output equilibrium map $y(h(l(\theta)))$ as close as possible to the maximum $y(h(l(\theta^*)))$ but without requiring the knowledge of either θ^* or the functions h and l.

We use a stochastic rather than deterministic perturbation signal and choose the parameter update law as (Fig. 5.4)

$$\frac{d\hat{\theta}}{dt} = k\xi, \tag{5.38}$$

$$\frac{d\xi}{dt} = -w_1\xi + w_1(y - \zeta)\sin(\eta), \tag{5.39}$$

$$\frac{d\zeta}{dt} = -w_2\zeta + w_2 y, \tag{5.40}$$

$$\varepsilon\, d\eta = -\eta\, dt + \sqrt{\varepsilon q}\, dW, \tag{5.41}$$

where $k > 0$, $w_1 > 0$, $w_2 > 0$, $\varepsilon > 0$, and $q > 0$ are design parameters and $(W(t), t \geq 0)$ is a 1-dimensional standard Brownian motion on some probability space $(\Omega, \mathcal{F}, P)$.

Remark 5.1 As in the deterministic case [6], the parameters k, w_1, w_2 need to be chosen as $O(\delta)$, where $0 < \delta \ll \varepsilon$. This yields a decomposition into three time scales (in contrast to two time scales encountered with the static map in Sect. 5.1). The fastest of the three time scales, the time scale associated with the plant $\dot{x} = f(x, \alpha(x, \theta))$, requires the employment of a singular perturbation argument, whereas averaging analysis is applied to the two lower time scales. Since we do not have a suitable infinite-time stochastic singular perturbation theorem at our disposal, we apply the singular perturbation reduction without invoking a formal theorem, though the reduced and boundary layer systems do satisfy the usual local exponential stability assumptions. In addition, the low-pass filter (5.39), together with the high-pass filter (5.8) in Sect. 5.1, is introduced for improved asymptotic performance but is not essential for achieving stability [137].

We define

$$\theta = \hat{\theta} + a \sin(\eta(t)) \tag{5.42}$$

with $a > 0$ and obtain the closed-loop system as

$$\frac{dx}{dt} = f\left(x, \alpha\left(x, \hat{\theta} + a \sin(\eta(t))\right)\right), \tag{5.43}$$

$$\frac{d\hat{\theta}}{dt} = k\xi, \tag{5.44}$$

$$\frac{d\xi}{dt} = -w_1\xi + w_1(y - \zeta) \sin(\eta(t)), \tag{5.45}$$

$$\frac{d\zeta}{dt} = -w_2\zeta + w_2 y, \tag{5.46}$$

$$\varepsilon \, d\eta(t) = -\eta(t) \, dt + \sqrt{\varepsilon} q \, dW(t). \tag{5.47}$$

Define $\chi(t) = \eta(\varepsilon t)$ and $B(t) = \frac{1}{\sqrt{\varepsilon}} W(\varepsilon t)$. Then with the error variables

$$\tilde{\theta} = \hat{\theta} - \theta^*, \tag{5.48}$$

$$\tilde{\zeta} = \zeta - h \circ l(\theta^*), \tag{5.49}$$

the closed-loop system is rewritten as

$$\frac{dx}{dt} = f\left(x, \alpha\left(\theta^* + \tilde{\theta} + a \sin(\chi(t/\varepsilon))\right)\right), \tag{5.50}$$

$$\frac{d}{dt} \begin{bmatrix} \tilde{\theta} \\ \xi \\ \tilde{\zeta} \end{bmatrix} = \tilde{E}, \tag{5.51}$$

where

$$\tilde{E} \triangleq \begin{bmatrix} k\xi \\ -w_1\xi + w_1(h(x) - h \circ l(\theta^*) - \tilde{\zeta}) \sin(\chi(t/\varepsilon)) \\ -w_2\tilde{\zeta} + w_2(h(x) - h \circ l(\theta^*)) \end{bmatrix} \tag{5.52}$$

and $d\chi(t) = -\chi(t) \, dt + q \, dB(t)$.

As indicated in Remark 5.1, we employ a singular perturbation reduction. Assuming ε to be large compared to the size of parameters in (5.33), we freeze x in (5.50) at its quasi-steady state value as

$$x = l\big(\theta^* + \tilde{\theta} + a\sin(\chi(t/\varepsilon))\big), \tag{5.53}$$

and substitute it into (5.51), and then get the reduced system

$$\frac{d}{dt}\begin{bmatrix} \tilde{\theta}_r \\ \xi_r \\ \tilde{\zeta}_r \end{bmatrix} = \tilde{E}_r, \tag{5.54}$$

where

$$\tilde{E}_r \triangleq \begin{bmatrix} k\xi_r \\ -w_1\xi_r + w_1(v(\tilde{\theta}_r + a\sin(\chi(t/\varepsilon))) - \tilde{\zeta}_r)\sin(\chi(t/\varepsilon)) \\ -w_2\tilde{\zeta}_r + w_2v(\tilde{\theta}_r + a\sin(\chi(t/\varepsilon))) \end{bmatrix} \tag{5.55}$$

and

$$v\big(\tilde{\theta}_r + a\sin(\chi(t/\varepsilon))\big) = h \circ l\big(\theta^* + \tilde{\theta}_r + a\sin(\chi(t/\varepsilon))\big) - h \circ l(\theta^*). \tag{5.56}$$

With Assumption 5.3, we have

$$v(0) = 0, \tag{5.57}$$

$$v'(0) = (h \circ l)'(\theta^*) = 0, \tag{5.58}$$

$$v''(0) = (h \circ l)''(\theta^*) < 0. \tag{5.59}$$

Now we use our stochastic averaging theorems to analyze system (5.54). According to (4.12), we obtain that the average system of (5.54) is

$$\frac{d}{dt}\begin{bmatrix} \tilde{\theta}_r^{\text{ave}} \\ \xi_r^{\text{ave}} \\ \tilde{\zeta}_r^{\text{ave}} \end{bmatrix} = \tilde{E}_r^{\text{ave}}, \tag{5.60}$$

where

$$\tilde{E}_r^{\text{ave}} \triangleq \begin{bmatrix} k\xi_r^{\text{ave}} \\ -w_1\xi_r^{\text{ave}} + w_1\frac{1}{\sqrt{\pi}q}\int_{-\infty}^{+\infty} v(\tilde{\theta}_r^{\text{ave}} + a\sin(y))\sin(y)e^{-y^2/q^2}\,dy \\ -w_2\tilde{\zeta}_r^{\text{ave}} + w_2\frac{1}{\sqrt{\pi}q}\int_{-\infty}^{+\infty} v(\tilde{\theta}_r^{\text{ave}} + a\sin(y))e^{-y^2/q^2}\,dy \end{bmatrix}. \tag{5.61}$$

First, we determine the average equilibrium $(\tilde{\theta}_r^{a,e}, \xi_r^{a,e}, \tilde{\eta}_r^{a,e})$ which satisfies

$$\xi_r^{a,e} = 0, \tag{5.62}$$

$$\int_{-\infty}^{+\infty} v(\tilde{\theta}_r^{a,e} + a\sin(y))\frac{\sin(y)}{\sqrt{\pi}q}e^{-y^2/q^2}\,dy = 0, \tag{5.63}$$

$$\tilde{\zeta}_r^{a,e} = \frac{1}{\sqrt{\pi}q}\int_{-\infty}^{+\infty} v(\tilde{\theta}_r^{a,e} + a\sin(y))e^{-y^2/q^2}\,dy. \tag{5.64}$$

Assume that $\tilde{\theta}_r^{a,e}$ has the form

$$\tilde{\theta}_r^{a,e} = b_1a + b_2a^2 + O(a^3), \tag{5.65}$$

and by (5.57)–(5.58), define

$$v(x) = \frac{v''(0)}{2}x^2 + \frac{v'''(0)}{3!}x^3 + O(x^4).$$

(5.66)

Then substituting (5.65) and (5.66) into (5.63), we have

$$\int_{-\infty}^{+\infty} v(b_1a + b_2a^2 + O(a^3) + a\sin(y))\sin(y)\frac{1}{\sqrt{\pi}q}e^{-y^2/q^2}\,dy$$

$$= \int_{-\infty}^{+\infty} \left[\frac{v''(0)}{2}(b_1a + b_2a^2 + O(a^3) + a\sin(y))^2\right.$$

$$+ \frac{v'''(0)}{3!}(b_1a + b_2a^2 + O(a^3) + a\sin(y))^3$$

$$\left.+ O((b_1a + b_2a^2 + O(a^3) + a\sin(y))^4)\right]\sin(y)\frac{1}{\sqrt{\pi}q}e^{-y^2/q^2}\,dy$$

$$= \int_{-\infty}^{\infty} \left[\frac{v''(0)}{2}(2b_1a^2 + 2b_2a^3 + O(a^4))\sin^2(y)\right.$$

$$\left.+ \frac{v'''(0)}{3!}(3b_1^2a^3 + O(a^4) + a^3\sin^2(y))\sin^2(y)\right]\frac{1}{\sqrt{\pi}q}e^{-y^2/q^2}\,dy + O(a^4)$$

$$= O(a^4) + v''(0)b_1\left(\frac{1}{2} - \frac{1}{2}e^{-q^2}\right)a^2$$

$$+ \left[\left(b_2v''(0) + \frac{v'''(0)}{2}b_1^2\right)\left(\frac{1}{2} - \frac{1}{2}e^{-q^2}\right)\right.$$

$$\left.+ \frac{v'''(0)}{6}\left(\frac{3}{8} - \frac{1}{2}e^{-q^2} + \frac{1}{8}e^{-4q^2}\right)\right]a^3$$

$$= 0,$$

(5.67)

where the following facts are used:

$$\frac{1}{\sqrt{\pi}q}\int_{-\infty}^{+\infty}\sin^{2k+1}(y)e^{-y^2/q^2}\,dy = 0, \quad k = 0, 1, 2, \ldots,$$

(5.68)

$$\frac{1}{\sqrt{\pi}q}\int_{-\infty}^{+\infty}\sin^2(y)e^{-y^2/q^2}\,dy = \frac{1}{2} - \frac{1}{2}e^{-q^2},$$

(5.69)

$$\frac{1}{\sqrt{\pi}q}\int_{-\infty}^{+\infty}\sin^4(y)e^{-y^2/q^2}\,dy = \frac{3}{8} - \frac{1}{2}e^{-q^2} + \frac{1}{8}e^{-4q^2}.$$

(5.70)

Comparing the coefficients of the powers of a on the right-hand and left-hand sides of (5.67), we have

$$b_1 = 0,$$

(5.71)

$$b_2 = -\frac{v'''(0)(3 - 4e^{-q^2} + e^{-4q^2})}{24v''(0)(1 - e^{-q^2})},$$

(5.72)

and thus by (5.65), we have

$$\tilde{\theta}_r^{a,e} = -\frac{v'''(0)(3 - 4e^{-q^2} + e^{-4q^2})}{24v''(0)(1 - e^{-q^2})} a^2 + O(a^3).$$ (5.73)

From this equation, together with (5.64), we have

$$\begin{aligned}
\zeta_r^{a,e} &= \int_{-\infty}^{+\infty} v(\tilde{\theta}_r^{a,e} + a\sin(y)) \frac{1}{\sqrt{\pi}q} e^{-y^2/q^2} \, dy \\
&= \int_{-\infty}^{+\infty} v(b_2 a^2 + O(a^3) + a\sin(y)) \frac{e^{-y^2/q^2}}{\sqrt{\pi}q} \, dy \\
&= \int_{-\infty}^{+\infty} \left[\frac{v''(0)}{2} (b_2 a^2 + O(a^3) + a\sin(y))^2 \right. \\
&\quad + \frac{v'''(0)}{3!} (b_2 a^2 + O(a^3) + a\sin(y))^3 \\
&\quad \left. + O((b_2 a^2 + O(a^3) + a\sin(y))^4) \right] \frac{e^{-y^2/q^2}}{\sqrt{\pi}q} \, dy \\
&= \frac{a^2 v''(0)}{2} \int_{-\infty}^{+\infty} \sin^2(y) \frac{1}{\sqrt{\pi}q} e^{-y^2/q^2} \, dy + O(a^3) \\
&= \frac{v''(0)(1 - e^{-q^2})}{4} a^2 + O(a^3).
\end{aligned}$$ (5.74)

Thus the equilibrium of the average system (5.60) is

$$\begin{bmatrix} \tilde{\theta}_r^{a,e} \\ \xi_r^{a,e} \\ \zeta_r^{a,e} \end{bmatrix} = \begin{bmatrix} -\frac{v'''(0)(3-4e^{-q^2}+e^{-4q^2})}{24v''(0)(1-e^{-q^2})} a^2 + O(a^3) \\ 0 \\ \frac{v''(0)(1-e^{-q^2})}{4} a^2 + O(a^3) \end{bmatrix}.$$ (5.75)

The Jacobian matrix of the average system (5.60) at the equilibrium $(\tilde{\theta}_r^{a,e}, \xi_r^{a,e}, \zeta_r^{a,e})$ is

$$J_r^a = \begin{bmatrix} 0 & k & 0 \\ J_{r21}^a & -w_1 & 0 \\ J_{r31}^a & 0 & -w_2 \end{bmatrix},$$ (5.76)

where

$$J_{r21}^a = \frac{w_1}{\sqrt{\pi}q} \int_{-\infty}^{+\infty} v'(\tilde{\theta}_r^{a,e} + a\sin(y)) \sin(y) e^{-y^2/q^2} \, dy,$$ (5.77)

$$J_{r31}^a = \frac{w_2}{\sqrt{\pi}q} \int_{-\infty}^{+\infty} v'(\tilde{\theta}_r^{a,e} + a\sin(y)) e^{-y^2/q^2} \, dy.$$ (5.78)

Since J_r^a is block-lower triangular, we see that it will be Hurwitz if and only if

$$\int_{-\infty}^{+\infty} v'(\tilde{\theta}_r^{a,e} + a\sin(y)) \sin(y) e^{-y^2/q^2} \, dy < 0.$$ (5.79)

With Taylor expansion and by calculating the integral, we get

$$\int_{-\infty}^{+\infty} v'\left(\tilde{\theta}_r^{a,e} + a\sin(y)\right)\sin(y)e^{-y^2/q^2}\,dy$$

$$= a\sqrt{\pi}qv''(0)\left(\frac{1}{2} - \frac{1}{2}e^{-q^2}\right) + O\left(a^2\right). \tag{5.80}$$

By substituting (5.80) into (5.76), we get

$$\det\left(\lambda I - J_r^a\right)$$

$$= \left(\lambda^2 + w_1\lambda - \frac{w_1 k}{2}v''(0)a\left(1 - e^{-q^2}\right) + O\left(a^2\right)\right)(\lambda + w_2), \tag{5.81}$$

which proves that J_r^a is Hurwitz for sufficiently small a. This implies that the equilibrium of the average system is exponentially stable for sufficiently small a. Then according to Theorems 4.3 and 4.4, we have the following result for stochastic extremum seeking algorithm in Fig. 5.4.

Theorem 2 *Consider system* (5.54) *under Assumption 5.3. Then there exists a constant $a^* > 0$ such that for any $0 < a < a^*$ there exist constants $r > 0$, $c > 0$, $\gamma > 0$, and a function $T(\varepsilon) : (0, \varepsilon_0) \to \mathbb{N}$ such that, for any initial condition $|\Delta^{\varepsilon,a}(0)| < r$ and any $\delta > 0$,*

$$\lim_{\varepsilon \to 0} \inf\left\{t \geq 0 : \left|\Delta^{\varepsilon,a}(t)\right| > c\left|\Delta^{\varepsilon,a}(0)\right|e^{-\gamma t} + \delta\right\} = \infty \quad \text{a.s.} \tag{5.82}$$

and

$$\lim_{\varepsilon \to 0} P\left\{\left|\Delta^{\varepsilon,a}(t)\right| \leq c\left|\Delta^{\varepsilon,a}(0)\right|e^{-\gamma t} + \delta, \forall t \in \left[0, T(\varepsilon)\right]\right\} = 1 \quad \text{with}$$

$$\lim_{\varepsilon \to 0} T(\varepsilon) = \infty, \tag{5.83}$$

where

$$\Delta^{\varepsilon,a}(t) \triangleq \left(\tilde{\theta}_r(t), \xi_r(t), \tilde{\zeta}_r(t)\right)$$

$$-\left(-\frac{v'''(0)(3 - 4e^{-q^2} + e^{-4q^2})}{24v''(0)(1 - e^{-q^2})}a^2 + O\left(a^3\right), 0,\right.$$

$$\left.\frac{v''(0)(1 - e^{-q^2})}{4}a^2 + O\left(a^3\right)\right). \tag{5.84}$$

These results imply that the norm of the error vector $\Delta^{\varepsilon,a}(t)$ exponentially converges, both almost surely and in probability, to below an arbitrarily small residual value δ over an arbitrary large time interval, which tends to infinity as the perturbation parameter ε goes to zero. In particular, the $\tilde{\theta}^\varepsilon(t)$-component of the error vector converges to below δ. To quantify the output convergence to the extremum, we define a stopping time

$$\tau_\varepsilon^\delta = \inf\left\{t \geq 0 : \left|\Delta^{\varepsilon,a}(t)\right| > c\left|\Delta^{\varepsilon,a}(0)\right|e^{-\gamma t} + \delta\right\}.$$

Then by (5.82) and the definition of $\Delta^{\varepsilon,a}(t)$, we know that $\lim_{\varepsilon \to 0} \tau_\varepsilon^\delta = \infty$ a.s., and for all $t \leq \tau_\varepsilon^\delta$,

$$\left|\tilde{\theta}_r(t) - \left(-\frac{v'''(0)(3 - 4e^{-q^2} + e^{-4q^2})}{24v''(0)(1 - e^{-q^2})}a^2 + O(a^3)\right)\right|$$
$$\le c|\Delta^{\varepsilon,a}(0)|e^{-\gamma t} + \delta, \tag{5.85}$$

which implies that

$$|\tilde{\theta}_r(t)| \le O(a^2) + c|\Delta^{\varepsilon,a}(0)|e^{-\gamma t} + \delta \quad \forall t \le \tau_\varepsilon^\delta. \tag{5.86}$$

Since $y(t) = h(l(\theta^* + \tilde{\theta}_r(t) + a\sin(\eta(t))))$ and $(h \circ l)'(\theta^*) = 0$, we have

$$y(t) - h \circ l(\theta^*)$$
$$= \frac{(h \circ l)''(\theta^*)}{2}(\tilde{\theta}_r(t) + a\sin(\eta(t)))^2 + O((\tilde{\theta}_r + a\sin(\eta(t)))^3). \tag{5.87}$$

Thus by (5.86), it holds that

$$|y(t) - h \circ l(\theta^*)| \le O(a^2) + O(\delta^2) + C|\Delta^{\varepsilon,a}(0)|^2 e^{-2\gamma t} \quad \forall t \le \tau_\varepsilon^\delta \tag{5.88}$$

for some positive constant C. Similarly, by (5.83),

$$\lim_{\varepsilon \to 0} P\{|y(t) - h \circ l(\theta^*)| \le O(a^2) + O(\delta^2) + C|\Delta^{\varepsilon,a}(0)|^2 e^{-2\gamma t},$$
$$\forall t \in [0, T(\varepsilon)]\} = 1, \tag{5.89}$$

where $T(\varepsilon)$ is a deterministic function with $\lim_{\varepsilon \to 0} T(\varepsilon) = \infty$.

5.3 Notes and References

This chapter is based on our results in [89]. The first design of a stochastic extremum seeking algorithm was proposed in discrete time in [98]. In this chapter, we introduced continuous-time extremum seeking algorithms that employ stochastic excitation signals instead of deterministic periodic signals. In the subsequent chapters, we explore more specific applications of the stochastic extremum seeking algorithms.

The stochastic extremum seeking approach introduced in [89] has inspired a design of an algorithm for maximization of endurance of aircraft in [69] by exploring the stochastic character of the air turbulence disturbance, which affects the airspeed-dependent drag force. The algorithm in [69] is a nontrivial modification of the algorithm in this chapter because the air turbulence is not a disturbance that is introduced by the user and hence it enters the feedback system differently.

Chapter 6
Stochastic Source Seeking for Nonholonomic Vehicles

Steering mobile robots in concentration fields with an unknown spatial distribution, and without position (GPS) measurements available to the robots, has become a very active field in recent years, with entire conference sessions dedicated to the topic and with many grants, in various countries, awarded to the study of this topic. The motivation comes from environmental (tracking of oil spill plumes) to homeland security (contaminants released into the atmosphere via a "dirty bomb") to biology and medicine (understanding the feedback mechanism that underlies chemotaxis of bacteria and cancer cells).

We refer to the problem of steering of vehicles in GPS-denied environments, with unknown spatially distributed concentration fields, as *source seeking*. In this chapter, we investigate a stochastic version of source seeking by navigating a unicycle robot with the help of a random perturbation. Our vehicle has no knowledge of its position, nor of the distribution of the signal field. To find the source, we employ a stochastic extremum seeking approach and provide a stability analysis based on stochastic averaging theorems that we developed in Chap. 4. The key challenge is that we cannot directly control the two-dimensional position vector of the robot but can control its scalar angular velocity input, for steering. With a controller that we design in the chapter, the vehicle is driven to approach a small neighborhood of the source in a manner that seems partly random but is convergent in a suitable sense. We present a stability proof for the scheme with a static source and simulation results for both static and moving sources. Convergence is proved both in the "almost sure" sense and "in probability".

The chapter is organized as follows. In Sect. 6.1, we present the vehicle model and state the problem. In Sect. 6.2, we present our stochastic source seeking controller. In Sect. 6.3, we prove local exponential convergence for circular level sets, namely, where the signal depends only on the distance from the source and decays quadratically. In Sect. 6.4, we calculate the convergence speed, for particular parameter choices for which it is possible to do so explicitly, and characterize the best achievable convergence speed. In Sect. 6.5, we present simulations and discussions about dependence on design parameters. In Sect. 6.6, we discuss the dependence on damping term. In Sect. 6.7, we discuss the effect of constraints of the angular ve-

S.-J. Liu, M. Krstic, *Stochastic Averaging and Stochastic Extremum Seeking*,
Communications and Control Engineering,
DOI 10.1007/978-1-4471-4087-0_6, © Springer-Verlag London 2012

Fig. 6.1 The notation used in
the model of vehicle sensor
and center dynamics

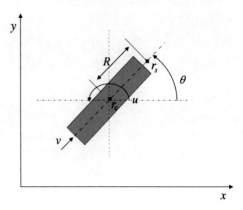

locity and design alternatives. In Sect. 6.8, we consider signal fields with elliptical
level sets.

6.1 Vehicle Model and Problem Statement

As in [28], we consider a mobile agent modeled as a unicycle with a sensor mounted
at its front end, a distance R from the center. Figure 6.1 depicts the position, heading,
angular and forward velocities for the center and sensor. The equations of motion
for the vehicle center are

$$\dot{r}_c = v e^{j\theta}, \tag{6.1}$$

$$\dot{\theta} = u, \tag{6.2}$$

where r_c is the vehicle center, θ is the orientation, v, u are the forward and angular
velocity inputs, respectively, and j is the imaginary unit. The sensor is located at
$r_s = r_c + R e^{j\theta}$.

The task of the vehicle is to seek a source that emits a spatially distributed signal
$J = f(r(x, y))$ which has an isolated local maximum $f^* = f(r^*)$, where r^* is the
location of the local maximum. We achieve local convergence to r^*, in a particular
probabilistic sense, without the knowledge of the shape of $f(\cdot)$, and without the
measurement of r_c, using only the measurement of $J(t)$ at the vehicle sensor.

6.2 Stochastic Source Seeking Controller

We employ the scheme depicted by the block diagram in Fig. 6.2. The forward
velocity of the vehicle is set to $v(t) = V_c \equiv$ const., whereas the angular velocity $\dot{\theta}$
is tuned by the extremum seeking control law

$$\dot{\theta} = a\dot{\eta} + c\xi \sin(\eta) - d_0 \xi^2 \sin(\eta), \tag{6.3}$$

Fig. 6.2 Block diagram of stochastic source seeking via tuning of angular velocity of the vehicle

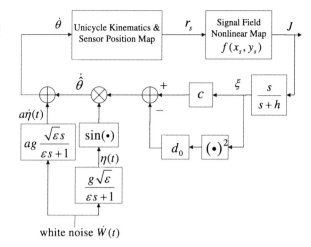

white noise $\dot{W}(t)$

where $\xi = \frac{s}{s+h}[J]$ is the output of the washout filter for the sensor reading J, $\eta = \frac{g\sqrt{\varepsilon}}{\varepsilon s + 1}[\dot{W}]$ ($\varepsilon \in (0, \varepsilon_0)$) is colored noise used as a perturbation in stochastic extremum seeking, and $V_c, a, c, d_0, g, \varepsilon, h > 0$ are design parameters which (along with parameter R) influence the performance. The signal $(W(t), t \geq 0)$ is a standard Brownian motion defined on a complete probability space $(\Omega, \mathcal{F}, P)$ with the sample space Ω, σ-field $\mathcal{F}$, and probability measure P.

With the observation that the transfer function from white noise $\dot{W}$ to $\dot{\eta}$ has relative degree zero, giving

$$\dot{\eta} = \frac{g\sqrt{\varepsilon}s}{\varepsilon s + 1}[\dot{W}] = \frac{1}{\sqrt{\varepsilon}}\frac{g\varepsilon s + g - g}{\varepsilon s + 1}[\dot{W}] = \frac{g}{\sqrt{\varepsilon}}\dot{W} - \frac{1}{\varepsilon}\eta, \qquad (6.4)$$

the control law is rewritten as

$$d\theta = -\frac{a}{\varepsilon}\eta\, dt + \left(c\xi - d_0\xi^2\right)\sin(\eta)\, dt + \frac{ag}{\sqrt{\varepsilon}}\, dW, \qquad (6.5)$$

$$d\eta = -\frac{1}{\varepsilon}\eta\, dt + \frac{g}{\sqrt{\varepsilon}}\, dW. \qquad (6.6)$$

Compared with the deterministic case in [28], where $\sin(\omega t)$ was used as the probing signal, we use the stochastic signal $\sin(\eta(t))$ to develop a gradient estimate. It is not essential to choose the sinusoidal nonlinearity $\sin(\eta)$ in the stochastic design. This choice is primarily made for the ease of deriving the average system in the stability analysis. We can replace $\sin(\eta)$ with other bounded and odd functions, such as $\eta e^{-\eta^2}$, however, the integrals in calculating the expectations in the derivation of the average system become more complicated. In fact, the boundedness of the perturbation (such as $\sin\eta$ or $\eta e^{-\eta^2}$) is only needed in the analysis, whereas in the simulations, successful convergence is achieved even when $\sin(\eta)$ is replaced by η.

We refer to the term $-d_0 \xi^2 \sin(\eta)$ as the "d_0-term" or the damping term. This term is not needed in the basic stochastic extremum seeking algorithm for a static map in Chap. 5. This term is essential for achieving exponential stability in source seeking problems with a vehicle employing constant forward velocity.

6.3 Stability Analysis

We assume that the nonlinear map defining the distribution of the signal field is quadratic and takes the form

$$J = f(r_s) = f^* - q_r |r_s - r^*|^2, \tag{6.7}$$

where r^* is the unknown maximizer, $f^* = f(r^*)$ is the unknown maximum, and q_r is an unknown positive constant. We define an output error variable

$$e = \frac{h}{s+h}[J] - f^*, \tag{6.8}$$

which allows us to express the signal ξ after the washout filter, as

$$\xi = \frac{s}{s+h}[J] = J - \frac{h}{s+h}[J] = J - f^* - e, \tag{6.9}$$

and thus we have $\dot{e} = h\xi$.

We now use our general stochastic averaging theory from Chap. 4 to analyze the stability of the closed-loop system.

Theorem 6.1 *Consider the closed-loop system*

$$dr_c = V_c e^{j\theta} \, dt, \tag{6.10}$$

$$d\theta = \frac{-a}{\varepsilon} \eta \, dt + \left(c\xi - d_0 \xi^2 \right) \sin(\eta) \, dt + \frac{ag}{\sqrt{\varepsilon}} \, dW, \tag{6.11}$$

$$de = h\xi \, dt, \tag{6.12}$$

$$\xi = -\left(q_r |r_s - r^*|^2 + e \right), \tag{6.13}$$

$$r_s = r_c + Re^{j\theta}, \tag{6.14}$$

$$d\eta = -\frac{1}{\varepsilon} \eta \, dt + \frac{g}{\sqrt{\varepsilon}} \, dW, \tag{6.15}$$

where $c, d_0, h, R, V_c, q_r > 0$, and the parameters $h, V_c, a, g > 0$ are chosen such that

$$\frac{1}{h} > \frac{R}{2V_c} \left(2 - \frac{I_2(2a, g)}{I_1(a, g)I_2(a, g)} \right), \tag{6.16}$$

where $I_1(a, g) = e^{-a^2 g^2/4}$, $I_2(a, g) = \frac{1}{2}[e^{-(a-1)^2 g^2/4} - e^{-(a+1)^2 g^2/4}]$. *(Condition (6.16) is satisfied for any $h > 0$ and $V_c > 0$, provided g is chosen as $g = \sqrt{\frac{\beta}{a}}$ and a is chosen as $0 < a < a^*(\beta) \triangleq \frac{2}{\beta} \ln \frac{e^{2\beta}-1}{2\sqrt{e^\beta}(e^\beta-1)}$ for any $\beta > 0$. For example, for $\beta = 1$, $a^*(1) \approx 0.24$.) If the initial conditions $r_c(0)$, $\theta(0)$, $e(0)$ are such that the following quantities are sufficiently small,*

$$\left| \left| r_c(0) - r^* \right| - \rho \right|, \qquad \left| e(0) + q_r \left(R^2 + \rho^2 \right) \right|, \tag{6.17}$$

either

$$\left| \theta(0) - \arg\left(r^* - r_c(0) \right) + \frac{\pi}{2} \right| \quad or \tag{6.18}$$

$$\left| \theta(0) - \arg\left(r^* - r_c(0) \right) - \frac{\pi}{2} \right|, \tag{6.19}$$

where

$$\rho = \sqrt{\frac{V_c I_1(a, g)}{2 q_r c R I_2(a, g)}}, \tag{6.20}$$

then there exist constants $C_0, \gamma_0 > 0$, and a function $T(\varepsilon) : (0, \varepsilon_0) \to \mathbb{N}$ such that, for any $\delta > 0$,

$$\lim_{\varepsilon \to 0} \inf \left\{ t \geq 0 : \left| \left| r_c(t) - r^* \right| - \rho \right| > C_0 e^{-\gamma_0 t} + \delta \right\} = \infty \quad a.s. \tag{6.21}$$

and

$$\lim_{\varepsilon \to 0} P \left\{ \left| \left| r_c(t) - r^* \right| - \rho \right| \leq C_0 e^{-\gamma_0 t} + \delta, \forall t \in \left[0, T(\varepsilon) \right] \right\} = 1 \tag{6.22}$$

with $\lim_{\varepsilon \to 0} T(\varepsilon) = \infty$, where the constant C_0 is dependent on the initial condition $(r_c(0), \theta(0), e(0))$ and on the parameters $a, c, d_0, h, R, V_c, q_r, g$, and the constant γ_0 is dependent on the parameters $a, c, d_0, h, R, V_c, q_r, g$.

Proof We start by defining the shifted variables

$$\hat{r}_c = r_c - r^*, \tag{6.23}$$

$$\hat{\theta} = \theta - a\eta, \tag{6.24}$$

and a map between $\hat{r}_c$ and a new quantity θ^* given by

$$-\hat{r}_c = |\hat{r}_c| e^{j\theta^*}, \tag{6.25}$$

$$\theta^* = \arg(-\hat{r}_c) = \arg\left(r^* - r_c \right)$$

$$= \begin{cases} -\pi - \frac{j}{2} \ln(\frac{\hat{r}_c}{\bar{\hat{r}}_c}) & \text{if } \theta^* \in (-\pi, -\frac{\pi}{2}], \\ -\frac{j}{2} \ln(\frac{\hat{r}_c}{\bar{\hat{r}}_c}) & \text{if } \theta^* \in (-\frac{\pi}{2}, \frac{\pi}{2}], \\ \pi - \frac{j}{2} \ln(\frac{\hat{r}_c}{\bar{\hat{r}}_c}) & \text{if } \theta^* \in (\frac{\pi}{2}, \pi], \end{cases} \tag{6.26}$$

where θ^* represents the heading angle toward the source located at r^* when the vehicle is at r_c. Using these definitions, the expression for ξ is

$$\xi = -\left(q_r\left(R^2 + |\hat{r}_c|^2 - 2R|\hat{r}_c|\cos(\hat{\theta} - \theta^* + a\eta)\right) + e\right). \tag{6.27}$$

Since

$$d\hat{\theta} = d\theta - a\,d\eta$$

$$= \frac{-a}{\varepsilon}\eta\,dt + \left(c\xi - d_0\xi^2\right)\sin(\eta)\,dt + \frac{ag}{\sqrt{\varepsilon}}dW + \frac{a}{\varepsilon}\eta\,dt - \frac{ag}{\sqrt{\varepsilon}}dW$$

$$= \left(c\xi - d_0\xi^2\right)\sin(\eta)\,dt, \tag{6.28}$$

we obtain the dynamics of the shifted system as

$$\frac{d\hat{r}_c}{dt} = \frac{dr_c}{dt} = V_c e^{j(\hat{\theta} + a\eta)}, \tag{6.29}$$

$$\frac{d\hat{\theta}}{dt} = \left(c\xi - d_0\xi^2\right)\sin(\eta), \tag{6.30}$$

$$\frac{de}{dt} = -hq_r\left(R^2 + |\hat{r}_c|^2 - 2R|\hat{r}_c|\cos(\hat{\theta} - \theta^* + a\eta)\right) - he. \tag{6.31}$$

Similar to Sect. 5.1, define

$$B(t) = \frac{1}{\sqrt{\varepsilon}}W(\varepsilon t), \qquad \chi(t) = \eta(\varepsilon t). \tag{6.32}$$

Then we have

$$d\chi(t) = -\chi(t)\,dt + g\,dB(t), \tag{6.33}$$

where $B(t)$ is a standard Brownian motion and the process $\chi(t)$ is an Ornstein–Uhlenbeck (OU) process which is ergodic with invariant distribution

$$\mu(dy) = \frac{1}{\sqrt{\pi}g}e^{-y^2/g^2}\,dy. \tag{6.34}$$

Now we define error variables $\tilde{r}_c$ and $\tilde{\theta}$ which represent the distance to the source, and the difference between the vehicle's heading and the optimal heading, respectively,

$$\tilde{r}_c = |\hat{r}_c| = |r_c - r^*|, \tag{6.35}$$

$$\tilde{\theta} = \hat{\theta} - \theta^*. \tag{6.36}$$

Thus we obtain the following dynamics for the error variables

$$\frac{d\tilde{r}_c}{dt} = \frac{d|\hat{r}_c|}{dt} = \frac{d\sqrt{\hat{r}_c\bar{\hat{r}}_c}}{dt} = \frac{1}{2|\hat{r}_c|}\left(\frac{d\hat{r}_c}{dt}\bar{\hat{r}}_c + \hat{r}_c\frac{d\bar{\hat{r}}_c}{dt}\right)$$

$$= -V_c\cos(\tilde{\theta} + a\chi(t/\varepsilon)), \tag{6.37}$$

$$\frac{d\tilde{\theta}}{dt} = \frac{d\hat{\theta}}{dt} - \frac{d\theta^*}{dt} = \frac{d\hat{\theta}}{dt} + \frac{j}{2|\hat{r}_c|^2}\left(\frac{d\hat{r}_c}{dt}\bar{\hat{r}}_c - \hat{r}_c\frac{d\bar{\hat{r}}_c}{dt}\right)$$

$$= (c - d_0\xi)\xi\,\sin(\chi(t/\varepsilon)) + \frac{V_c}{\tilde{r}_c}\sin(\tilde{\theta} + a\chi(t/\varepsilon)), \tag{6.38}$$

$$\frac{de}{dt} = h\xi, \tag{6.39}$$

$$\xi = -\left(q_r\left(R^2 + \tilde{r}_c^2 - 2R\tilde{r}_c\cos(\tilde{\theta} + a\chi(t/\varepsilon))\right) + e\right), \tag{6.40}$$

$$d\chi(t) = -\chi(t)\,dt + g\,dB(t). \tag{6.41}$$

We use general stochastic averaging presented in Chap. 4 to analyze this error system.

First we calculate the average system of (6.37), (6.38), and (6.39). Since

$$\int_{\mathbb{R}}\sin(ay)\mu(dy) = \int_{-\infty}^{+\infty}\sin(ay)\frac{1}{\sqrt{\pi}g}e^{-y^2/g^2}\,dy = 0, \tag{6.42}$$

$$\int_{\mathbb{R}}\cos(ay)\sin(ay)\mu(dy) = \int_{\mathbb{R}}\cos(2ay)\sin(ay)\mu(dy) = 0, \tag{6.43}$$

$$\int_{\mathbb{R}}\cos(ay)\mu(dy) = \int_{-\infty}^{+\infty}\cos(ay)\frac{1}{\sqrt{\pi}g}e^{-y^2/g^2}\,dy$$

$$= e^{-a^2g^2/4} \triangleq I_1(a, g), \tag{6.44}$$

$$\int_{\mathbb{R}}\sin(ay)\sin(y)\mu(dy) = \int_{-\infty}^{+\infty}\sin(ay)\sin(y)\frac{1}{\sqrt{\pi}g}e^{-y^2/g^2}\,dy$$

$$= \frac{1}{2}\left[e^{-(a-1)^2g^2/4} - e^{-(a+1)^2g^2/4}\right]$$

$$\triangleq I_2(a, g), \tag{6.45}$$

by (4.12), we obtain that the average error system is

$$\frac{d\tilde{r}_c^{\text{ave}}}{dt} = -V_c I_1(a, g)\cos(\tilde{\theta}^{\text{ave}}), \tag{6.46}$$

$$\frac{d\tilde{\theta}^{\text{ave}}}{dt} = -\left(2c + 4d_0\left(q_r\left(R^2 + \tilde{r}_c^{\text{ave}^2}\right) + e^{\text{ave}}\right)\right) \times q_r R\tilde{r}_c^{\text{ave}}\sin(\tilde{\theta}^{\text{ave}})I_2(a, g)$$

$$+ 2d_0q_r^2 R^2\tilde{r}_c^{\text{ave}^2}\sin(2\tilde{\theta}^{\text{ave}})I_2(2a, g) + \frac{V_c}{\tilde{r}_c^{\text{ave}}}\sin(\tilde{\theta}^{\text{ave}})I_1(a, g), \tag{6.47}$$

$$\frac{de^{\text{ave}}}{dt} = -hq_r R^2 - hq_r \tilde{r}_c^{\text{ave}^2} + 2hRq_r \tilde{r}_c^{\text{ave}} \cos\left(\tilde{\theta}^{\text{ave}}\right) I_1(a, g) - he^{\text{ave}}. \tag{6.48}$$

The average error system has two equilibria defined by

$$\left[\tilde{r}_c^{\text{ave}^{\text{eq1}}}, \tilde{\theta}^{\text{ave}^{\text{eq1}}}, e^{\text{ave}^{\text{eq1}}}\right] = \left[\rho, +\frac{\pi}{2}, -q_r\left(R^2 + \rho^2\right)\right], \tag{6.49}$$

$$\left[\tilde{r}_c^{\text{ave}^{\text{eq2}}}, \tilde{\theta}^{\text{ave}^{\text{eq2}}}, e^{\text{ave}^{\text{eq2}}}\right] = \left[\rho, -\frac{\pi}{2}, -q_r\left(R^2 + \rho^2\right)\right], \tag{6.50}$$

where ρ is given by (6.20). The above two equilibria have the following Jacobians, respectively,

$$A^{\text{eq1}} = - \begin{bmatrix} 0 & -V_c I_1(a, g) & 0 \\ A_{21}^{\text{eq1}} & 4d_0 \gamma^2 \rho^2 I_2(2a, g) & 4d_0 \gamma \rho I_2(a, g) \\ 2hq_r \rho & 2h\gamma \rho I_1(a, g) & h \end{bmatrix} \tag{6.51}$$

and

$$A^{\text{eq2}} = \begin{bmatrix} 0 & -V_c I_1(a, g) & 0 \\ A_{21}^{\text{eq2}} & -4d_0 \gamma^2 \rho^2 I_2(2a, g) & 4d_0 \gamma \rho I_2(a, g) \\ -2hq_r \rho & 2h\gamma \rho I_1(a, g) & -h \end{bmatrix}, \tag{6.52}$$

where $A_{21}^{\text{eq1}} = A_{21}^{\text{eq2}} = 4\gamma(c + 2d_0 q_r \rho^2) I_2(a, g)$, $\gamma \triangleq q_r R$. The characteristic polynomial for both Jacobians is

$$0 = \lambda^3 + h\lambda^2 + \frac{2V_c^2 I_1^2(a, g)}{\rho^2} \lambda + h\frac{2V_c^2 I_1^2(a, g)}{\rho^2}$$
$$+ 4d_0 \rho^2 q_r^2 R\left[R I_2(2a, g)\lambda^2 + \left(2V_c I_1(a, g) I_2(a, g)\right.\right.$$
$$\left.\left. + hR\left(I_2(2a, g) - 2I_1(a, g) I_2(a, g)\right)\right)\lambda\right]. \tag{6.53}$$

Since $a > 0$, we have $I_2(a, g) > 0$ and $I_1(a, g) > 0$. For the roots of the polynomial (6.53) to be in the left half-plane, all of its three coefficients need to be positive and the product of the coefficients associated with λ^2 and λ^1 needs to be greater than the coefficient associated with λ^0. All of these conditions are satisfied whenever

$$2V_c I_1(a, g) I_2(a, g) + hR\left(I_2(2a, g) - 2I_1(a, g) I_2(a, g)\right) > 0, \tag{6.54}$$

which is equivalent to the condition (6.16). When the condition (6.16) is satisfied, the Jacobians (6.51) and (6.52) are Hurwitz, which implies that both average equilibria (6.49) and (6.50) are exponentially stable. Thus by Theorems 4.3 and 4.4, there exist constants $c_0^{(i)} > 0, r_0^{(i)} > 0, \gamma_0^{(i)} > 0$, and functions $T^{(i)}(\varepsilon) : (0, \varepsilon_0) \to \mathbb{N}$, $i = 1, 2$, such that, for any $\delta > 0$ and any initial condition $|\Lambda_\varepsilon^{(i)}(0)| < r_0^{(i)}$,

$$\lim_{\varepsilon \to 0} \inf\left\{t \geq 0 : \left|\Lambda_\varepsilon^{(i)}(t)\right| > c_0^{(i)} \left|\Lambda_\varepsilon^{(i)}(0)\right| e^{-\gamma_0^{(i)} t} + \delta\right\} = \infty \quad \text{a.s.} \tag{6.55}$$

and

$$\lim_{\varepsilon \to 0} P\{|\Lambda_\varepsilon^{(i)}(t)| \le c_0^{(i)}|\Lambda_\varepsilon^{(i)}(0)|e^{-\gamma_0^{(i)}t} + \delta, t \in [0, T^{(i)}(\varepsilon)]\} = 1 \qquad (6.56)$$

with $\lim_{\varepsilon \to 0} T^{(i)}(\varepsilon) = \infty$, where $\Lambda_\varepsilon^{(1)}(t) = (\tilde{r}_c(t) - \rho, \tilde{\theta}(t) - \frac{\pi}{2}, e(t) + q_r(R^2 + \rho^2))$ and $\Lambda_\varepsilon^{(2)}(t) = (\tilde{r}_c(t) - \rho, \tilde{\theta}(t) + \frac{\pi}{2}, e(t) + q_r(R^2 + \rho^2))$. By the results (6.55), (6.56), together with the fact $|\tilde{r}_c(t) - \rho| < |(\tilde{r}_c(t) - \rho, \tilde{\theta}(t) \pm \frac{\pi}{2}, e(t) + q_r(R^2 + \rho^2))|$ and the definition of $\tilde{r}_c$, we have

$$\lim_{\varepsilon \to 0} \inf\{t \ge 0 : \left| |r_c(t) - r^*| - \rho \right| > C_0^{(i)}e^{-\gamma_0^{(i)}t} + \delta\} = \infty \quad \text{a.s.} \qquad (6.57)$$

and

$$\lim_{\varepsilon \to 0} P\{\left| |r_c(t) - r^*| - \rho \right| \le C_0^{(i)}e^{-\gamma_0^{(i)}t} + \delta, \forall t \in [0, T^{(i)}(\varepsilon)]\} = 1 \qquad (6.58)$$

with $\lim_{\varepsilon \to 0} T^{(i)}(\varepsilon) = \infty$, where $C_0^{(1)} = c_0^{(1)}|(\tilde{r}_c(0) - \rho, \tilde{\theta}(0) - \frac{\pi}{2}, e(0) + q_r(R^2 + \rho^2))|$ and $C_0^{(2)} = c_0^{(2)}|(\tilde{r}_c(0) - \rho, \tilde{\theta}(0) + \frac{\pi}{2}, e(0) + q_r(R^2 + \rho^2))|$. This completes the proof. $\qquad \square$

6.4 Convergence Speed

Theorem 6.1 establishes exponential convergence, however, the convergence rate is determined by the complicated cubic polynomial (6.53), whose roots are in general hard to find analytically. However, for particular parameter choice, they can be found explicitly, as given in the next proposition.

Proposition 6.1 *Let the vehicle speed V_c and the parameter h of the washout filter be chosen according to the following relation:*

$$V_c = hR. \qquad (6.59)$$

Then the exponential convergence rate of the source seeking system in Theorem 6.1 is determined by the eigenvalues

$$\lambda_1 = -h, \qquad (6.60)$$

$$\lambda_2 = -\frac{d_0 q_r R^2 h I_1(a, g) I_2(2a, g)}{c I_2(a, g)}(1 - \sqrt{1 - \psi}), \qquad (6.61)$$

$$\lambda_3 = -\frac{d_0 q_r R^2 h I_1(a, g) I_2(2a, g)}{c I_2(a, g)}(1 + \sqrt{1 - \psi}), \qquad (6.62)$$

where

$$\psi = \frac{4c^3 I_2^3(a, g)}{d_0^2 q_r h R^2 I_1(a, g) I_2^2(2a, g)} > 0, \tag{6.63}$$

and the radius of the residual annulus is

$$\rho = \sqrt{\frac{h I_1(a, g)}{2 q_r c I_2(a, g)}}. \tag{6.64}$$

Proof With $V_c = hR$, the stability condition (6.16) becomes

$$0 > -\frac{I_2(2a, g)}{2 I_1(a, g) I_2(a, g)}, \tag{6.65}$$

which is satisfied for all parameters $a, g, h, R > 0$. Thus the characteristic polynomial (6.53) has all three roots with negative real parts. Let

$$H \triangleq 4 d_0 \rho^2 q_r^2 R^2 I_2(2a, g), \tag{6.66}$$

$$M \triangleq \frac{2 V_c^2 I_1^2(a, g)}{\rho^2}, \tag{6.67}$$

$$Q \triangleq 4 d_0 \rho^2 q_r^2 R \big[2 V_c I_1(a, g) I_2(a, g) \\ + h R \big(I_2(2a, g) - 2 I_1(a, g) I_2(a, g) \big) \big]. \tag{6.68}$$

Then we write the characteristic polynomial compactly as

$$\lambda^3 + (h + H)\lambda^2 + (M + Q)\lambda + hM = 0. \tag{6.69}$$

Denote by λ_i, $i = 1, 2, 3$, the roots of the polynomial (6.69). Then by the relation between the roots and the coefficients in the polynomial, we have

$$h + H = -\lambda_1 - \lambda_2 - \lambda_3, \tag{6.70}$$

$$hM = -\lambda_1 \lambda_2 \lambda_3, \tag{6.71}$$

$$M + Q = \lambda_1 \lambda_2 + \lambda_2 \lambda_3 + \lambda_3 \lambda_1. \tag{6.72}$$

At this point one can just verify (6.60)–(6.62) by direct substitution into (6.70)–(6.72), however, we explain how we have arrived at (6.60)–(6.62). Let $\lambda_1 = -h$. We shall show that this choice satisfies (6.70)–(6.72) by also finding λ_2 and λ_3 which satisfy (6.70)–(6.72). With $\lambda_1 = -h$, (6.70)–(6.72) become

$$H = -\lambda_2 - \lambda_3, \tag{6.73}$$

$$M = \lambda_2 \lambda_3, \tag{6.74}$$

$$Q = hH. \tag{6.75}$$

Substituting $V_c = hR$ into (6.68), we immediately see that (6.75) is verified. From (6.73) and (6.74) we see that we only need to solve the quadratic equation $\lambda^2 + H\lambda + M = 0$. Applying the formula for the roots of a quadratic equation, we arrive at

$$\lambda_2 = -2d_0\rho^2 q_r^2 R^2 I_2(2a, g) + \frac{1}{\rho}\sqrt{4d_0^2\rho^6 q_r^4 R^4 I_2^2(2a, g) - 2V_c^2 I_1^2(a, g)}$$

$$= -d_0 q_r R V_c \frac{I_1(a, g) I_2(2a, g)}{c I_2(a, g)}$$

$$+ \frac{R}{c}\sqrt{d_0^2 q_r^2 V_c^2 \frac{I_1^2(a, g) I_2^2(2a, g)}{I_2^2(a, g)} - 4h q_r c^3 I_1(a, g) I_2(a, g)}, \qquad (6.76)$$

$$\lambda_3 = -2d_0\rho^2 q_r^2 R^2 I_2(2a, g) - \frac{1}{\rho}\sqrt{4d_0^2\rho^6 q_r^4 R^4 I_2^2(2a, g) - 2V_c^2 I_1^2(a, g)}$$

$$= -d_0 q_r R V_c \frac{I_1(a, g) I_2(2a, g)}{c I_2(a, g)}$$

$$- \frac{R}{c}\sqrt{d_0^2 q_r^2 V_c^2 \frac{I_1^2(a, g) I_2^2(2a, g)}{I_2^2(a, g)} - 4h q_r c^3 I_1(a, g) I_2(a, g)}, \qquad (6.77)$$

which, with some simplifications, gives (6.60)–(6.63). This completes the proof. $\square$

Of the three eigenvalues in Proposition 6.1 one is real and can be placed arbitrarily far to the left by choosing h large, whereas the other two can either be real or conjugate complex. The optimal choice is where the eigenvalues λ_2 and λ_3 are equal, because otherwise, either one or both of these eigenvalues are closer to the imaginary axis then when $\lambda_2 = \lambda_3$. Unfortunately, this optimal eigenvalue placement cannot be achieved by intent, since the design parameters would have to depend on the unknown q_r, however, in the next corollary we state this result in order to note what the best achievable convergence speed is.

Corollary 6.1 *Let $V_c = hR$ and let the damping parameter be chosen as*

$$d_0 = \frac{1}{\sqrt{q_r}} \frac{2}{R h I_2(2a, g)} \sqrt{\frac{c^3 I_2^2(a, g)}{I_1(a, g)}}. \qquad (6.78)$$

Then the exponential convergence rate of the source seeking system in Theorem 6.1 is determined by the eigenvalues

$$\lambda_1 = -h, \qquad (6.79)$$

$$\lambda_2 = \lambda_3 = -2R\sqrt{q_r h c I_1(a, g) I_2(a, g)}, \qquad (6.80)$$

whereas the residual annulus is as in (6.64).

From Corollary 6.1 we note that the optimizing damping coefficient d_0 grows, whereas the convergence rate $\lambda_2 = \lambda_3$ decays, with a decrease of the parameter q_r, namely, with the flattening of the extremum, as should be expected. Not surprisingly, the residual annulus (6.64) also grows with the flattening of the extremum. The convergence speed grows, whereas the annulus size shrinks, with the tuning gain c.

Proposition 6.2 *For a fixed a, the optimal convergence speed* (6.80) *has a non-monotonic dependence on the noise intensity g, with the maximal convergence speed achieved for*

$$g^* = \sqrt{\frac{1}{a} \ln \frac{2a^2 + 1 + 2a}{2a^2 + 1 - 2a}}. \tag{6.81}$$

Proof It is enough to consider (6.80) and maximize

$$I_1(a, g) I_2(a, g) = \frac{1}{2} e^{-a^2 g^2/4} \left[e^{-(a-1)^2 g^2/4} - e^{-(a+1)^2 g^2/4} \right] \tag{6.82}$$

with respect to g^2. This completes the proof. $\square$

The non-monotonic dependence of the convergence speed on the noise intensity g is intuitive. If the noise is low, the gradient exploration is insufficient and the tuning process is ineffective. Too much noise, and the perturbation takes the trajectories too far from the average trajectory, slowing the approach to the annulus.

Proposition 6.3 *For $a \geq 1/2$ the annulus radius ρ defined in* (6.64) *is a decreasing function of the noise intensity g. For $a \in (0, 1/2)$, the radius ρ has a non-monotonoic dependence on g, with the minimal ρ achieved for*

$$g^\circ = \sqrt{\frac{1}{a} \ln \frac{1 + 2a}{1 - 2a}}. \tag{6.83}$$

Proof By considering (6.64) and minimizing

$$\frac{I_2(a, g)}{I_1(a, g)} = \frac{2 e^{(1/4) g^2}}{(e^{(a/2) g^2} - e^{-(a/2) g^2})} \tag{6.84}$$

with respect to g^2, we complete the proof. $\square$

Since we want to operate with a relatively small perturbation parameter a, the annulus-minimizing value of g in (6.83) is of interest. Both very large and very small intensities of perturbation noise result in a large annulus, whereas a medium range of g is optimal. It is worth comparing the optimizing g for convergence speed in (6.81) with the optimizing g for the annulus in (6.83). For small a, they are similar, which is very fortunate.

Fig. 6.3 (**a**) The trajectory of
the vehicle center for the case
of source with circular level
sets. The trajectory converges
to an annulus.
(**b**) A zoomed-in section of
the vehicle trajectory,
displaying the vehicle motion
more clearly. For both
simulations: $V_c = 0.1$,
$c = 10000$, $d_0 = 10$, $a = 0.1$,
$g = 1$, $\varepsilon = 0.01$, $R = 0.1$,
$f^* = 0$, $h = 1$, $q_r = 1.5$. The
source is at $r^* = (0, 0)$

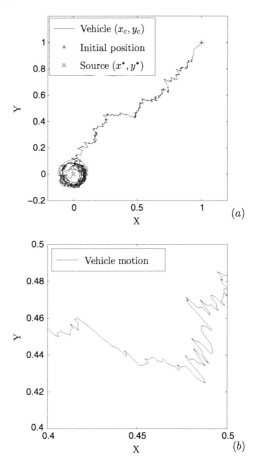

6.5 Simulations and Dependence on Design Parameters

6.5.1 Basic Simulations

Without loss of generality, we let the *unknown* location of the source be at the origin $r^* = (0, 0)$. We pick the design parameters as $V_c = 0.1$, $c = 10000$, $d_0 = 10$, $a = 0.1$, $h = 1$, $g = 1$, $\varepsilon = 0.01$, $R = 0.1$, and take the parameters of the map as $f^* = 0$, $q_r = 1.5$. The simulation results are given in Fig. 6.3. We observe that the trajectories of the vehicle center go to a small neighborhood of the source and the vehicle motion involves a random perturbation component, instead of a sinusoidal perturbation employed in the deterministic case [28]. In the simulations, we use band-limited white noise to approximate the white noise.

The stochastic source seeking approach can also be used for pursuit of nonstationary sources. For the case where the source is performing a "figure eight" motion, unknown to the pursuing vehicle, the simulation result is shown in Fig. 6.4.

Fig. 6.4 Vehicle following a moving source with circular level sets. The simulation parameters are $V_c = 0.1$, $c = 10000$, $d_0 = 10$, $a = 0.1$, $g = 1$, $\varepsilon = 0.01$, $R = 0.1$, $f^* = 0$, $h = 1$, $q_r = 1.5$. The source moves according to $x_{\text{src}}(t) = 0.5 \sin(0.13t)$, $y_{\text{src}}(t) = 0.5 \sin(0.26t)$

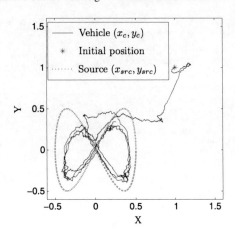

6.5.2 *Dependence of Annulus Radius* ρ *on Parameters*

From (6.20), we see that the radius ρ of the attractive annulus is dependent on the model parameters q_r, R and design parameters V_c, c, a, g, and that it can be made as small as desired. Hence, by (6.21) and (6.22), by making ρ as small as desired, the vehicle can converge as closely to the source as desired.

The dependence of ρ on the noise intensity is characterized by Proposition 6.3. Figure 6.5 shows some of this dependence. For a fixed small $a = 0.1$, the radius for $g = 2$ is $\rho = 0.021$, which is smaller than the radius $\rho = 0.029$ for $g = 1$.

6.6 Dependence on Damping Term d_0

Similar to the deterministic case in [28], the damping term $-d_0\xi^2 \sin(\eta)$ in the control law (6.3) for $\dot{\theta}$ plays a crucial role in achieving convergence of the vehicle to an annulus of radius ρ and arbitrarily small thickness δ near the source, cf. (6.21) and (6.22), and long-term retention (in a probabilistic sense) in that annulus. To analyze the effect of the damping term in the stochastic setting, we consider two cases.

6.6.1 *No Damping* ($d_0 = 0$)

From (6.49) and (6.50), the location of the equilibria are independent of d_0. Let $d_0 = 0$. Then the average error system (6.46)–(6.47) simplifies to

$$\frac{d\tilde{r}_c^{\text{ave}}}{dt} = -V_c I_1(a, g) \cos(\tilde{\theta}^{\text{ave}}), \tag{6.85}$$

$$\frac{d\tilde{\theta}^{\text{ave}}}{dt} = \sin(\tilde{\theta}^{\text{ave}}) \left(\frac{V_c}{\tilde{r}_c^{\text{ave}}} I_1(a, g) - 2cq_r R \tilde{r}_c^{\text{ave}} I_2(a, g) \right), \tag{6.86}$$

Fig. 6.5 The radius of the attractive annulus of the vehicle center for the case of source with circular level sets: (**a**) $g = 1$; (**b**) $g = 2$. The other simulation parameters are $V_c = 0.1$, $c = 10000$, $d_0 = 10$, $a = 0.1$, $\varepsilon = 0.01$, $R = 0.1$, $f^* = 0$, $h = 1$, $q_r = 1.5$. The source is at $r^* = (0, 0)$

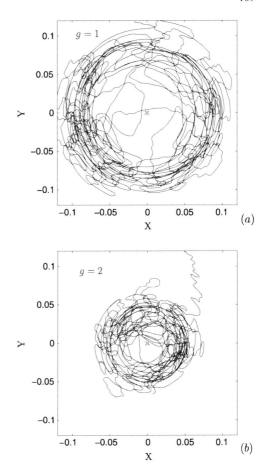

$$\frac{de^{\text{ave}}}{dt} = -h\left(q_r R^2 + q_r \tilde{r}_c^{\text{ave}^2} - 2R q_r \tilde{r}_c^{\text{ave}} \cos\left(\tilde{\theta}^{\text{ave}}\right) I_1(a, g)\right)$$
$$- h e^{\text{ave}}, \tag{6.87}$$

and the corresponding characteristic polynomial becomes

$$0 = \lambda^3 + h\lambda^2 + \frac{2V_c^2 I_1^2(a, g)}{\rho^2}\lambda + h\frac{2V_c^2 I_1^2(a, g)}{\rho^2}$$
$$= (\lambda + h)\left(\lambda^2 + \frac{2V_c^2 I_1^2(a, g)}{\rho^2}\right), \tag{6.88}$$

which means that the average error system has one stable eigenvalue and two purely imaginary eigenvalues, rendering it neutrally stable instead of exponentially stable. Thus the locally exponential convergence result in Theorem 6.1 does not hold, neither in almost sure sense, nor in the sense of convergence in probability. We add the

Fig. 6.6 Phase portrait of the average error system: (**a**) $d_0 = 0$ and the average error system is not exponentially stable but only marginally stable; (**b**) $d_0 = 10000$ and the average error system is exponentially stable. The simulation parameters are $V_c = 0.1$, $c = 1000$, $a = 0.1$, $g = 1$, $\varepsilon = 0.01$, $R = 0.1$, $f^* = 0$, $h = 1$, $q_r = 1.5$. The source is at $\tilde{r}_c^{ave} = 0$ (which is the $\tilde{\theta}^{ave}$-axis manifold in the state space of the average system)

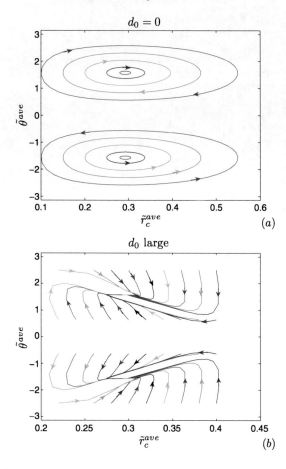

term $-d_0 \xi^2 \sin(\eta)$ to the stochastic extremum seeking control law (6.3) precisely in order to achieve local exponential stability of the average error system without changing its equilibria, and thus to obtain the local exponential convergence of the original error system. Figure 6.6 depicts the phase portrait of the average error system, from which we see for $d_0 > 0$ that the two equilibria are exponentially stable, each one with a region of attraction being exactly one half of the state space of the average error system $(\tilde{r}_c^{ave}, \tilde{\theta}^{ave}) \in \mathbb{R}_+ \times \{-\pi, \pi\}$, whereas for $d_0 = 0$ the two average equilibria are only neutrally stable.

6.6.2 Effect of Damping ($d_0 > 0$)

Figure 6.7 displays two distinct behaviors of the source seeking scheme. For large d_0, the vehicle undergoes a "roundabout" transient but settles quickly into a small neighborhood of the source (see Fig. 6.7(a)). From Fig. 6.7(b), we see that, for

Fig. 6.7 (**a**) The trajectories of the vehicle center for different values of d_0. (**b**) A zoomed-in section of attractive annulus with different values of d_0. The simulation parameters are $V_c = 0.1, c = 1000, a = 0.1, g = 1, \varepsilon = 0.01, R = 0.1, f^* = 0, h = 1, q_r = 1.5$. The source is at $r^* = (0, 0)$

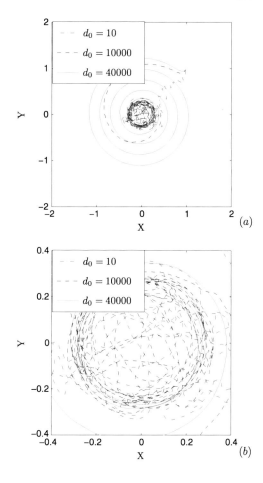

smaller d_0, the vehicle manifests distinct overshoot phenomenon before setting to the attractive annulus, and owing to the use of stochastic perturbation, the vehicle moves randomly and even turns around in the small neighborhood of the source. From Fig. 6.3(b), we see that, for a small d_0, the vehicle makes sharp turns, particularly during the transient motion toward the annulus. However, by taking a larger d_0, the vehicle avoids sharp turns and moves in a smoother way (see Fig. 6.7(a)). By smoothing the trajectories, the damping term actually steers the vehicle faster to the origin, even though the *average* trajectory appears more roundabout. To see this point, consider Fig. 6.8 which shows three complete trajectories over the time interval $[0, T]$, for $T = 60$. For a small value of the damping coefficient, $d_0 = 10$, the vehicle has not even reached the annulus over the time interval considered, whereas for the large damping value, $d_0 = 40000$, the vehicle has long arrived to the annulus and has moved several times around the source in the same time interval.

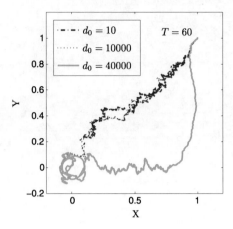

Fig. 6.8 The trajectories of the vehicle center, for different values of d_0, showing that the vehicle arrives faster to the source for larger values of d_0, due to smoothing, or "straightening out", of its trajectory. For $d_0 = 10000$, the trajectory just makes it to the annulus in $T = 60$ seconds. Over the same time interval, the trajectory does not make it yet to the annulus for a smaller d_0, but arrives much sooner to the annulus, and makes a couple of revolutions around it, for a larger d_0. The simulation parameters are $V_c = 0.1$, $c = 10000$, $a = 0.1$, $g = 1$, $\varepsilon = 0.01$, $R = 0.1$, $f^* = 0$, $h = 1$, $q_r = 1.5$. The source is at $r^* = (0, 0)$

6.7 Effect of Constraints of the Angular Velocity and Design Alternatives

6.7.1 Effect of Constraints of the Angular Velocity

A physical vehicle always has a steering constraint, namely, a limit on the angular velocity $\dot{\theta}$. This type of a unicycle model is commonly referred to as the Dubins vehicle. Figure 6.9 depicts the trajectories of the vehicle center when the angular velocity is restricted to a symmetric interval, $[-u_{max}, +u_{max}]$, for several values of u_{max}. We observe that, for u_{max} as small as 20, our control law successfully steers the vehicle to the annulus, and keeps the vehicle near the source, see Fig. 6.9(a). In addition, the vehicle moves more smoothly for smaller u_{max}, see Fig. 6.9(b). However, if the actuator constraint u_{max} is too small, for example, $u_{max} = 10$, the algorithm cannot keep the vehicle very near the source, as observed in Fig. 6.10.

6.7.2 Alternative Designs

In the standard extremum seeking algorithm (see [6]), the probing signal and the de-modulation signal are the same, typically $\sin(\omega t)$. Looking at the probing equation (6.24) and the demodulation equation (6.28) in the present work, the reader should note that the probing and demodulation signals are different. They are η and $\sin(\eta)$,

Fig. 6.9 (a) The trajectories of the vehicle center for different constraints of angular velocity. (b) A zoomed-in section of vehicle motion for different constraints u_{max} on angular velocity. The simulation parameters are $V_c = 0.1$, $c = 10000$, $a = 0.1$, $g = 1$, $\varepsilon = 0.01$, $R = 0.1$, $f^* = 0$, $h = 1$, $q_r = 1.5$. The source is at $r^* = (0, 0)$

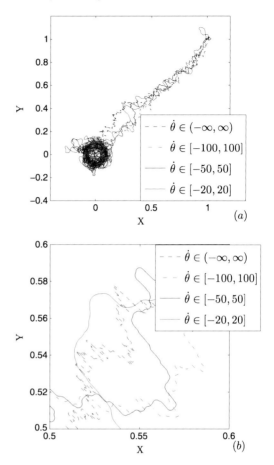

respectively. In this chapter, we make such a choice for the sake of simplicity of calculating the average error system in the stability analysis—the integrals in the expectations are easier to obtain analytically with such a choice. If η is replaced by $\sin(\eta)$ as the stochastic perturbation in (6.24), the extremum seeking control (6.3) is replaced by

$$\dot{\theta} = a\cos(\eta)\dot{\eta} - \frac{ag^2}{2\varepsilon}\sin(\eta) + c\xi\sin(\eta) - d_0\xi^2\sin(\eta) \qquad (6.89)$$

and thus (6.11) in the closed-loop system changes to

$$d\theta = \left[\frac{-a}{\varepsilon}\cos(\eta)\eta - \frac{ag^2}{2\varepsilon}\sin(\eta)\right]dt$$
$$+ \left(c\xi - d_0\xi^2\right)\sin(\eta)\,dt + \frac{ag}{\sqrt{\varepsilon}}\cos(\eta)\,dW, \qquad (6.90)$$

Fig. 6.10 The trajectories of the vehicle center under a severe constraint on the angular velocity input ($u_{max} = 10$). The simulation parameters are $V_c = 0.1$, $c = 10000$, $a = 0.1$, $g = 1$, $\varepsilon = 0.01$, $R = 0.1$, $f^* = 0$, $h = 1$, $q_r = 1.5$. The source is at $r^* = (0, 0)$

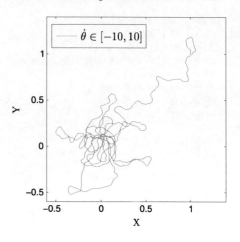

Fig. 6.11 The trajectory of the vehicle center under the control law (6.89). The simulation parameters are $V_c = 0.1$, $c = 10000$, $d_0 = 10$, $a = 0.1$, $g = 1$, $\varepsilon = 0.01$, $R = 0.1$, $f^* = 0$, $h = 1$, $q_r = 1.5$. The source is at $r^* = (0, 0)$

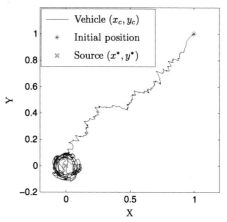

where the additional term $-\frac{ag^2}{2\varepsilon}\sin(\eta)$ results from the Ito formula. Consequently, the two terms $\cos(\tilde{\theta} + a\chi(t/\varepsilon))$ and $\sin(\tilde{\theta} + a\chi(t/\varepsilon))$ in the error system (6.37), (6.38), and (6.39) should be replaced by $\cos(\tilde{\theta} + a\sin(\chi(t/\varepsilon)))$ and $\sin(\tilde{\theta} + a\sin(\chi(t/\varepsilon)))$, respectively. It is hard to obtain the corresponding analytical average error system because we need to calculate two integrals: $\int_{-\infty}^{+\infty}\cos(a\sin(y))e^{-y^2/g^2}\,dy$ and $\int_{-\infty}^{+\infty}\sin(a\sin(y))\sin(y)e^{-y^2/g^2}\,dy$ and it is hard to obtain the analytical results though we can obtain numerical results. Figure 6.11 depicts the trajectory of the vehicle center when the control law (6.89) is used. From the simulation, there is no noticeable difference relative to the trajectory in Fig. 6.3(a).

Now we analyze the radius of the annulus for three alternative perturbation signals. Let $V_c = 0.1$, $c = 10000$, $d_0 = 10$, $a = 0.1$, $g = 1$, $\varepsilon = 0.01$, $R = 0.1$, $f^* = 0$, $q_r = 1.5$. Then

1. For the probing signal η in (6.24) and demodulation signal $\sin(\eta)$ in (6.28), we obtain the radius of the annulus as $\rho^1 = 0.0293$.

2. If we use $\sin(\eta)$ to replace η as the probing signal in (6.24), the expressions $I_1(a, g)$ and $I_2(a, g)$ are replaced by $I_1^*(a, g)$ and $I_2^*(a, g)$, where

$$I_1^*(a, g) \triangleq \int_{\mathbb{R}} \cos(a\sin(y))\mu(dy)$$

$$= \int_{-\infty}^{+\infty} \cos(a\sin(y))\frac{1}{\sqrt{\pi}g}e^{-y^2/g^2}\,dy$$

and

$$I_2^*(a, g) \triangleq \int_{\mathbb{R}} \sin(a\sin(y))\sin(y)\mu(dy)$$

$$= \int_{-\infty}^{+\infty} \sin(a\sin(y))\sin(y)\frac{1}{\sqrt{\pi}g}e^{-y^2/g^2}\,dy.$$

By calculating the integrals numerically, we obtain $I_1^*(0.1, 1) = 0.9984$ and $I_2^*(0.1, 1) = 0.0316$. Thus, we get the radius of the annulus as $\rho^{II} = \sqrt{\frac{V_c I_1^*(a,g)}{2q_r cR I_2^*(a,g)}} = 0.0325$, which is a little larger than ρ^{I}.

3. If we use the bounded function $\eta e^{-\eta^2}$ to replace both η as the probing signal in (6.24) and $\sin(\eta)$ as the demodulating signal in (6.28), by numerical calculation we obtain

$$\int_{\mathbb{R}} \cos(0.1ye^{-y^2})\mu(dy) = \int_{-\infty}^{+\infty} \cos(0.1ye^{-y^2})\frac{1}{\sqrt{\pi}g}e^{-y^2/g^2}\,dy$$

$$= 0.9995,$$

$$\int_{\mathbb{R}} \sin(0.1ye^{-y^2})ye^{-y^2}\mu(dy) = \int_{-\infty}^{+\infty} \sin(0.1ye^{-y^2})ye^{-y^2}\frac{1}{\sqrt{\pi}g}e^{-y^2/g^2}\,dy$$

$$= 0.0096.$$

Thus the radius is $\rho^{III} = 0.0588$, which is considerably larger than both ρ^{I} and ρ^{II}.

Therefore, from the point of view of the annulus radius, our choice η as the probing signal in (6.24) and $\sin(\eta)$ as the demodulation signal in (6.28) achieves the best performance, in addition to facilitating the analysis.

If OU process $(\eta(t), t \geq 0)$ is used not only as the probing signal, but also as the demodulation signal in (6.28), the extremum seeking control law (6.3) is replaced by

$$\dot{\theta} = a\dot{\eta} + (c\xi - d_0\xi^2)\eta. \tag{6.91}$$

With $\sin(\eta)$ replaced by η as a demodulation signal, where the latter signal is not uniformly bounded, the local Lipschitz condition (Assumption 4.1) is not satisfied uniformly in the perturbation process for the resulting closed-loop system. For this reason, we cannot use general stochastic averaging theorem to analyze stability.

Fig. 6.12 The trajectory of the vehicle center under the control law (6.91). The simulation parameters are $V_c = 0.1, c = 10000,$ $d_0 = 10, a = 0.1, g = 1,$ $\varepsilon = 0.01, R = 0.1, f^* = 0,$ $h = 1, q_r = 1.5$. The source is at $r^* = (0, 0)$

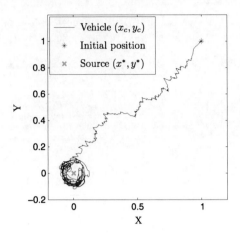

However, from simulation results given by Fig. 6.12, we observe that the vehicle achieves convergence to a an annulus near the source under the control law (6.91).

6.8 System Behavior for Elliptical Level Sets

Our analysis is limited to circular level sets, namely, to fields that depend on the distance from the source only. In this section, we present simulation results for elliptical level sets. Without loss of generality, we assume the source is at $r^* = (0, 0)$, and the signal distribution in space is given (at the sensor location) by

$$J = f(r_s) = f^* - q_r |r_s|^2 - q_p (r_s^2 + \bar{r}_s^2)$$
$$= f^* - (q_r + 2q_p)x_s^2 - (q_r - 2q_p)y_s^2$$
$$= f^* - q_r |r_c + Re^{j\theta}|^2 - q_p ((r_c + Re^{j\theta})^2 + (\bar{r}_c + Re^{-j\theta})^2), \quad (6.92)$$

where $q_r > 0, q_r \pm 2q_p > 0$.

Figure 6.13 depicts the trajectory of the vehicle center for a signal field with elliptical level sets. The vehicle reaches a small neighborhood of the source, however, the average motion is not circular revolution around the source, nor elliptical revolution, but a motion biased to one of the flatter sides of the ellipse. More than one such attractor exists. It depends on the initial condition and on the noise sequence which of the average attractors the trajectory will converge to.

Figure 6.14 depicts the trajectories of the vehicle center with different d_0-values in the control law. From Fig. 6.14(a), we see that for larger d_0 the vehicle undergoes a "roundabout" behavior and then moves into a small neighborhood of the source. This is no different than the situation for circular level sets, with either stochastic or deterministic source seeking algorithms. However, from Fig. 6.14(b), we observe a difference relative to the results obtained for elliptical level sets in the deterministic case in [28]. The value of d_0 does not affect the shape and size of the system attractors—the motion near the source is limited to an elliptical shape.

Fig. 6.13 The trajectory of
the vehicle center for signal
field with elliptical level sets.
The simulation parameters
are $V_c = 0.1$, $c = 10000$,
$d_0 = 10$, $a = 0.1$, $g = 1$,
$\varepsilon = 0.01$, $R = 0.1$, $f^* = 0$,
$h = 1$, $q_r = 1.5$, $q_p = 0.25$.
The source is located at
$r^* = (0, 0)$

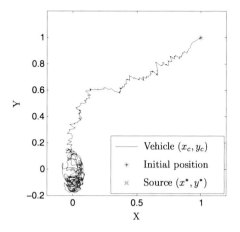

6.9 Notes and References

The research on GPS-denied source seeking has been initiated by the second author
and his students in [147] (for fully actuated point mass vehicles in 2D), [28, 46, 146]
(for nonholonomic unicycles), [29] (for underactuated vehicles in 3D), and [30] (for
models of fish locomotion involving multi-link structures in ideal fluid and flexible
hydrofoils in vertical flows).

In this chapter, we have investigated a stochastic version of source seeking by
navigating the unicycle with the help of a random perturbation, achieving a behavior
that mimics the chemotaxis-like motion observed in the bacterium *Escherichia coli*
(*E. coli*). *E. coli* is a single celled organism consisting of a cell body with multiple
trailing flagella used for propulsion. In [18] and [19], it is observed that the bacterium is able to move up chemical gradients toward higher densities of nutrients by
switching between alternate behaviors known as "run" and "tumble". The behavior
"run" means that the bacterium moves in essentially a straight line by rotating the
flagella counter-clockwise as viewed from behind the cell and the behavior "tumble"
means that the bacterium ceases forward motion and spins by turning some flagella
in a clockwise direction. It is also observed that the tumble behavior displays ap-
parent random nature, although the net motion of the bacterium is not completely
random but is in the direction of higher nutrient concentrations.

Motivated by the chemotactic behavior of *E. coli*, in this chapter we have considered the problem of stochastic source seeking for a nonholonomic unicycle. The
analogy is appropriate since neither the unicycle nor *E. coli* can exhibit sideways
motions, though they can be steered. The unicycle vehicle that we considered in this
chapter has no knowledge of its own position, or of the position of the source. It is
only able to sense a scalar signal which emanates from the source. In an application
to autonomous vehicles, the signal could be the concentration of a chemical or biological agent, or it could be an electromagnetic, acoustic, thermal or radar signal.
The strength of the signal is assumed to decay away from the source through diffusion or other physical processes, however, the spatial distribution of the signal is not
available to the vehicle.

Fig. 6.14 Signal field with elliptical level sets. (**a**) The trajectories of the vehicle center for different d_0-values. (**b**) A zoomed-in section of attractors for different d_0-values. The simulation parameters are $V_c = 0.1$, $c = 10000$, $a = 0.1$, $g = 1$, $\varepsilon = 0.01$, $R = 0.1$, $f^* = 0$, $h = 1$, $q_r = 1.5$, $q_p = 0.25$. The source is located at $r^* = (0, 0)$

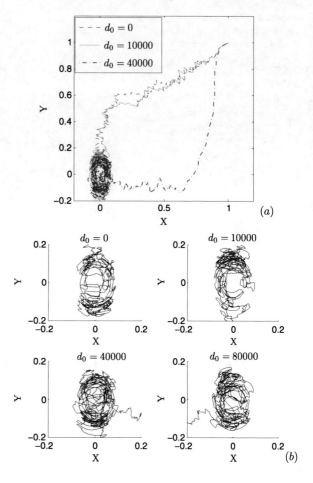

To find the source, we employed a stochastic extremum seeking approach and provided a stability analysis based on stochastic averaging theorems that we developed in Chap. 4. With a controller that we designed in this chapter, the vehicle is driven to approach a small neighborhood of the source in a manner that seems partly random but is convergent in a suitable sense. We presented a stability proof for the scheme with a static source and simulation results for both static and moving sources. Convergence was proved both in the "almost sure" sense and "in probability".

It is important to consider the relative merits of the deterministic solution to the source seeking problem in [28] and the stochastic solution presented here. As expected, the steering inputs in the stochastic approach are less smooth, which is a disadvantage of the stochastic approach from the viewpoint of actuator wear. However, the nearly random motion of the stochastic seeker has its advantage in applications where the seeker itself may be pursued by another pursuer. A seeker, which successfully performs the source finding task but with an unpredictable, nearly random trajectory, is a more challenging target, and is hence less vulnerable, than a deterministic seeker.

Motivated by *E. coli* chemotaxis, in [100] the authors consider a similar problem of seeking the maximum of a scalar signal, using a swarm of autonomous vehicles, and propose a control design which induces the vehicles to perform a biased random walk, with a net motion of the swarm toward the maximum, and achieving higher vehicle densities near the maximum at the end of the search. Besides the difference in the algorithms presented in [100] and in the present work, different results are proved. The result in [100] guarantees that the probability density function of the positions of the vehicles evolves toward a specified function of the spatial profile of the measured signal, whereas in this chapter we proved convergence (in probability and almost surely), for any single vehicle, to a specific small neighborhood of the source.

Another significant difference is that we establish exponential convergence, and in fact characterize the best achievable value and the worst-case value of the exponential convergence rate as a function of the design parameters. In contrast, in [100] exponential convergence is not shown, or formally claimed. A considerable difference in performance is also observed in simulations. The algorithm in [100] at best matches the convergence of the deterministic algorithm in [146], whereas the present algorithm has superior convergence to that in [146] as it does not employ motions that would, in the absence of a gradient, keep a vehicle in place on the average (such as random walk, or the triangle and diamond-shaped gaits in [146]), but employs a strategy that keeps the vehicle moving in some average direction even when the gradient is zero, as is the case with the design in [28]. However, it is important to note that the results we proved here are only for signal fields that have circular level sets, whereas in [100] such a restriction is not present.

The results of this chapter are not difficult to extend to 3D source seeking, as in [29], for underwater vehicle applications, or even to source seeking for fish models, as in [30].

Chapter 7
Stochastic Source Seeking with Tuning of Forward Velocity

In this chapter, we investigate the same source seeking problem as in Chap. 6, but by controlling the forward velocity of the vehicle instead of the angular velocity.

The chapter is organized as follows. In Sect. 7.1, we give the description of the vehicle model and state the problem. In Sect. 7.2, we present our stochastic source seeking controller and prove local exponential convergence to a small neighborhood for the case where the signal field has circular level sets, namely, where the signal depends only on the distance from the source and decays quadratically. In Sect. 7.3, we present simulations. Section 7.4 contains some notes and references.

7.1 The Model of Autonomous Vehicle

We consider a unicycle model of a mobile robot with sensor that is collocated at the center of the vehicle. A diagram depicting the position, heading, angular and forward velocities, and the sensor location on the autonomous vehicle is shown in Fig. 7.1. The equations of motion for the vehicle center are

$$\dot{x}_c = v\cos(\theta), \tag{7.1}$$

$$\dot{y}_c = v\sin(\theta), \tag{7.2}$$

$$\dot{\theta} = \omega_0, \tag{7.3}$$

where $(x_c, y_c) = z_c$ is the center of the vehicle, θ is the orientation, and v, ω_0 are the forward and angular velocity inputs. Our stochastic extremum seeking algorithm will tune only the forward velocity input v, while keeping the angular velocity input ω_0 constant.

Different from the case in Chap. 6 and [90] of tuning the angular velocity, the sensor is collocated with the vehicle center. For the non-collocated sensor case, where the sensor is mounted some distance away from the center, there is no essential difference but the calculations are more complex.

S.-J. Liu, M. Krstic, *Stochastic Averaging and Stochastic Extremum Seeking*,
Communications and Control Engineering,
DOI 10.1007/978-1-4471-4087-0_7, © Springer-Verlag London 2012

Fig. 7.1 The notation used in the vehicle model where $z_c = (x_c, y_c)$ is the center of the vehicle

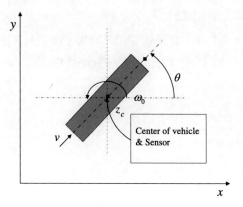

Fig. 7.2 Stochastic extremum seeking scheme for unicycle with forward velocity tuning

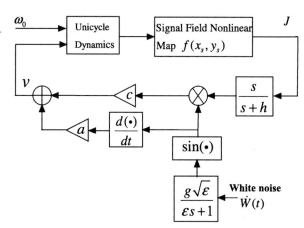

7.2 Search Algorithm and Convergence Analysis

We assume that the signal source being tracked is distributed according to an unknown nonlinear map $J = f(x, y)$, which has an isolated local maximum $f^* = f(x^*, y^*)$ at (x^*, y^*). Our purpose is to control the autonomous vehicle to achieve local convergence to the maximizer (x^*, y^*) without knowledge of the shape of $f(x, y)$ and using only the measurements of its value at the vehicle's position. A block diagram of the stochastic extremum seeking scheme is shown in Fig. 7.2.

Consider the controller for the forward velocity,

$$v = cJ \sin(\eta) + \frac{a}{\varepsilon} \left[-\left(\eta \cos(\eta) + \frac{1}{2} g^2 \sin(\eta) \right) + g \sqrt{\varepsilon} \cos(\eta) \dot{W} \right], \quad (7.4)$$

where $d\eta = -\frac{1}{\varepsilon} \eta \, dt + \frac{g}{\sqrt{\varepsilon}} dW$, $0 < \varepsilon < \varepsilon_0$ is a small parameter for fixed $\varepsilon_0 > 0$, $\dot{W}$ denotes the white noise, and $c, a, q > 0$ are design parameters. For our analysis, we assume that the nonlinear map is quadratic:

$$J = J^* + \frac{1}{2}(z_c - z^*)^T H(z_c - z^*),$$ (7.5)

where $H = H^T < 0$, $z_c = [x_c, y_c]^T$, and $z^* = (x^*, y^*)$.

After defining

$$x_c - x^* = \tilde{x} + a\cos(\theta)\sin(\eta),$$ (7.6)
$$y_c - y^* = \tilde{y} + a\sin(\theta)\sin(\eta),$$ (7.7)

we can rewrite (7.5) as

$$J = J^* + \frac{1}{2}\left(\tilde{Z}^\varepsilon + a\begin{bmatrix} \cos(\theta) \\ \sin(\theta) \end{bmatrix}\sin(\eta)\right)^T H\left(\tilde{Z}^\varepsilon + a\begin{bmatrix} \cos(\theta) \\ \sin(\theta) \end{bmatrix}\sin(\eta)\right),$$ (7.8)

where $\tilde{Z}^\varepsilon$ denotes the error variables

$$\tilde{Z}^\varepsilon = \begin{bmatrix} \tilde{x} \\ \tilde{y} \end{bmatrix}.$$ (7.9)

Here we use the superscript ε to make the dependence on the small parameter ε more clear. Then by (7.1), (7.2), (7.3), (7.4), (7.6), (7.7), and Ito formula, we have the following error dynamics

$$\frac{d\tilde{Z}^\varepsilon}{dt} = \left(-a\omega_0\begin{bmatrix} -\sin(\theta) \\ \cos(\theta) \end{bmatrix} + cJ\begin{bmatrix} \cos(\theta) \\ \sin(\theta) \end{bmatrix}\right)\sin(\eta)$$

$$= \begin{bmatrix} -\sin(\theta) \\ \cos(\theta) \end{bmatrix}(-a\omega_0)\sin(\eta)$$

$$+ c\begin{bmatrix} \cos(\theta) \\ \sin(\theta) \end{bmatrix}\sin(\eta)\left\{J^* + \frac{1}{2}\tilde{Z}^{\varepsilon^T} H\tilde{Z}^\varepsilon + a\begin{bmatrix} \cos(\theta) \\ \sin(\theta) \end{bmatrix}^T\sin(\eta)H\tilde{Z}^\varepsilon$$

$$+ \frac{a^2}{2}\begin{bmatrix} \cos(\theta) \\ \sin(\theta) \end{bmatrix}^T H\begin{bmatrix} \cos(\theta) \\ \sin(\theta) \end{bmatrix}\sin^2(\eta)\right\},$$ (7.10)

$$\frac{d\theta}{dt} = \omega_0.$$ (7.11)

By the definition of Ito stochastic differential equation, we have

$$\eta(t) = \eta(0) - \int_0^t \frac{1}{\varepsilon}\eta(s)\,ds + \int_0^t \frac{g}{\sqrt{\varepsilon}}\,dW(s).$$ (7.12)

Thus, it holds that

$$\eta(\varepsilon t) = \eta(0) - \int_0^t \eta(\varepsilon u)\,du + \int_0^t \frac{g}{\sqrt{\varepsilon}}\,dW(\varepsilon u).$$ (7.13)

Now define

$$B(t) = \frac{1}{\sqrt{\varepsilon}} W(\varepsilon t), \tag{7.14}$$

$$\chi(t) = \eta(\varepsilon t). \tag{7.15}$$

Then, the error dynamics (7.10)–(7.11) are transformed to

$$\frac{d\tilde{Z}^\varepsilon}{dt} = \begin{bmatrix} -\sin(\theta) \\ \cos(\theta) \end{bmatrix} (-a\omega_0) \sin(\chi(t/\varepsilon))$$

$$+ c \begin{bmatrix} \cos(\theta) \\ \sin(\theta) \end{bmatrix} \sin(\chi(t/\varepsilon)) \left\{ J^* + \frac{1}{2} \tilde{Z}^{\varepsilon T} H \tilde{Z}^\varepsilon \right.$$

$$+ a \begin{bmatrix} \cos(\theta) \\ \sin(\theta) \end{bmatrix}^T \sin(\chi(t/\varepsilon)) H \tilde{Z}^\varepsilon$$

$$+ \frac{a^2}{2} \begin{bmatrix} \cos(\theta) \\ \sin(\theta) \end{bmatrix}^T H \begin{bmatrix} \cos(\theta) \\ \sin(\theta) \end{bmatrix} \sin^2(\chi(t/\varepsilon)) \right\}, \tag{7.16}$$

$$\frac{d\theta}{dt} = \omega_0, \tag{7.17}$$

where

$$d\chi(t) = -\chi(t)\,dt + g\,dB(t), \tag{7.18}$$

$B(t)$ is a standard Brownian motion and the process $\chi(t)$ is an Ornstein–Uhlenbeck (OU) process, which is ergodic with invariant distribution

$$\mu(dy) = \frac{1}{\sqrt{\pi}\,g} e^{-y^2/g^2}\,dy. \tag{7.19}$$

Since

$$\int_{\mathbb{R}} \sin^{2k+1}(y)\mu(dy) = \int_{-\infty}^{+\infty} \sin^{2k+1}(y)\frac{1}{\sqrt{\pi}\,g} e^{-y^2/g^2}\,dy = 0, \quad k = 0, 1,$$

$$\int_{\mathbb{R}} \sin^2(y)\mu(dy) = \int_{-\infty}^{+\infty} \sin^2(y)\frac{1}{\sqrt{\pi}\,g} e^{-y^2/g^2}\,dy = \frac{1}{2}(1 - e^{-g^2}), \tag{7.20}$$

by (4.12), we obtain the average system of (7.16)–(7.17),

$$\frac{d\tilde{Z}^{\mathrm{ave}}}{dt} = \frac{ca}{2}(1 - e^{-g^2}) \begin{bmatrix} \cos(\theta^{\mathrm{ave}}) \\ \sin(\theta^{\mathrm{ave}}) \end{bmatrix} \begin{bmatrix} \cos(\theta^{\mathrm{ave}}) \\ \sin(\theta^{\mathrm{ave}}) \end{bmatrix}^T H \tilde{Z}^{\mathrm{ave}}, \tag{7.21}$$

$$\frac{d\theta^{\mathrm{ave}}}{dt} = \omega_0. \tag{7.22}$$

The average system (7.21)–(7.22), is not exponentially stable, however, since

$$\theta^{\mathrm{ave}}(t) = \theta_0 + \omega_0 t, \quad \forall t \geq 0, \tag{7.23}$$

where $\theta_0 = \theta(0)$, (7.21) can be rewritten as the linear time-varying system

$$\frac{d\tilde{Z}^{\mathrm{ave}}}{dt} = \frac{ca}{2}(1 - e^{-g^2})\Psi(t)H\tilde{Z}^{\mathrm{ave}}, \tag{7.24}$$

where

$$\Psi(t) = \begin{bmatrix} \cos(\theta_0 + \omega_0 t) \\ \sin(\theta_0 + \omega_0 t) \end{bmatrix} \begin{bmatrix} \cos(\theta_0 + \omega_0 t) \\ \sin(\theta_0 + \omega_0 t) \end{bmatrix}^T. \tag{7.25}$$

By the property of persistency of excitation (PE) of $[\cos(\theta_0 + \omega_0 t), \sin(\theta_0 + \omega_0 t)]^T$, (7.23) can be shown to be exponentially stable for any θ_0 and any $\omega_0 \neq 0$ (see Lemma 3.4, [138]). Thus, there exist $M, \gamma > 0$ such that the following holds for the average system (7.21)–(7.22):

$$\left| \tilde{Z}^{\mathrm{ave}}(t) \right| \leq M e^{-\gamma t} \left| \tilde{Z}_0^{\mathrm{ave}} \right|, \tag{7.26}$$

$$\theta^{\mathrm{ave}}(t) = \theta_0 + \omega_0 t, \quad \forall t \geq 0, \tag{7.27}$$

for all $(\tilde{Z}^{\mathrm{ave}}(0), \theta^{\mathrm{ave}}(0)) = (\tilde{Z}_0^{\mathrm{ave}}, \theta_0) \in \mathbb{R}^3$ and all $\omega_0 \neq 0$.

If the error dynamics (7.16)–(7.17) have a unique continuous solution on $[0, \infty)$ and $\tilde{Z}^{\varepsilon}(0) = \tilde{Z}^{\mathrm{ave}}(0)$, then by the approximate result Theorem 4.1, we have for any $\delta > 0$

$$\liminf_{\varepsilon \to 0} \left\{ t \geq 0 : \left| \begin{bmatrix} \tilde{Z}^{\varepsilon}(t) \\ \theta(t) \end{bmatrix} - \begin{bmatrix} \tilde{Z}^{\mathrm{ave}}(t) \\ \theta^{\mathrm{ave}}(t) \end{bmatrix} \right| > \delta \right\} = +\infty \quad \text{a.s.}, \tag{7.28}$$

and there exists a function $T(\varepsilon) : (0, \varepsilon_0) \to \mathbb{N}$ such that for any $\delta > 0$

$$\lim_{\varepsilon \to 0} P \left\{ \sup_{0 \leq t \leq T(\varepsilon)} \left| \begin{bmatrix} \tilde{Z}^{\varepsilon}(t) \\ \theta(t) \end{bmatrix} - \begin{bmatrix} \tilde{Z}^{\mathrm{ave}}(t) \\ \theta^{\mathrm{ave}}(t) \end{bmatrix} \right| > \delta \right\} = 0, \tag{7.29}$$

where

$$\lim_{\varepsilon \to 0} T(\varepsilon) = +\infty. \tag{7.30}$$

Noting (7.26) and (7.27), we have

$$\liminf_{\varepsilon \to 0} \left\{ t \geq 0 : \left| \begin{bmatrix} \tilde{Z}^{\varepsilon}(t) \\ \theta(t) - \theta_0 - \omega_0 t \end{bmatrix} \right| \leq M e^{-\gamma t} |\tilde{Z}_0^{\varepsilon}| + \delta \right\} = +\infty \quad \text{a.s.} \tag{7.31}$$

and

$$\lim_{\varepsilon \to 0} P \left\{ \left| \begin{bmatrix} \tilde{Z}^{\varepsilon}(t) \\ \theta(t) - \theta_0 - \omega_0 t \end{bmatrix} \right| \leq M e^{-\gamma t} |\tilde{Z}_0^{\varepsilon}| + \delta, \forall t \in [0, T(\varepsilon)] \right\} = 1. \tag{7.32}$$

Finally, recall that

$$\tilde{Z}^{\varepsilon} = \begin{bmatrix} x - x^* \\ y - y^* \end{bmatrix} + a \begin{bmatrix} \cos(\theta) \\ \sin(\theta) \end{bmatrix} \sin(\eta), \tag{7.33}$$

which leads us to the following result.

Theorem 7.1 *Consider the vehicle (7.1)–(7.3) under the controller (7.4). If the error dynamics (7.16)–(7.17) have a unique continuous solution on $[0, \infty)$, then there exist constants $r, c, \gamma > 0$, and a function $T(\varepsilon) : (0, \varepsilon_0) \to \mathbb{N}$ such that, for any initial condition $|z_c(0) - z^*| < r$ and any $\delta > 0$,*

$$
\liminf_{\varepsilon \to 0} \left\{ t \geq 0 : \left\| \begin{bmatrix} x^\varepsilon(t) - x^* \\ y^\varepsilon(t) - y^* \\ \theta(t) - \theta_0 - \omega_0 t \end{bmatrix} \right\| > c \left\| \begin{bmatrix} x_0 - x^* \\ y_0 - y^* \end{bmatrix} \right\| e^{-\gamma t} + \delta + O(a) \right\}
$$

$$
= +\infty \quad a.s. \tag{7.34}
$$

and

$$
\lim_{\varepsilon \to 0} P \left\{ \left\| \begin{bmatrix} x^\varepsilon(t) - x^* \\ y^\varepsilon(t) - y^* \\ \theta(t) - \theta_0 - \omega_0 t \end{bmatrix} \right\| \leq c \left\| \begin{bmatrix} x_0 - x^* \\ y_0 - y^* \end{bmatrix} \right\| e^{-\gamma t} + \delta + O(a), \right.
$$

$$
\left. \forall t \in \left[0, T(\varepsilon) \right] \right\} = 1 \tag{7.35}
$$

with $\lim_{\varepsilon \to 0} T(\varepsilon) = +\infty$.

Remark 7.1 In summary, for the stability analysis of the error dynamics (7.16)–(7.17), we compute the average system (7.21)–(7.22) and solve (7.22) directly. Substituting the solution of (7.22) into (7.21) yields the time-varying system (7.23), which is exponentially stable since $\Psi(t)$ is persistently exciting. Thus, by the approximation results of stochastic averaging, convergence of (7.16)–(7.17) is proved. We do not substitute the solution of (7.17) into (7.16) before computing the average system since the stochastic averaging theory in Chap. 4 does not apply to systems that depend explicitly on time.

7.3 Simulation

For a numerical example, we employ the forward velocity controller (7.4), with parameters $\varepsilon = 0.05$, $a = 0.025$, $c = 25$, $g = 0.6$, to steer the unicycle with $\omega_0 = 5$ rad/s in an unknown signal field given by

$$
J = 1 - 0.5(x_c - x^*)^2 - 0.25(y_c - y^*)^2,
$$

where $(x^*, y^*) = (0, 0)$. Hence,

$$
H = \begin{bmatrix} -0.5 & 0 \\ 0 & -0.25 \end{bmatrix}.
$$

The vehicle's initial position is $(x_c(0), y_c(0)) = (1, 1.5)$.

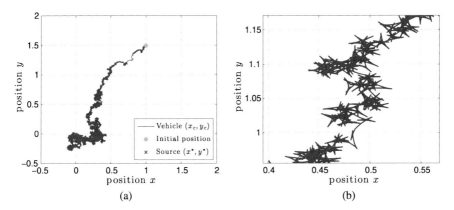

Fig. 7.3 (**a**) The trajectory of the vehicle center, and (**b**) a zoomed-in section of the vehicle trajectory, displaying the vehicle motion more clearly. The source is at $(0, 0)$, and the vehicle is initialized at $(1, 1.5)$

Fig. 7.4 Time history of the signal field measured at the vehicle's position

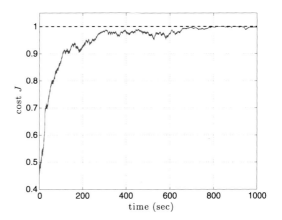

Figure 7.3 shows the trajectory of the vehicle as it converges toward a small neighborhood of the source. It is interesting to note the star-shaped trajectory that occurs on average in Fig. 7.3(b) and the star-pattern that occurs when deterministic extremum seeking is employed [146]. Figure 7.4 depicts the time history of the measured signal field, which converges to a neighborhood of $J^* = 1$. The time histories of the x- and y-positions are shown in Fig. 7.5.

7.4 Notes and References

This chapter is the stochastic version of the deterministic result in [146]. Owing to the co-occurrence of a sinusoidal signal and stochastic perturbation, the deterministic/stochastic averaging theorems are not applicable. In this chapter, we replace the

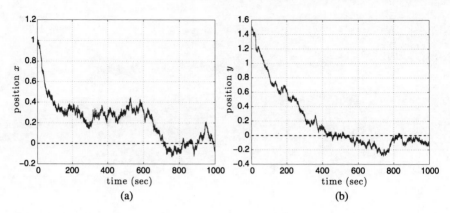

Fig. 7.5 Time history of the vehicle's position (**a**) x-position and (**b**) y-position

time-varying average error system (the "time-varying" character is caused by the sinusoidal evolution of vehicle heading) by a time-invariant average system with one
more state and supply convergence analysis by the approximation results developed
in Chap. 4.

Chapter 8
Multi-parameter Stochastic Extremum Seeking and Slope Seeking

In Chaps. 5, 6, and 7, we considered single-input ES problems, even though the physical space had more than one dimension, such as in the source seeking problems in Chaps. 6 and 7. In this chapter, we introduce multivariable (multi-parameter) ES algorithms in which a distinct white noise signal is used for each channel of the input vector.

Numerous applications motivate the development of multivariable extremum seeking: formation flight for drag minimization, source seeking with fully actuated vehicles, locomotion of fish with elongated bodies that can be approximated by a mechanism with more than two links, beam matching in charged particle accelerators, and many other problems where performance is shaped using more than one parameter.

In this chapter, we first develop the tools for a theoretical analysis of multi-parameter stochastic ES algorithms. These tools are multi-input stochastic averaging theorems in Sect. 8.1. Then, we introduce multi-parameter stochastic ES algorithms for static maps in Sect. 8.2. Finally, in Sect. 8.3, we present stochastic gradient-seeking algorithms in which the input to a plant is tuned to a value at which the gradient of the map approximately equals the commanded gradient.

8.1 Multi-input Stochastic Averaging

Consider the following system

$$\begin{cases} \frac{dX^\varepsilon(t)}{dt} = a(X^\varepsilon(t), Y_1(t/\varepsilon_1), Y_2(t/\varepsilon_2), \dots, Y_l(t/\varepsilon_l)), \\ X^\varepsilon(0) = x, \end{cases} \tag{8.1}$$

where $X^\varepsilon(t) \in \mathbb{R}^n$, $Y_i(t) \in \mathbb{R}^{m_i}$, $1 \leq i \leq l$, are time homogeneous continuous Markov processes defined on a complete probability space $(\Omega, \mathcal{F}, P)$, where Ω is the sample space, $\mathcal{F}$ is a σ-field, and P is a probability measure. The initial condition $X(0) = x$ is deterministic. The small parameters ε_i, $i = 1, 2, \dots, l$, are in $(0, \varepsilon_0)$ with fixed $\varepsilon_0 > 0$, $\varepsilon = [\varepsilon_1, \dots, \varepsilon_l]^T$. Let $S_{Y_i} \subset \mathbb{R}^{m_i}$ be the living space of the

S.-J. Liu, M. Krstic, *Stochastic Averaging and Stochastic Extremum Seeking*,
Communications and Control Engineering,
DOI 10.1007/978-1-4471-4087-0_8, © Springer-Verlag London 2012

perturbation process $(Y_i(t), t \geq 0)$ and note that S_{Y_i} may be a proper (e.g., compact) subset of $\mathbb{R}^{m_i}$.

Assume that

$$\varepsilon_i = \frac{\varepsilon_1}{c_i}, \quad i = 2, \ldots, l, \tag{8.2}$$

for some positive real constants c_i. Denote

$$Z_1(t) = Y_1(t), \qquad Z_2(t) = Y_2(c_2 t), \qquad \ldots, \qquad Z_l(t) = Y_l(c_l t). \tag{8.3}$$

Then (8.1) becomes

$$\begin{cases} \frac{dX^{\varepsilon_1}(t)}{dt} = a(X^{\varepsilon_1}(t), Z_1(t/\varepsilon_1), Z_2(t/\varepsilon_1), \ldots, Z_l(t/\varepsilon_1)), \\ X^{\varepsilon_1}(0) = x, \end{cases} \tag{8.4}$$

where $X^{\varepsilon_1}(t) \triangleq X^{\varepsilon}(t)$.

We have the following lemma about the ergodicity of the processes $(Y_i(t), t \geq 0)$ and $(Z_i(t), t \geq 0)$.

Lemma 8.1 *For $i = 1, \ldots, l$, if the process $(Y_i(t), t \geq 0)$ is ergodic with invariant distribution $\mu_i(dx_i)$ (i.e., if, for any x in the living space of $(Y_i(t), t \geq 0)$, we have that $\| P_i(x, t, \cdot) - \mu_i \|_{\text{var}} \to 0$ as $t \to \infty$, where $P_i(x, t, \cdot)$ is the distribution of $Y_i(t)$ when $Y_i(0) = x$, and $\| \cdot \|_{\text{var}}$ is the total variation norm), then the process $(Z_i(t), t \geq 0)$ is ergodic with the same invariant distribution $\mu_i(dx_i)$.*

Proof Since $Z_1 \equiv Y_1$, we only need to prove the claim for $i = 2, \ldots, l$. For any $i = 2, \ldots, l$, denote by $Q_i(z_i, t, \cdot)$ the distribution of $Z_i(t)$ when $Z_i(0) = Y_i(0) = z_i$. Then by the definition of $Z_i(t)$, we have that $Q_i(z_i, t, \cdot) = P_i(z_i, c_i t, \cdot)$, and thus

$$\left\| Q_i(z_i, t, \cdot) - \mu_i \right\|_{\text{var}} = \left\| P_i(z_i, c_i t, \cdot) - \mu_i \right\|_{\text{var}} \to 0 \quad \text{as } t \to \infty. \tag{8.5}$$

The proof is complete. $\qquad\square$

Denote $Z(t) = [Z_1(t)^T, Z_2(t)^T, \ldots, Z_l(t)^T]^T$. Then for the vector-valued process we have the following result:

Lemma 8.2 *If the process $(Y_i(t), t \geq 0)$ is ergodic with invariant distribution $\mu_i(dx_i)$, and the processes $(Y_1(t), t \geq 0), \ldots, (Y_l(t), t \geq 0)$ are independent, then the process $(Z(t), t \geq 0)$ is ergodic with the invariant distribution $\mu_1(dx_1) \times \cdots \times \mu_l(dx_l)$.*

Proof By the independence of $\{Y_1, \ldots, Y_l\}$, we can assume that the process $(Z(t), t \geq 0)$ lives in the product space of $S_{Y_1} \times \cdots \times S_{Y_l}$. Denote the distribution of $Z_i(t)$ when $Z_i(0) = z_i$, $i = 1, \ldots, l$, by $Q_i(z_i, t, \cdot)$ and the distribution of $Z(t)$ when $Z(0) = z = (z_1, \ldots, z_l)$ by $Q(z, t, \cdot)$. Then by independence, we have that

$$Q(z, t, \cdot) = Q_1(z_1, t, \cdot) \times \cdots \times Q_l(z_l, t, \cdot). \tag{8.6}$$

And thus, by Lemma 8.1, we get

$$\left\| Q(z,t,\cdot) - \mu_1 \times \cdots \times \mu_l \right\|_{\mathrm{var}}$$

$$= \left\| Q_1(z_1,t,\cdot) \times \cdots \times Q_l(z_l,t,\cdot) - \mu_1 \times \mu_2 \times \cdots \times \mu_l \right\|_{\mathrm{var}}$$

$$\leq \left\| Q_1(z_1,t,\cdot) \times \cdots \times Q_l(z_l,t,\cdot) - \mu_1 \times Q_2(z_2,t,\cdot) \times \cdots \times Q_l(z_l,t,\cdot) \right\|_{\mathrm{var}}$$

$$+ \left\| \mu_1 \times Q_2(z_2,t,\cdot) \times \cdots \times Q_l(z_l,t,\cdot) \right.$$

$$\left. - \mu_1 \times \mu_2 \times Q_3(z_3,t,\cdot) \times \cdots \times Q_l(z_l,t,\cdot) \right\|_{\mathrm{var}}$$

$$+ \cdots + \left\| \mu_1 \times \cdots \times \mu_{l-1} \times Q_l(z_l,t,\cdot) - \mu_1 \times \cdots \times \mu_{l-1} \times \mu_l \right\|_{\mathrm{var}}$$

$$\leq \left\| Q_1(z_1,t,\cdot) - \mu_1 \right\|_{\mathrm{var}} + \cdots + \left\| Q_l(z_l,t,\cdot) - \mu_l \right\|_{\mathrm{var}} \to 0, \quad t \to \infty. \quad (8.7)$$

The proof is complete. □

So we obtain the average system of system (8.4) as follows:

$$\frac{d\bar{X}(t)}{dt} = \bar{a}\big(\bar{X}(t)\big), \quad \bar{X}_0 = x, \qquad (8.8)$$

where

$$\bar{a}(x) = \int_{S_{Y_1} \times \cdots \times S_{Y_l}} a(x, z_1, \ldots, z_l) \mu_1(dz_1) \times \cdots \times \mu_l(dz_l). \qquad (8.9)$$

To develop a multi-input stochastic averaging theorem, we make the following assumptions:

Assumption 8.1 The vector field $a(x, y_1, y_2, \ldots, y_l)$ is a continuous function of $(x, y_1, y_2, \ldots, y_l)$, and for any $x \in \mathbb{R}^n$, it is a bounded function of $y = [y_1^T, y_2^T, \ldots, y_l^T]^T$. Further it satisfies the local Lipschitz condition in $x \in \mathbb{R}^n$ uniformly in $y \in S_{Y_1} \times S_{Y_2} \times \cdots \times S_{Y_l}$, i.e., for any compact subset $D \subset \mathbb{R}^n$, there is a constant k_D such that for all $x_1, x_2 \in D$ and all $y \in S_{Y_1} \times S_{Y_2} \times \cdots \times S_{Y_l}$,

$$\left| a(x_1, y) - a(x_2, y) \right| \leq k_D |x_1 - x_2|. \qquad (8.10)$$

Assumption 8.2 The perturbation processes $(Y_i(t), t \geq 0)$, $i = 1, \ldots, l$, are ergodic with invariant distributions μ_i, respectively, and independent.

By the same method as in Chap. 4 for single input stochastic averaging theorem, we obtain the following multi-input averaging theorem:

Theorem 8.1 *Consider system* (8.1) *under Assumptions* 8.1 *and* 8.2. *If the equilibrium* $\bar{X}(t) \equiv 0$ *of the average system* (8.8) *is exponentially stable, then*

(i) *The solution of system* (8.1) *is weakly stochastically exponentially stable, i.e., there exist constants* $r > 0$, $c > 0$, *and* $\gamma > 0$ *such that, for any initial condition* $x \in \{ \check{x} \in \mathbb{R}^n : |\check{x}| < r \}$ *and any* $\delta > 0$, *the solution* $X^\varepsilon(t) = X^{\varepsilon_1}(t)$ *of system* (8.1) *satisfies*

$$\lim_{\varepsilon_1 \to 0} \inf \left\{ t \geq 0 : \left| X^{\varepsilon_1}(t) \right| > c|x|e^{-\gamma t} + \delta \right\} = +\infty \quad a.s. \qquad (8.11)$$

(ii) *There exists a function* $T(\varepsilon_1) : (0, \varepsilon_0) \to \mathbb{N}$ *such that*

$$\lim_{\varepsilon_1 \to 0} P \left\{ \sup_{0 \le t \le T(\varepsilon_1)} \left\{ \left| X^{\varepsilon_1}(t) \right| - c|x| e^{-\gamma t} \right\} > \delta \right\} = 0 \quad \text{with}$$

$$\lim_{\varepsilon_1 \to 0} T(\varepsilon_1) = \infty. \tag{8.12}$$

Furthermore, (8.12) is equivalent to

$$\lim_{\varepsilon_1 \to 0} P \left\{ \left| X^{\varepsilon_1}(t) \right| \le c|x| e^{-\gamma t} + \delta, \forall t \in \left[0, T(\varepsilon_1)\right] \right\} = 1 \quad \text{with}$$

$$\lim_{\varepsilon_1 \to 0} T(\varepsilon_1) = \infty. \tag{8.13}$$

8.2 Multi-parameter Stochastic ES for Static Maps

8.2.1 Algorithm for Multi-parameter Stochastic ES

Let $f(\theta)$ be a function of the form

$$f(\theta) = f^* + \left(\theta - \theta^*\right)^T P \left(\theta - \theta^*\right), \tag{8.14}$$

where $P = (p_{ij})_{l \times l} \in \mathbb{R}^{l \times l}$ is an unknown symmetric matrix, f^* is an unknown constant, $\theta = [\theta_1, \dots, \theta_l]^T$, and $\theta^* = [\theta_1^*, \dots, \theta_l^*]^T$. Any $C^2(\mathbb{R}^l)$ function $f(\theta)$ with an extremum at $\theta = \theta^*$ and with $\nabla^2 f \ne \mathbf{0}$ can be locally approximated by (8.14). Without loss of generality, we assume the matrix P is positive definite.

The objective is to design an algorithm to make $|\theta - \theta^*|$ as small as possible, so that the output $y = f(\theta)$ is driven to its minimum f^*.

Denote $\hat{\theta}_j(t)$ as the estimate of the unknown optimal input θ_j^* and let

$$\tilde{\theta}_j(t) = \theta_j^* - \hat{\theta}_j(t) \tag{8.15}$$

denote the estimation error.

We use a stochastic perturbation to develop a gradient estimate for every parameter. Let

$$\theta_j(t) = \hat{\theta}_j(t) + a_j \sin\left(\eta_j(t)\right), \tag{8.16}$$

where $a_j > 0$ is the perturbation amplitude and $(\eta_j(t), t \ge 0)$ is an OU process which is given by

$$\varepsilon_j \, d\eta_j(t) = -\eta_j(t) \, dt + \sqrt{\varepsilon_j} q_j \, dW_j(t), \tag{8.17}$$

where ε_j, $j = 1, \dots, l$, are small parameters.

By (8.15) and (8.16), we have

$$\theta_j(t) - \theta_j^* = a_j \sin\left(\eta_j(t)\right) - \tilde{\theta}_j(t). \tag{8.18}$$

Substituting (8.18) into (8.14), we have the output

$$y(t) = f^* + \left(\theta(t) - \theta^*\right)^T P \left(\theta(t) - \theta^*\right), \tag{8.19}$$

where $\theta(t) - \theta^* = [a_1 \sin(\eta_1(t)) - \tilde{\theta}_1(t), \ldots, a_l \sin(\eta_l(t)) - \tilde{\theta}_l(t)]^T$.

We design the parameter update law as follows:

$$\frac{d\hat{\theta}_j(t)}{dt} = -k_j a_j \sin(\eta_j(t))(y(t) - \xi_i(t)), \tag{8.20}$$

$$\frac{d\xi_j(t)}{dt} = -h_j \xi_j(t) + h_j y(t), \tag{8.21}$$

$$\varepsilon_j \, d\eta_j(t) = -\eta_j(t) \, dt + \sqrt{\varepsilon_j} q_j \, dW_j(t), \tag{8.22}$$

where h_j, k_j, $j = 1, \ldots, l$, are scalar design parameters. To improve the performance, here we use a washout filter $\frac{s}{s+h_j}$ for each parameter, and the gradient estimation for each parameter is based on the output $\frac{s}{s+h_j}[y] = y(t) - \xi_j(t)$ of this filter.

Define $\chi_j(t) = \eta_j(\varepsilon_j t)$ and $B_j(t) = \frac{1}{\sqrt{\varepsilon_j}} W_j(\varepsilon_j t)$. Then we have

$$d\chi_j(t) = -\chi_j(t) \, dt + q_j \, dB_j(t), \tag{8.23}$$

where $B_j(t)$ is a 1-dimensional standard Brownian motion defined on the complete probability space $(\Omega, \mathcal{F}, P)$, while $[B_1(t), \ldots, B_l(t)]^T$ is an l-dimensional independent standard Brownian motion on the same space.

Define the output error variable

$$e_j(t) = \xi_j(t) - f^*, \quad j = 1, \ldots, l. \tag{8.24}$$

Therefore, it follows from (8.15), (8.19), (8.20), and (8.21) that we have the error dynamics

$$\begin{aligned}
\frac{d\tilde{\theta}_j(t)}{dt} &= -\frac{d\hat{\theta}_j(t)}{dt} \\
&= -k_j a_j \sin(\eta_j(t))\left((\theta(t) - \theta^*)^T P(\theta(t) - \theta^*) - e_j(t)\right) \\
&= -k_j a_j \sin(\chi_j(t/\varepsilon_j))\left((\theta(t) - \theta^*)^T P(\theta(t) - \theta^*) - e_j(t)\right), \quad (8.25)
\end{aligned}$$

$$\begin{aligned}
\frac{de_j(t)}{dt} &= h_j\left(y(t) - f^* - e_j(t)\right) \\
&= h_j\left((\theta(t) - \theta^*)^T P(\theta(t) - \theta^*) - e_j(t)\right), \quad j = 1, \ldots, l. \quad (8.26)
\end{aligned}$$

Denote $\tilde{\theta}(t) = [\tilde{\theta}_1(t), \ldots, \tilde{\theta}_l(t)]^T$ and $e(t) = [e_1(t), \ldots, e_l(t)]^T$, which are dependent on the small parameter ε_1. Then we have the following result:

Theorem 8.2 *Consider the static map* (8.14) *under the parameter update law* (8.20)–(8.22). *Then the error system* (8.25)–(8.26) *is weakly stochastically exponentially stable, i.e., there exist constants $r > 0$, $c > 0$, and $\gamma > 0$ such that, for any initial condition $|\Lambda_1^{\varepsilon_1}(0)| < r$ and any $\delta > 0$,*

$$\lim_{\varepsilon_1 \to 0} \inf\{t \geq 0 : |\Lambda_1^{\varepsilon_1}(t)| > c|\Lambda_1^{\varepsilon_1}(0)|e^{-\gamma t} + \delta\} = +\infty \quad a.s. \tag{8.27}$$

Moreover, there exists a function $T(\varepsilon_1) : (0, \varepsilon_0) \to \mathbb{N}$ such that

$$\lim_{\varepsilon_1 \to 0} P\left\{ \sup_{0 \le t \le T(\varepsilon_1)} \left\{ \left| \Lambda_1^{\varepsilon_1}(t) \right| - c \left| \Lambda_1^{\varepsilon_1}(0) \right| e^{-\gamma t} \right\} > \delta \right\} = 0 \quad with$$

$$\lim_{\varepsilon_1 \to 0} T(\varepsilon_1) = \infty, \tag{8.28}$$

where $\Lambda_1^{\varepsilon_1}(t) = (\tilde{\theta}(t)^T, e(t)^T) - (0_{l \times 1}^T, \sum_{i=1}^{l} p_{ii} a_i^2 G_0(q_i) I_1^T)$, $I_1 = [1, 1, \ldots,$
$1]_{1 \times l}^T$. *Furthermore,* (8.28) *is equivalent to*

$$\lim_{\varepsilon_1 \to 0} P\left\{ \left| \Lambda_1^{\varepsilon_1}(t) \right| \le c \left| \Lambda_1^{\varepsilon_1}(0) \right| e^{-\gamma t} + \delta, \forall t \in \left[0, T(\varepsilon_1) \right] \right\} = 1 \quad with$$

$$\lim_{\varepsilon_1 \to 0} T(\varepsilon_1) = \infty. \tag{8.29}$$

8.2.2 Convergence Analysis

We rewrite the error dynamics (8.25)–(8.26) as

$$\frac{d\tilde{\theta}_j(t)}{dt} = k_j a_j \sin\left(\chi_j(t/\varepsilon_j)\right)$$

$$\times \left(\left[a_1 \sin\left(\chi_1(t/\varepsilon_1)\right) - \tilde{\theta}_1(t), \ldots, a_l \sin\left(\chi_l(t/\varepsilon_l)\right) - \tilde{\theta}_l(t) \right]^T \right.$$

$$\times P\left[a_1 \sin\left(\chi_1(t/\varepsilon_1)\right) - \tilde{\theta}_1(t), \ldots, a_l \sin\left(\chi_l(t/\varepsilon_l)\right) - \tilde{\theta}_l(t) \right] \right)$$

$$= k_j a_j \sin\left(\chi_j(t/\varepsilon_j)\right)$$

$$\times \left(\sum_{i,k=1}^{l} p_{ik} \left(a_i \sin\left(\chi_i(t/\varepsilon_i)\right) - \tilde{\theta}_i(t) \right) \right.$$

$$\times \left(a_k \sin\left(\chi_k(t/\varepsilon_k)\right) - \tilde{\theta}_k(t) \right) - e_j(t) \right), \tag{8.30}$$

$$\frac{de_i(t)}{dt} = h_j \left(\sum_{i,k=1}^{l} p_{ik} \left(a_i \sin\left(\chi_i(t/\varepsilon_i)\right) - \tilde{\theta}_i(t) \right) \right.$$

$$\times \left(a_k \sin\left(\chi_k(t/\varepsilon_k)\right) - \tilde{\theta}_k(t) \right) - e_j(t) \right), \quad j = 1, \ldots, l. \tag{8.31}$$

Now we calculate the average system of the error system. Assume that

$$\varepsilon_i = \frac{\varepsilon_1}{c_i}, \quad i = 2, \ldots, l, \tag{8.32}$$

for some positive real constants c_i. Denote

$$Z_1(t) = \chi_1(t), \qquad Z_2(t) = \chi_2(c_2 t), \qquad \ldots, \qquad Z_l(t) = \chi(c_l t). \tag{8.33}$$

Then the error dynamics become

$$\frac{d\tilde{\theta}_j(t)}{dt} = k_j a_j \sin\left(Z_j(t/\varepsilon_1)\right)$$

$$\times \left(\sum_{i,k=1}^{l} p_{ik} \left(a_i \sin\left(Z_i(t/\varepsilon_1)\right) - \tilde{\theta}_i(t)\right) \right.$$

$$\left. \times \left(a_k \sin\left(Z_k(t/\varepsilon_1)\right) - \tilde{\theta}_k(t)\right) - e_j(t) \right), \tag{8.34}$$

$$\frac{de_j(t)}{dt} = h_j\left(y(t) - f^* - e_j(t)\right)$$

$$= h_j\left(\sum_{i,k=1}^{l} p_{ik} \left(a_i \sin\left(Z_i(t/\varepsilon_1)\right) - \tilde{\theta}_i(t)\right) \right.$$

$$\left. \times \left(a_k \sin\left(Z_k(t/\varepsilon_1)\right) - \tilde{\theta}_k(t)\right) - e_j(t) \right), \quad j = 1, \ldots, l. \tag{8.35}$$

It is known that for given $j = 1, \ldots, l$, the stochastic process $(\chi_j(t), t \geq 0)$ is ergodic and has invariant distribution $\mu_j(dx_j) = \frac{1}{\sqrt{\pi}q_j} e^{-x_j^2/q_j^2} dx_j$. Thus by Lemma 8.2, the vector-valued process $[Z_1(t), Z_2(t), \ldots, Z_l(t)]^T$ is also ergodic with invariant distribution $\mu_1(dx_1) \times \cdots \times \mu_l(dx_l)$.

To calculate the average system of system (8.34)–(8.35), we need to consider the following terms

$$\sin\left(Z_j(t/\varepsilon_1)\right) \sin\left(Z_i(t/\varepsilon_1)\right) \sin\left(Z_k(t/\varepsilon_1)\right), \quad i \neq j, j \neq k, k \neq i, \tag{8.36}$$

$$\sin^3\left(Z_j(t/\varepsilon_1)\right), \tag{8.37}$$

$$\sin\left(Z_j(t/\varepsilon_1)\right) \sin^2\left(Z_i(t/\varepsilon_1)\right), \quad i \neq j, \tag{8.38}$$

$$\sin^2\left(Z_j(t/\varepsilon_1)\right), \tag{8.39}$$

$$\sin\left(Z_j(t/\varepsilon_1)\right) \sin\left(Z_i(t/\varepsilon_1)\right), \quad i \neq j. \tag{8.40}$$

Averaging calculation gives

$$\int_{\mathbb{R}^3} \sin(x_i) \sin(x_j) \sin(x_k) \mu_i(dx_i) \times \mu_j(dx_j) \times \mu_k(dx_k)$$

$$= \int_{-\infty}^{+\infty} \int_{-\infty}^{+\infty} \int_{-\infty}^{+\infty} \sin(x_i) \sin(x_j) \sin(x_k)$$

$$\times \frac{1}{\sqrt{\pi}q_i} e^{-x_i^2/q_i^2} \frac{1}{\sqrt{\pi}q_j} e^{-x_j^2/q_j^2} \frac{1}{\sqrt{\pi}q_k} e^{-x_k^2/q_k^2} dx_i\, dx_j\, dx_k$$

$$= 0, \tag{8.41}$$

$$\int_{\mathbb{R}} \sin^{2k+1}(x_i) \mu_i(dx_i)$$

$$= \int_{-\infty}^{+\infty} \sin^{2k+1}(x_i) \frac{1}{\sqrt{\pi}q_i} e^{-x_i^2/q_i^2} dx_i = 0, \quad k = 0, 1, 2, \ldots, \tag{8.42}$$

$$\int_{\mathbb{R}^2} \sin^2(x_i)\sin(x_j)\mu_i(dx_i) \times \mu_j(dx_j)$$

$$= \int_{-\infty}^{+\infty}\int_{-\infty}^{+\infty} \sin^2(x_i)\sin(x_j)\frac{1}{\sqrt{\pi}q_i}e^{-x_i^2/q_i^2}\frac{1}{\sqrt{\pi}q_j}e^{-x_j^2/q_j^2}\,dx_i\,dx_j$$

$$= 0, \tag{8.43}$$

$$\int_{\mathbb{R}} \sin^2(x_i)\mu_i(dx_i)$$

$$= \int_{-\infty}^{+\infty} \sin^2(x_i)\frac{1}{\sqrt{\pi}q_i}e^{-x_i^2/q_i^2}\,dx_i = \frac{1}{2}\left(1-e^{-q_i^2}\right) \triangleq G_0(q_i), \tag{8.44}$$

$$\int_{\mathbb{R}^2} \sin(x_i)\sin(x_j)\mu_i(dx_i) \times \mu_j(dx_j)$$

$$= \int_{-\infty}^{+\infty}\int_{-\infty}^{+\infty} \sin(x_i)\sin(x_j)\frac{1}{\sqrt{\pi}q_i}e^{-x_i^2/q_i^2}\frac{1}{\sqrt{\pi}q_j}e^{-x_j^2/q_j^2}\,dx_i\,dx_j$$

$$= 0. \tag{8.45}$$

Then we get the average error system as follows:

$$\frac{d\tilde{\theta}_j^{\text{ave}}(t)}{dt} = -a_j^2 k_j\left(1-e^{-q_j^2}\right)\sum_{i=1}^{l} p_{ji}\tilde{\theta}_i^{\text{ave}}(t), \tag{8.46}$$

$$\frac{de_j^{\text{ave}}(t)}{dt} = h_j\left(\sum_{i=1}^{l} p_{ii}a_i^2\frac{1}{2}\left(1-e^{-q_i^2}\right) + \sum_{i,k=1}^{l} p_{ik}\tilde{\theta}_i^{\text{ave}}\tilde{\theta}_k^{\text{ave}} - e_j^{\text{ave}}(t)\right),$$

$$j = 1,\ldots,l. \tag{8.47}$$

In the matrix form, the average error system is

$$\frac{d\tilde{\theta}^{\text{ave}}(t)}{dt} = -\Pi P\tilde{\theta}^{\text{ave}}(t), \tag{8.48}$$

$$\frac{de^{\text{ave}}(t)}{dt} = H\left(\sum_{i=1}^{l} p_{ii}a_i^2 G_0(q_i)I_1 - e^{\text{ave}}(t) + Q\left(\tilde{\theta}^{\text{ave}}(t)\right)\right), \tag{8.49}$$

where

$$\Pi = \begin{bmatrix} a_1^2 k_1(1-e^{-q_1^2}) & 0 & \cdots & 0 \\ 0 & a_2^2 k_2(1-e^{-q_2^2}) & \cdots & 0 \\ \vdots & \vdots & \ddots & \vdots \\ 0 & 0 & \cdots & a_l^2 k_l(1-e^{-q_l^2}) \end{bmatrix},$$

$$\tilde{\theta}^{\text{ave}}(t) = \left[\tilde{\theta}_1^{\text{ave}}(t),\ldots,\tilde{\theta}_l^{\text{ave}}(t)\right]^T, \qquad e^{\text{ave}}(t) = \left[e_1^{\text{ave}}(t),\ldots,e_l^{\text{ave}}(t)\right]^T,$$

$$H = \begin{bmatrix} h_1 & 0 & \cdots & 0 \\ 0 & h_2 & \cdots & 0 \\ \vdots & \vdots & \ddots & \vdots \\ 0 & 0 & \cdots & h_l \end{bmatrix},$$

$$Q\left(\tilde{\theta}^{\text{ave}}(t)\right) = \tilde{\theta}^{\text{ave}^T}(t)P\tilde{\theta}^{\text{ave}}(t)I_1, \qquad I_1 = [1,1,\ldots,1]_{1\times l}^T.$$

The average error system has equilibrium $(\tilde{\theta}^{e^T}, ee^T) = (0^T_{l \times 1}, \sum^l_{i=1} p_{ii} a^2_i \times G_0(q_i) I^T_1)$. The corresponding Jacobian matrix at this equilibrium is

$$\Xi_1 = \begin{bmatrix} -\Pi P & 0 \\ 0 & -H \end{bmatrix}. \tag{8.50}$$

Since Π and P are positive definite and Π is diagonal, all eigenvalues of the matrix ΠP are positive, i.e., the eigenvalues of the matrix $-\Pi P$ are negative. Furthermore, from the fact $h_i > 0$, $i = 1, \ldots, l$, it follows that the matrix Ξ_1 is Hurwitz and hence the equilibrium is exponentially stable. Thus by Theorem 8.1, the convergence results (8.27) and (8.29) hold. The proof is complete.

To quantify the output convergence to the extremum, for any $\varepsilon_1 > 0$, define a stopping time

$$\tau^\delta_{\varepsilon_1} = \inf\{t \geq 0 : |\Lambda^{\varepsilon_1}_1(t)| > c|\Lambda^{\varepsilon_1}_1(0)|e^{-\gamma t} + \delta\}.$$

Then by (8.27), we know that $\lim_{\varepsilon_1 \to 0} \tau^\delta_{\varepsilon_1} = \infty$ a.s., and

$$|\tilde{\theta}(t)| \leq c|\Lambda^{\varepsilon_1}_1(0)|e^{-\gamma t} + \delta, \quad \forall t \leq \tau^\delta_{\varepsilon_1}. \tag{8.51}$$

Denote $\hat{\theta}(t) = [\hat{\theta}_1(t), \ldots, \hat{\theta}_l(t)]^T$, $a\sin(\eta(t)) = [a_1 \sin(\eta_1(t)), \ldots, a_l \sin(\eta_l(t))]^T$. Then the output $y(t) = f(\theta) = f(\theta^* + \tilde{\theta}(t) + a\sin(\eta(t)))$, for $\nabla f(\theta^*) = 0$, we have

$$y(t) - f(\theta^*) = (\tilde{\theta}(t) + a\sin(\eta(t)))^T P(\tilde{\theta}(t) + a\sin(\eta(t)))$$
$$+ O(|\tilde{\theta}(t) + a\sin(\eta(t))|^3). \tag{8.52}$$

Thus by (8.51), it holds that

$$|y(t) - f(\theta^*)| \leq O(|a|^2) + O(\delta^2) + C|\Lambda^{\varepsilon_1}_1(0)|^2 e^{-2\gamma t} \quad \forall t \leq \tau^\delta_{\varepsilon_1} \tag{8.53}$$

for some positive constant C, where $|a| = \sqrt{a^2_1 + a^2_2 + \cdots + a^2_l}$. Similarly, by (8.29), we have

$$\lim_{\varepsilon_1 \to 0} P\{|y(t) - f(\theta^*)| \leq O(|a|^2) + O(\delta^2) + C|\Lambda^{\varepsilon_1}_1(0)|^2 e^{-2\gamma t},$$
$$\forall t \in [0, T(\varepsilon_1)]\} = 1, \tag{8.54}$$

where $T(\varepsilon_1)$ is a deterministic function with $\lim_{\varepsilon_1 \to 0} T(\varepsilon_1) = \infty$.

Figure 8.1 displays the simulation results with $f^* = 1$, $\theta^* = [0, 1]^T$, $P = \begin{bmatrix} 1 & 1 \\ 1 & 2 \end{bmatrix}$ in the static map (8.14) and $a_1 = 0.8$, $a_2 = 0.6$, $k_1 = 1.25$, $k_2 = 5/3$, $h_1 = 1$, $h_2 = 2$, $q_1 = q_2 = 1$, $\varepsilon_1 = 0.25$, $\varepsilon_2 = 0.01$ in the parameter update law (8.20)–(8.22) and initial condition $\tilde{\theta}_1(0) = 1$, $\tilde{\theta}_2(0) = -1$, $\hat{\theta}_1(0) = -1$, $\hat{\theta}_2(0) = 2$, $\xi_1(0) = \xi_2(0) = 0$.

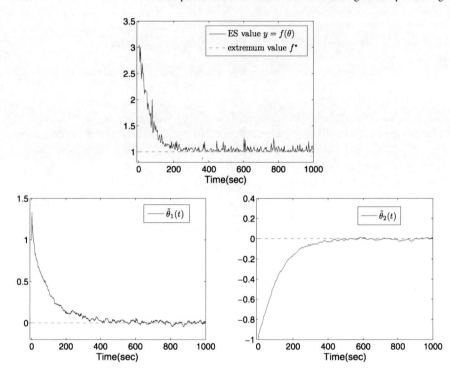

Fig. 8.1 Stochastic extremum seeking with an OU process perturbation. *Top*: output and extremum values. *Bottom*: solutions of the error system

8.3 Stochastic Gradient Seeking

8.3.1 Single-parameter Stochastic Slope Seeking

Let f'_{ref} denote the commanded slope on an unknown single-input quadratic operating map. Let the map be parameterized as

$$f(\theta) = f^* + f'_{\text{ref}}(\theta - \theta^*) + \frac{f''}{2}(\theta - \theta^*)^2, \qquad (8.55)$$

which means that θ^* is the unknown input that produces the slope f'_{ref}, f^* is the value of the output at $\theta = \theta^*$, and $f'' > 0$ is the second derivative of $f(\theta)$ at the point $\theta = \theta^*$ where the slope is f'_{ref}. The object is to design an algorithm to make $\theta(t) - \theta^*$ as small as possible, so that the slope $f'(\theta)$ is driven to f'_{ref}.

Denote $\hat{\theta}(t)$ as the estimate of the unknown optimal input θ^*. Let

$$\tilde{\theta}(t) = \theta^* - \hat{\theta}(t) \qquad (8.56)$$

denote the estimation error. Let

$$\theta(t) = \hat{\theta}(t) + a\sin(\eta(t)), \qquad (8.57)$$

where $a > 0$ and $(\eta(t), t \geq 0)$ is a stochastic process satisfying

$$\eta = \frac{\sqrt{\varepsilon q}}{\varepsilon s + 1}[\dot{W}], \quad \text{or} \quad \varepsilon \, d\eta = -\eta \, dt + \sqrt{\varepsilon q} \, dW, \tag{8.58}$$

where $q > 0$ and $(W(t), t \geq 0)$ is a 1-dimensional standard Brownian motion defined on some complete probability space $(\Omega, \mathcal{F}, P)$. Thus, by (8.56) and (8.57), we have

$$\theta - \theta^* = a \sin(\eta) - \tilde{\theta}. \tag{8.59}$$

Substituting (8.59) into (8.14), we have the output

$$y = f^* + f'_{\text{ref}}(a \sin(\eta) - \tilde{\theta}) + \frac{f''}{2}(a \sin(\eta) - \tilde{\theta})^2. \tag{8.60}$$

Now, we design the parameter update law as follows:

$$\frac{d\hat{\theta}}{dt} = -k\left[\sin(\eta)(y - \xi) + r\left(f'_{\text{ref}}\right)\right], \tag{8.61}$$

$$\frac{d\xi}{dt} = -h\xi + hy, \tag{8.62}$$

$$\varepsilon \, d\eta = -\eta \, dt + \sqrt{\varepsilon q} \, dW, \tag{8.63}$$

where $k > 0$, $h > 0$ are scalar design parameters, and r is a function to be designed. Define $\chi(t) = \eta(\varepsilon t)$ and $B(t) = \frac{1}{\sqrt{\varepsilon}} W(\varepsilon t)$. Then we have

$$d\chi(t) = -\chi(t) \, dt + q \, dB(t), \tag{8.64}$$

where $B(t)$ is a 1-dimensional standard Brownian motion.
Define the output error variable

$$e = \frac{h}{s + h}[y] - f^*. \tag{8.65}$$

Then we have the following error dynamics

$$\begin{aligned}
\frac{d\tilde{\theta}(t)}{dt} &= -\frac{d\hat{\theta}(t)}{dt} \\
&= k \sin(\chi(t/\varepsilon)) \\
&\quad \cdot \left(f'_{\text{ref}}(a \sin(\chi(t/\varepsilon)) - \tilde{\theta}) + \frac{f''}{2}(a \sin(\chi(t/\varepsilon)) - \tilde{\theta})^2 - e \right) \\
&\quad + kr\left(f'_{\text{ref}}\right),
\end{aligned} \tag{8.66}$$

$$\begin{aligned}
\frac{de(t)}{dt} &= h\bigg(f'_{\text{ref}}(a \sin(\chi(t/\varepsilon)) - \tilde{\theta}) \\
&\quad + \frac{f''}{2}(a \sin(\chi(t/\varepsilon)) - \tilde{\theta})^2 - e \bigg).
\end{aligned} \tag{8.67}$$

Now we calculate the average system. By (5.18), (5.19), and (4.11), we obtain that the average system of (8.66)–(8.67) is

$$\frac{d\tilde{\theta}^{\text{ave}}(t)}{dt} = \frac{akf'_{\text{ref}}}{2}\left(1 - e^{-q^2}\right) - \frac{akf''}{2}\left(1 - e^{-q^2}\right)\tilde{\theta}^{\text{ave}} + kr\left(f'_{\text{ref}}\right), \quad (8.68)$$

$$\frac{de^{\text{ave}}(t)}{dt} = h\left(-f'_{\text{ref}}\tilde{\theta}^{\text{ave}} + \frac{f''a^2}{4}\left(1 - e^{-q^2}\right) + \frac{f''}{2}\tilde{\theta}^{\text{ave}^2} - e^{\text{ave}}\right). \quad (8.69)$$

We choose

$$r\left(f'_{\text{ref}}\right) = -\frac{af'_{\text{ref}}}{2}\left(1 - e^{-q^2}\right). \quad (8.70)$$

Then by simple calculation, we get the following equilibrium of the above average system at $\tilde{\theta}^{a,e} = 0$, $e^{a,e} = \frac{a^2 f''}{4}(1 - e^{-q^2})$ with the corresponding Jacobian matrix

$$\begin{bmatrix} -\frac{akf''}{2}(1 - e^{-q^2}) & 0 \\ -hf'_{\text{ref}} & -h \end{bmatrix}. \quad (8.71)$$

Noticing that $f'' > 0$, $k > 0$, $a > 0$, and $h > 0$, the above Jacobian is Hurwitz, i.e., the equilibrium $(0, \frac{a^2 f''}{4}(1 - e^{-q^2}))$ of the average system is exponentially stable. Then by stochastic averaging Theorems 4.3 and 4.4, we have the following result.

Theorem 8.3 *Consider the static map* (8.55) *with the commanded slope under the parameter update law* (8.61)–(8.63). *Then there exist constants* $r > 0$, $c > 0$, $\gamma > 0$, *and a function* $T(\varepsilon) : (0, \varepsilon_0) \to \mathbb{N}$ *such that, for any initial condition* $|\Lambda_2^\varepsilon(0)| < r$ *and any* $\delta > 0$,

$$\lim_{\varepsilon \to 0} \inf\left\{t \geq 0 : \left|\Lambda_2^\varepsilon(t)\right| > c\left|\Lambda_2^\varepsilon(0)\right|e^{-\gamma t} + \delta\right\} = \infty \quad a.s. \quad (8.72)$$

and

$$\lim_{\varepsilon \to 0} P\left\{\left|\Lambda_2^\varepsilon(t)\right| \leq c\left|\Lambda_2^\varepsilon(0)\right|e^{-\gamma t} + \delta, \forall t \in \left[0, T(\varepsilon)\right]\right\} = 1 \quad with$$

$$\lim_{\varepsilon \to 0} T(\varepsilon) = \infty, \quad (8.73)$$

where $\Lambda_2^\varepsilon(t) \triangleq (\tilde{\theta}^\varepsilon(t), e^\varepsilon(t)) - (0, \frac{a^2 f''}{4}(1 - e^{-q^2}))$.

These two results imply that the norm of the error vector $\Lambda_2^\varepsilon(t)$ exponentially converges, both almost surely and in probability, to below an arbitrarily small residual value δ, over an arbitrarily long time interval, which tends to infinity as ε goes to zero.

In particular, the $\tilde{\theta}^\varepsilon(t)$-component of the error vector converges to below δ. To quantify the output convergence to the optimum, for any $\varepsilon > 0$, define a stopping time

$$\tau_\varepsilon^\delta = \inf\left\{t \geq 0 : \left|\Lambda_2^\varepsilon(t)\right| > c\left|\Lambda_2^\varepsilon(0)\right|e^{-\gamma t} + \delta\right\}. \quad (8.74)$$

Then by (8.72) and the definition of $\Lambda_2^\varepsilon(t)$, we know that $\lim_{\varepsilon \to 0} \tau_\varepsilon^\delta = \infty$ a.s. and

$$\left|\tilde{\theta}^\varepsilon(t)\right| \leq c\left|\Lambda_2^\varepsilon(0)\right|e^{-\gamma t} + \delta \quad \forall t \leq \tau_\varepsilon^\delta. \quad (8.75)$$

Since $y(t) = f(\theta^* + \tilde{\theta}^\varepsilon(t) + a\sin(\eta(t)))$, we have

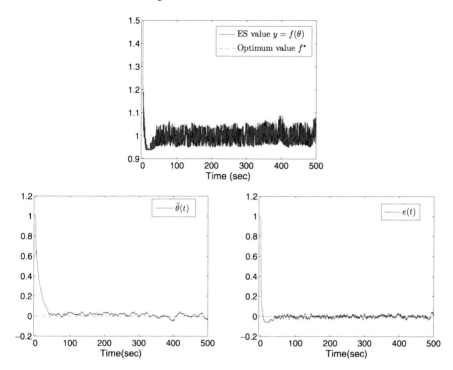

Fig. 8.2 Single-parameter stochastic slope seeking with an OU process perturbation. *Top*: output and optimum values. *Bottom*: solutions of the error system

$$y(t) - f\left(\theta^*\right) = f'\left(\theta^*\right)\left(\tilde{\theta}^\varepsilon(t) + a\sin(\eta(t))\right) + \frac{f''(\theta^*)}{2}\left(\tilde{\theta}^\varepsilon(t) + a\sin(\eta(t))\right)^2$$
$$+ O\left(\left(\tilde{\theta}^\varepsilon(t) + a\sin(\eta(t))\right)^3\right). \tag{8.76}$$

Thus by (8.75), it holds that

$$\left|y(t) - f\left(\theta^*\right)\right| \le O(a) + O(\delta)$$
$$+ c\left|\Lambda_2^\varepsilon(0)\right|e^{-\gamma t} + C\left|\Lambda_2^\varepsilon(0)\right|^2 e^{-2\gamma t} \quad \forall t \le \tau_\varepsilon^\delta \tag{8.77}$$

for some positive constant C. Similarly, by (8.73),

$$\lim_{\varepsilon \to 0} P\left\{\left|y(t) - f\left(\theta^*\right)\right| \le O(a) + O(\delta) + c\left|\Lambda_2^\varepsilon(0)\right|e^{-\gamma t} + C\left|\Lambda_2^\varepsilon(0)\right|^2 e^{-2\gamma t}\right.$$
$$\left. \forall t \in \left[0, T(\varepsilon)\right]\right\} = 1, \tag{8.78}$$

where $T(\varepsilon)$ is a deterministic function with $\lim_{\varepsilon \to 0} T(\varepsilon) = \infty$.

Figure 8.2 displays the simulation results with $f^* = 1$, $f'_{\text{ref}} = 0.5$, $\theta^* = 0$ in the static map (8.55) and $a = 0.1, k = 1, q = 1, \varepsilon = 0.25, h = 1$ in the parameter update law (8.61)–(8.62) and initial condition $\tilde{\theta}(0) = 1, \hat{\theta}(0) = -1, \xi(0) = 1.99$.

8.3.2 Multi-parameter Stochastic Gradient Seeking

Analogous to the single parameter case, we let $f(\theta)$ be a function of the form:

$$f(\theta) = f^* + J^T(\theta - \theta^*) + (\theta - \theta^*)^T P_1(\theta - \theta^*), \tag{8.79}$$

where $P_1 = P_1^T = (p_{1ij})_{l \times l} \in \mathbb{R}^{l \times l}$, $\theta = [\theta_1, \ldots, \theta_l]^T$, $\theta^* = [\theta_1^*, \ldots, \theta_l^*]^T$, and $J = [J_1, J_2, \ldots, J_l]$ is the commanded gradient.

According to the multi-parameter stochastic extremum seeking algorithm, we denote $\hat{\theta}_j(t)$ as the estimate of the unknown optimal input θ_j^* and let

$$\tilde{\theta}_j(t) = \theta_j^* - \hat{\theta}_j(t) \tag{8.80}$$

denote the estimation error.

We use stochastic perturbation to develop a gradient estimate for every parameter. Let

$$\theta_j(t) = \hat{\theta}_j(t) + a_j \sin(\eta_j(t)), \tag{8.81}$$

where $a_j > 0$ is the perturbation amplitude and $(\eta_j(t), t \geq 0)$ is an OU process which is given by

$$\varepsilon_j \, d\eta_j(t) = -\eta_j(t) \, dt + \sqrt{\varepsilon_j} q_j \, dW_j(t), \tag{8.82}$$

where ε_j, $j = 1, \ldots, l$, are small parameters.

By (8.80) and (8.81), we have

$$\theta_j(t) - \theta_j^* = a_j \sin(\eta_j(t)) - \tilde{\theta}_j(t). \tag{8.83}$$

Substituting (8.83) into (8.79), we have the output

$$y(t) = f^* + J^T(\theta(t) - \theta^*) + (\theta(t) - \theta^*)^T P_1(\theta(t) - \theta^*), \tag{8.84}$$

where $\theta(t) - \theta^* = [a_1 \sin(\eta_1(t)) - \tilde{\theta}_1(t), \ldots, a_l \sin(\eta_l(t)) - \tilde{\theta}_l(t)]^T$.

We design the parameter update law as follows:

$$\frac{d\hat{\theta}_j(t)}{dt} = -k_j[a_j \sin(\eta_j(t))(y(t) - \xi_j(t)) + r_j(J_j)], \tag{8.85}$$

$$\frac{d\xi_j(t)}{dt} = -h_j \xi_j(t) + h_j y(t), \tag{8.86}$$

$$\varepsilon_j \, d\eta_j(t) = -\eta_j(t) \, dt + \sqrt{\varepsilon_j} q_j \, dW_j(t), \tag{8.87}$$

where h_j, k_j, $j = 1, \ldots, l$, are scalar design parameters, and r_j are functions to be designed. We use a washout filter $\frac{s}{s+h_j}$ for each parameter and the gradient estimation for each parameter is based on the output $\frac{s}{s+h_j}[y] = y(t) - \xi_j(t)$ of this filter.

Define $\chi_j(t) = \eta_j(\varepsilon_j t)$ and $B_j(t) = \frac{1}{\sqrt{\varepsilon_j}} W_j(\varepsilon_j t)$. Then we have

$$d\chi_j(t) = -\chi_j(t) \, dt + q_j \, dB_j(t), \tag{8.88}$$

where $B_j(t)$ is a 1-dimensional standard Brownian motion defined on the complete probability space $(\Omega, \mathcal{F}, P)$, while $[B_1(t), \ldots, B_l(t)]^T$ is an l-dimensional independent standard Brownian motion on the same space.

Define the output error variable

$$e_j(t) = \xi_j(t) - f^*, \quad j = 1, \ldots, l. \tag{8.89}$$

Therefore, it follows from (8.80), (8.84), (8.85), and (8.86) that we have the error dynamics

$$\frac{d\tilde{\theta}_j(t)}{dt} = -\frac{d\hat{\theta}_j(t)}{dt}$$

$$= k_j \big[a_j \sin(\eta_j(t)) \big(J^T(\theta(t) - \theta^*) + (\theta(t) - \theta^*)^T P_1(\theta(t) - \theta^*) - e_j(t) \big)$$
$$+ r_j(J_j) \big]$$

$$= k_j \big[a_j \sin(\chi_j(t/\varepsilon_j))$$
$$\times \big(J^T(\theta(t) - \theta^*) + (\theta(t) - \theta^*)^T P_1(\theta(t) - \theta^*) - e_j(t) \big)$$
$$+ r_j(J_j) \big],$$

$$= k_j \Bigg[a_j \sin(\eta_j(t)) \Bigg(\sum_{k=1}^{l} J_k \big(a_k \sin(\chi_k(t/\varepsilon_k)) - \tilde{\theta}_k \big)$$
$$+ \sum_{i,k=1}^{l} p_{1ik} \big(a_i \sin(\chi_i(t/\varepsilon_i)) - \tilde{\theta}_i \big) \big(a_k \sin(\chi_k(t/\varepsilon_k)) - \tilde{\theta}_k \big) - e_j(t) \Bigg)$$
$$+ r_j(J_j) \Bigg], \tag{8.90}$$

$$\frac{de_j(t)}{dt} = h_j \big(y(t) - f^* - e_j(t) \big)$$

$$= h_j \big(J^T(\theta(t) - \theta^*) + (\theta(t) - \theta^*)^T P_1(\theta(t) - \theta^*) - e_j(t) \big)$$

$$= h_j \Bigg[\sum_{k=1}^{l} J_k \big(a_k \sin(\chi_k(t/\varepsilon_k)) - \tilde{\theta}_k \big)$$
$$+ \sum_{i,k=1}^{l} p_{1ik} \big(a_i \sin(\chi_i(t/\varepsilon_i)) - \tilde{\theta}_i \big) \big(a_k \sin(\chi_k(t/\varepsilon_k)) - \tilde{\theta}_k \big) - e_j(t) \Bigg],$$
$$j = 1, \ldots, l. \tag{8.91}$$

Thus the corresponding average system is

$$\frac{d\tilde{\theta}_j^{\text{ave}}(t)}{dt} = a_j^2 k_j J_j \frac{1 - e^{-q_j^2}}{2} - a_j^2 k_j \big(1 - e^{-q_j^2} \big) \sum_{i=1}^{l} p_{1ji} \tilde{\theta}_i^{\text{ave}}(t)$$
$$+ k_j r_j(J_j), \tag{8.92}$$

$$\frac{de_j^{\text{ave}}(t)}{dt} = h_j\left[-\sum_{k=1}^{l} J_k\tilde{\theta}_k^{\text{ave}} + \sum_{i=1}^{l} p_{1ii}a_i^2\frac{1 - e^{-q_i^2}}{2}\right.$$

$$\left. + \sum_{i,k=1}^{l} p_{1ik}\tilde{\theta}_i^{\text{ave}}\tilde{\theta}_k^{\text{ave}} - e_j^{\text{ave}}(t)\right]. \tag{8.93}$$

We choose

$$r_j(J_j) = -a_j^2 J_j \frac{1 - e^{-q_j^2}}{2}. \tag{8.94}$$

Then in matrix form, the average error system is

$$\frac{d\tilde{\theta}^{\text{ave}}(t)}{dt} = -\Pi P_1 \tilde{\theta}^{\text{ave}}(t), \tag{8.95}$$

$$\frac{de^{\text{ave}}(t)}{dt} = H\left(-J^T\tilde{\theta}^{\text{ave}} + \sum_{i=1}^{l} p_{1ii}a_i^2 G_0(q_i)I_1\right.$$

$$\left. - e^{\text{ave}}(t) + Q\big(\tilde{\theta}^{\text{ave}}(t)\big)\right), \tag{8.96}$$

where

$$\Pi = \begin{bmatrix} a_1^2 k_1(1 - e^{-q_1^2}) & 0 & \cdots & 0 \\ 0 & a_2^2 k_2(1 - e^{-q_2^2}) & \cdots & 0 \\ \vdots & \vdots & \ddots & \vdots \\ 0 & 0 & \cdots & a_l^2 k_l(1 - e^{-q_l^2}) \end{bmatrix},$$

$$\tilde{\theta}^{\text{ave}}(t) = \big[\tilde{\theta}_1^{\text{ave}}(t), \ldots, \tilde{\theta}_l^{\text{ave}}(t)\big]^T, \qquad e^{\text{ave}}(t) = \big[e_1^{\text{ave}}(t), \ldots, e_l^{\text{ave}}(t)\big]^T,$$

$$H = \begin{bmatrix} h_1 & 0 & \cdots & 0 \\ 0 & h_2 & \cdots & 0 \\ \vdots & \vdots & \ddots & \vdots \\ 0 & 0 & \cdots & h_l \end{bmatrix},$$

$$Q\big(\tilde{\theta}^{\text{ave}}(t)\big) = \tilde{\theta}^{\text{ave}^T}(t)P_1\tilde{\theta}^{\text{ave}}(t)I_1, \qquad I_1 = [1, 1, \ldots, 1]_{1 \times l}^T.$$

The average error system has equilibrium $(\tilde{\theta}^{e^T}, e^{e^T}) = (0_{l \times 1}^T, \sum_{i=1}^{l} p_{1ii}a_i^2 \times G_0(q_i)I_1^T)$. The corresponding Jacobi matrix at this equilibrium is

$$\Xi_2 = \begin{bmatrix} -\Pi P_1 & 0 \\ -HJ^T & -H \end{bmatrix}. \tag{8.97}$$

Since Π, P_1, H are positive definite and Π is diagonal, Ξ_2 is Hurwitz. By multi-input stochastic averaging theorem given in Theorem 8.1, we have the following result:

Theorem 8.4 *Consider the static map (8.79) under the parameter update law (8.85)–(8.87). Then the error system (8.90)–(8.91) is weakly stochastically expo-*

nentially stable, i.e., there exist constants $r > 0$, $c > 0$, and $\gamma > 0$ such that, for any initial condition $|\Lambda_3^{\varepsilon_1}(0)| < r$ and any $\delta > 0$,

$$\lim_{\varepsilon_1 \to 0} \inf\left\{t \geq 0 : \left|\Lambda_3^{\varepsilon_1}(t)\right| > c\left|\Lambda_3^{\varepsilon_1}(0)\right|e^{-\gamma t} + \delta\right\} = +\infty \quad a.s. \tag{8.98}$$

Moreover, there exists a function $T(\varepsilon_1) : (0, \varepsilon_0) \to \mathbb{N}$ such that

$$\lim_{\varepsilon_1 \to 0} P\left\{\sup_{0 \leq t \leq T(\varepsilon_1)} \left\{\left|\Lambda_3^{\varepsilon_1}(t)\right| - c\left|\Lambda_3^{\varepsilon_1}(0)\right|e^{-\gamma t}\right\} > \delta\right\} = 0 \quad with$$

$$\lim_{\varepsilon_1 \to 0} T(\varepsilon_1) = \infty, \tag{8.99}$$

where $\Lambda_3^{\varepsilon_1}(t) = (\tilde{\theta}(t)^T, e(t)^T) - (0_{l\times l}^T, \sum_{i=1}^{l} p_{1ii}a_i^2 G_0(q_i)I_1^T)$, $I_1 = [1, 1, \ldots, 1]_{1\times l}^T$. Furthermore, (8.99) is equivalent to

$$\lim_{\varepsilon_1 \to 0} P\left\{\left|\Lambda_3^{\varepsilon_1}(t)\right| \leq c\left|\Lambda_3^{\varepsilon_1}(0)\right|e^{-\gamma t} + \delta, \forall t \in \left[0, T(\varepsilon_1)\right]\right\} = 1 \quad with$$

$$\lim_{\varepsilon_1 \to 0} T(\varepsilon_1) = \infty. \tag{8.100}$$

To quantify the output convergence to the optimum, for any $\varepsilon_1 > 0$, define a stopping time

$$\tau_{\varepsilon_1}^{\delta} = \inf\left\{t \geq 0 : \left|\Lambda_3^{\varepsilon_1}(t)\right| > c\left|\Lambda_3^{\varepsilon_1}(0)\right|e^{-\gamma t} + \delta\right\}.$$

Then by (8.98), we know that $\lim_{\varepsilon_1 \to 0} \tau_{\varepsilon_1}^{\delta} = \infty$ a.s., and

$$\left|\tilde{\theta}(t)\right| \leq c\left|\Lambda_3^{\varepsilon_1}(0)\right|e^{-\gamma t} + \delta \quad \forall t \leq \tau_{\varepsilon_1}^{\delta}. \tag{8.101}$$

Denote $\hat{\theta}(t) = [\hat{\theta}_1(t), \ldots, \hat{\theta}_l(t)]^T$, $a\sin(\eta(t)) = [a_1\sin(\eta_1(t)), \ldots, a_l\sin(\eta_l(t))]^T$. Then the output $y(t) = f(\theta) = f(\theta^* + \tilde{\theta}(t) + a\sin(\eta(t)))$, we have

$$\begin{aligned}
y(t) - f(\theta^*) &= \nabla f(\theta^*)(\tilde{\theta}(t) + a\sin(\eta(t))) \\
&\quad + (\tilde{\theta}(t) + a\sin(\eta(t)))^T P_1(\tilde{\theta}(t) + a\sin(\eta(t))) \\
&\quad + O(|\tilde{\theta}(t) + a\sin(\eta(t))|^3).
\end{aligned} \tag{8.102}$$

Thus by (8.101), it holds that

$$\begin{aligned}
\left|y(t) - f(\theta^*)\right| \leq O(|a|) &+ O(\delta) \\
&+ \left|\Lambda_3^{\varepsilon_1}(0)\right|e^{-\gamma t} + C\left|\Lambda_3^{\varepsilon_1}(0)\right|^2 e^{-2\gamma t} \quad \forall t \leq \tau_{\varepsilon_1}^{\delta}
\end{aligned} \tag{8.103}$$

for some positive constant C, where $|a| = \sqrt{a_1^2 + a_2^2 + \cdots + a_l^2}$. Similarly, by (8.29), we have

$$\lim_{\varepsilon_1 \to 0} P\left\{\left|y(t) - f(\theta^*)\right| \leq O(|a|) + O(\delta) + \left|\Lambda_3^{\varepsilon_1}(0)\right|e^{-\gamma t} + C\left|\Lambda_3^{\varepsilon_1}(0)\right|^2 e^{-2\gamma t}\right.$$

$$\left.\forall t \in \left[0, T(\varepsilon_1)\right]\right\} = 1, \tag{8.104}$$

where $T(\varepsilon_1)$ is a deterministic function with $\lim_{\varepsilon_1 \to 0} T(\varepsilon_1) = \infty$.

Fig. 8.3 Multi-parameter
stochastic gradient seeking
with an OU process
perturbation. *Top*: output and
optimum values. *Bottom*:
solutions of the error system

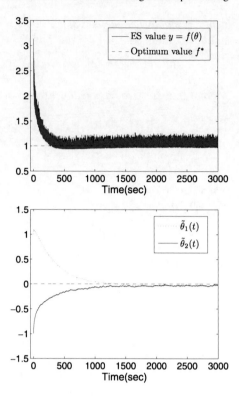

Figure 8.3 displays the simulation results with $f^* = 1$, $J = [0.5, 0.2]^T$, $\theta^* = [0, 1]^T$, $P_1 = \begin{bmatrix} 1 & 1 \\ 1 & 2 \end{bmatrix}$ in the static map (8.79) and $a_1 = 0.1$, $a_2 = 0.2$, $k_1 = 1.25$, $k_2 = 5/3$, $q_1 = 1$, $q_2 = 1$, $\varepsilon_1 = 0.25$, $\varepsilon_2 = 0.01$, $h_1 = 1$, $h_2 = 2$ in the parameter update law (8.85)–(8.86) and initial condition $\tilde{\theta}_1(0) = 1$, $\tilde{\theta}_2(0) = -1$, $\hat{\theta}_1(0) = -1$, $\hat{\theta}_2(0) = 2$, $\xi_1(0) = 0$, $\xi_2(0) = 0$.

8.4 Notes and References

Most of the existing stochastic averaging theory focuses on the systems with a single stochastic perturbation input [21, 39, 78, 88] or on two-time-scales systems with slow dynamics and fast dynamics [59, 129]. There are few result on stochastic averaging for systems with multi-scale stochastic perturbation inputs. With multi-input stochastic averaging theorems that we developed in this chapter, we designed multivariable stochastic extremum seeking and gradient seeking algorithms. Parts of this chapter are based on our results in [88].

Chapter 9
Stochastic Nash Equilibrium Seeking for Games with General Nonlinear Payoffs

Non-cooperative games have been a vibrant topic in mathematics and economics for decades. While work on control-theoretic problems in differential games has been conducted since at least the 1960s, the topic of games has, in recent years, been justifiably enjoying a renaissance in the field of engineering and, in particular, in the area of control systems. A comprehensive account of non-cooperative game theory is available in the seminal book [11].

It is inherent to the non-cooperative character of games that the opponents share as little information as possible. For example, in realistic games, players would not be inclined to inform their opponents about the functional form of their performance criteria (payoff functions). When a game is played iteratively (as the time evolves), the opposing players would not be inclined to share with others the information on the actions that they are taking and on the payoff values that they are obtaining. It is reasonable that each player is aware of the value of his own payoff achieved, but the player doesn't necessarily know the functional form of the payoff, namely, its dependence on the player's own action and on the actions of the opponents. This uncertainty in the functional form of the payoff functions comes from the uncertainty regarding the environment in which the game is played, such as the uncertainty that companies encounter in the marketplace when they play games in which pricing or production volumes are the actions, while the customer response is hard to predict.

Due to such a lack of modeling information and restricted measurements, the topic of learning in games has recently been popular [26, 37, 38, 42, 54, 83, 126, 127, 150]. Rather than seeking strategically optimal (Nash equilibrium) policies using iterative algorithms that employ the full modeling information and the measurement of the opponents' actions, learning-based algorithms attempt to achieve convergence to Nash policies using estimation and various other techniques.

In this chapter, a multi-parameter stochastic extremum seeking algorithm is developed for finding Nash equilibria in N-player noncooperative games. For an N-player noncooperative game, each player employs independently stochastic extremum seeking to attain a Nash equilibrium. The players are not required to know the mathematical model of the game (either their own or the opponents' payoff functions). The players are also not required to measure the opponents actions but only

S.-J. Liu, M. Krstic, *Stochastic Averaging and Stochastic Extremum Seeking*,
Communications and Control Engineering,
DOI 10.1007/978-1-4471-4087-0_9, © Springer-Verlag London 2012

measure their own payoff values. We prove that, under certain conditions, the actions of the players converge to a small neighborhood of a Nash equilibrium. The convergence result is local in the sense that convergence to any particular Nash equilibrium is assured only for initial conditions in a set around that specific stable Nash equilibrium. Moreover, convergence to a Nash equilibrium is biased in proportion to the third derivatives of the payoff functions and is dependent on the intensity of stochastic perturbation.

The chapter is organized as follows: we introduce the general problem formulation in Sect. 9.1, state the algorithm and convergence results in Sect. 9.2, and present the convergence proof in Sect. 9.3. We provide a numerical example for a two-player game in Sect. 9.4.

9.1 Problem Formulation

Consider an N-player noncooperative game where each player wishes to maximize its payoff function of the general nonlinear form. Assume the payoff function of player i is of the form

$$J_i = h_i(u_i, u_{-i}), \tag{9.1}$$

where u_i is player i's action, the action (strategy) space is the whole space $\mathbb{R}$, $u_{-i} = [u_1, \ldots, u_{i-1}, u_{i+1}, \ldots, u_N]$ represents the actions of other players, $h_i : \mathbb{R}^N \to \mathbb{R}$ is smooth, and $i \in \{1, \ldots, N\}$.

Our algorithm is based on the following assumptions.

Assumption 9.1 There exists at least one, possibly not unique, isolated stable Nash equilibrium $u^* = [u_1^*, \ldots, u_N^*]$ such that

$$\frac{\partial h_i}{\partial u_i}(u^*) = 0, \tag{9.2}$$

$$\frac{\partial^2 h_i}{\partial u_i^2}(u^*) < 0 \tag{9.3}$$

for all $i \in \{1, \ldots, N\}$.

Assumption 9.2 The matrix

$$\Xi = \begin{bmatrix} \frac{\partial^2 h_1(u^*)}{\partial u_1^2} & \frac{\partial^2 h_1(u^*)}{\partial u_1 \partial u_2} & \cdots & \frac{\partial^2 h_1(u^*)}{\partial u_1 \partial u_N} \\ \frac{\partial^2 h_2(u^*)}{\partial u_1 \partial u_2} & \frac{\partial^2 h_2(u^*)}{\partial u_2^2} & \cdots & \frac{\partial^2 h_2(u^*)}{\partial u_2 \partial u_N} \\ \vdots & \vdots & \ddots & \vdots \\ \frac{\partial^2 h_N(u^*)}{\partial u_1 \partial u_N} & \frac{\partial^2 h_N(u^*)}{\partial u_2 \partial u_N} & \cdots & \frac{\partial^2 h_N(u^*)}{\partial u_N^2} \end{bmatrix} \tag{9.4}$$

is strictly diagonally dominant and hence, nonsingular.

By Assumptions 9.1 and 9.2, Ξ is Hurwitz.

In our scheme, player i has no knowledge of others players' payoff h_j, $j \neq i$, and actions u_j ($j \neq i$). It can only measure its own payoff h_i. Our objective is to design a stochastic extremum seeking algorithm for each player to approximate Nash equilibrium.

9.2 Stochastic Nash Equilibrium Seeking Algorithm

In our algorithm, each player independently employs a stochastic seeking strategy to attain the stable Nash equilibrium of the game. Player i implements the following strategy:

$$u_i(t) = \hat{u}_i(t) + a_i f_i\big(\eta_i(t)\big), \tag{9.5}$$

$$\frac{d\hat{u}_i(t)}{dt} = k_i a_i f_i\big(\eta_i(t)\big) J_i(t), \tag{9.6}$$

where for any $i = 1, \ldots, N$, $a_i > 0$ is the perturbation amplitude, $k_i > 0$ is the adaptive gain, $J_i(t)$ is the measured payoff value for player i, and f_i is a bounded smooth function that player i chooses, e.g., a sine function. $\eta_i(t)$, $i = 1, \ldots, N$, are independent ergodic processes chosen by player i, e.g., the Ornstein–Uhlenbeck (OU) processes

$$\eta_i = \frac{\sqrt{\varepsilon_i} q_i}{\varepsilon_i s + 1}[\dot{W}_i], \quad \text{or} \quad \varepsilon_i \, d\eta_i(t) = -\eta_i(t)\, dt + \sqrt{\varepsilon_i} q_i \, dW_i(t), \tag{9.7}$$

$q_i > 0$, ε_i are small parameters satisfying $0 < \max_i \varepsilon_i < \varepsilon_0$ for fixed $\varepsilon_0 > 0$, and $W_i(t)$, $i = 1, \ldots, N$, are independent 1-dimensional standard Brownian motions on a complete probability space $(\Omega, \mathcal{F}, P)$ with the sample space Ω, σ-field $\mathcal{F}$, probability measure P.

Figure 9.1 depicts a noncooperative game played by two players implementing the stochastic extremum seeking strategy (9.5)–(9.6) to attain a Nash equilibrium.

To analyze the convergence of the algorithm, we denote the error relative to the Nash equilibrium as

$$\tilde{u}_i(t) = \hat{u}_i(t) - u_i^*. \tag{9.8}$$

Then, we obtain an error system as

$$\frac{d\tilde{u}_i(t)}{dt} = k_i \rho_i^{(1)}(t) h_i\big(u_i^* + \tilde{u}_i + \rho_i^{(1)}(t), u_{-i}^* + \tilde{u}_{-i} + \rho_{-i}^{(1)}(t)\big), \tag{9.9}$$

where $\rho_i^{(1)}(t) = a_i f_i(\eta_i(t))$, $\rho_{-i}^{(1)}(t) = [a_1 f_1(\eta_1(t)), \ldots, a_{i-1} f_{i-1}(\eta_{i-1}(t)), a_{i+1} f_{i+1}(\eta_{i+1}(t)), \ldots, a_N f_N(\eta_N(t))]$, $\tilde{u}_{-i}^* = [\tilde{u}_1^*, \ldots, \tilde{u}_{i-1}^*, \tilde{u}_{i+1}^*, \ldots, \tilde{u}_N^*]$, and $\tilde{u}_{-i} = [\tilde{u}_1, \ldots, \tilde{u}_{i-1}, \tilde{u}_{i+1}, \ldots, \tilde{u}_N]$.

If the players choose $f_i(x) = \sin x$ for all $i = 1, \ldots, N$, and η_i as OU processes (9.7), we have the following convergence result.

Fig. 9.1 Stochastic extremum seeking scheme for a two-player noncooperative game

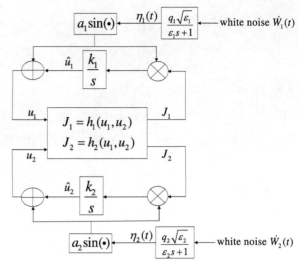

Theorem 9.1 *Consider the error system* (9.9) *for an N-player game under Assumptions 9.1 and 9.2. Then there exists a constant $a^* > 0$ such that for $\max_{1 \le i \le N} a_i \in (0, a^*)$ there exist constants $r > 0$, $c > 0$, $\gamma > 0$, and a function $T(\varepsilon_1): (0, \varepsilon_0) \to \mathbb{N}$ such that, for any initial condition $|\Lambda^{\varepsilon_1}(0)| < r$ and any $\delta > 0$,*

$$\lim_{\varepsilon_1 \to 0} \inf \left\{ t \ge 0 : \left| \Lambda^{\varepsilon_1}(t) \right| > c \left| \Lambda^{\varepsilon_1}(0) \right| e^{-\gamma t} + \delta + O\left(\max_i a_i^3 \right) \right\} = \infty \quad a.s. \quad (9.10)$$

and

$$\lim_{\varepsilon_1 \to 0} P \left\{ \left| \Lambda^{\varepsilon_1}(t) \right| \le c \left| \Lambda^{\varepsilon_1}(0) \right| e^{-\gamma t} + \delta + O\left(\max_i a_i^3 \right), \right.$$

$$\left. \forall t \in \left[0, T(\varepsilon_1) \right] \right\} = 1 \quad (9.11)$$

with

$$\lim_{\varepsilon_1 \to 0} T(\varepsilon_1) = \infty, \quad (9.12)$$

where

$$\Lambda^{\varepsilon_1}(t) = \left[\tilde{u}_1(t) - \sum_{j=1}^{N} d_{jj}^1 a_j^2, \ldots, \tilde{u}_N(t) - \sum_{j=1}^{N} d_{jj}^N a_j^2 \right], \quad (9.13)$$

$$
\begin{bmatrix} d^1_{jj} \\ \vdots \\ d^{j-1}_{jj} \\ d^j_{jj} \\ d^{j+1}_{jj} \\ \vdots \\ d^N_{jj} \end{bmatrix} = -\varXi^{-1} \begin{bmatrix} \frac{1}{2}G_0(q_j)\frac{\partial^3 h_1}{\partial u_1\,\partial u_j^2}(u^*) \\ \vdots \\ \frac{1}{2}G_0(q_j)\frac{\partial^3 h_{j-1}}{\partial u_{j-1}\,\partial u_j^2}(u^*) \\ \frac{1}{6}\frac{G_1(q_j)}{G_0(q_j)}\frac{\partial^3 h_j}{\partial u_j^3}(u^*) \\ \frac{1}{2}G_0(q_j)\frac{\partial^3 h_{j+1}}{\partial u_j^2\,\partial u_{j+1}}(u^*) \\ \vdots \\ \frac{1}{2}G_0(q_j)\frac{\partial^3 h_N}{\partial u_j^2\,\partial u_N}(u^*) \end{bmatrix}, \tag{9.14}
$$

and $G_0(q_j) = \frac{1}{2}(1 - e^{-q_j^2})$, $G_1(q_j) = \frac{3}{8} - \frac{1}{2}e^{-q_j^2} + \frac{1}{8}e^{-4q_j^2} = \frac{1}{8}(1 - e^{-q_j^2})^2(e^{-2q_j^2} + 2e^{-q_j^2} + 3)$.

Several remarks are needed in order to properly interpret Theorem 9.1. From (9.10) and the fact that $|\Lambda^{\varepsilon_1}(t)| \geq \max_i |\tilde{u}_i(t) - \sum_{j=1}^{N} d^i_{jj}a_j^2|$, we obtain

$$
\lim_{\varepsilon_1 \to 0} \inf \left\{ t \geq 0 : \max_i \left\| \tilde{u}_i(t) - \sum_{j=1}^{N} d^i_{jj}a_j^2 \right\| > c\left|\Lambda^{\varepsilon_1}(0)\right|e^{-\gamma t} + \delta + O\left(\max_i a_i^3\right) \right\}
$$
$$
= \infty \quad \text{a.s.}
$$

By taking all the a_i's small, $\max_i |\tilde{u}_i(t)|$ can be made arbitrarily small as $t \to \infty$.

The bias terms $\sum_{j=1}^{N} d^i_{jj}a_j^2$ defined by (9.14) appear complicated but have a simple physical interpretation. When the game's payoff functions are not quadratic (not symmetric), the extremum seeking algorithms which employ zero-mean (symmetric) perturbations will produce a bias. According to the formulas (9.14), the bias depends on the third derivatives of the payoff functions, namely, on the level of asymmetry in the payoff surfaces at the Nash equilibrium. In the trivial case of a single player, the interpretation is easy—extremum seeking settles on the flatter (more favorable) side of an asymmetric peak. In the case of multiple players, the interpretation is more difficult, as each player contributes both to his own bias and to the other players' biases. Though difficult to intuitively interpret in the multi-player case, the formula (9.14) is useful as it quantifies the biases.

The estimate of the region of attraction r can be conservatively taken as independent of the a_i's, for a_i's chosen sufficiently small. This fact can be only seen by going through the proof of the averaging theorem for the specific system (3.5). Hence, r is larger than the bias terms, which means that for small a_i's the algorithm reduces the distance to the Nash equilibrium for all initial conditions except for those within an $O(\max_i a_i^2)$ to the Nash equilibrium.

On the other hand, the convergence rate γ cannot be taken independently of the a_i's because the a_i's appear as factors on the entire right-hand side of (3.5). However, by letting the k_i's increase as the a_i's decrease, independence of γ from the a_i's can be ensured.

In the rare case where the error system (9.9) may be globally Lipschitz, we obtain global convergence using the global averaging theorem in Chap. 3.

9.3 Proof of the Algorithm Convergence

We apply the multi-input stochastic averaging theory presented in Chap. 8 to analyze the error system (9.9). First, we calculate the average system of (9.9).

Define $\chi_i(t) = \eta_i(\varepsilon_i t)$ and $B_i(t) = \frac{1}{\sqrt{\varepsilon_i}} W_i(\varepsilon_i t)$. Then, by (9.7), we have

$$d\chi_i(t) = -\chi_i(t)\,dt + q_i\,dB_i(t), \tag{9.15}$$

where $[B_1(t), \ldots, B_N(t)]^T$ is an N-dimensional standard Brownian motion on the space $(\Omega, \mathcal{F}, P)$.

Thus we can rewrite the error system (9.9) as

$$\frac{d\tilde{u}_i(t)}{dt} = k_i \rho_i^{(2)}(t/\varepsilon_i) h_i\big(u_i^* + \tilde{u}_i + \rho_i^{(2)}(t/\varepsilon_i), u_{-i}^* + \tilde{u}_{-i} + \rho_{-i}^{(2)}(t/\varepsilon_{-i})\big), \tag{9.16}$$

where $\rho_i^{(2)}(t) = a_i \sin(\chi_i(t))$, $\rho_{-i}^{(2)}(t/\varepsilon_{-i}) = [a_1 \sin(\chi_1(t/\varepsilon_1)), \ldots, a_{i-1} \sin(\chi_{i-1}(t/\varepsilon_{i-1})), a_{i+1} \sin(\chi_{i+1}(t/\varepsilon_{i+1})), \ldots, a_N \sin(\chi_N(t/\varepsilon_N))]$.

Denote

$$\varepsilon_i = \frac{\varepsilon_1}{c_i}, \quad i = 2, \ldots, N, \tag{9.17}$$

for some positive real constants c_i's and consider the change of variable

$$Z_1(t) = \chi_1(t), \qquad Z_2(t) = \chi_2(c_2 t), \qquad \cdots, \qquad Z_N(t) = \chi(c_N t). \tag{9.18}$$

Then the error system (9.16) can be transformed into one with a single small parameter ε_1:

$$\frac{d\tilde{u}_i(t)}{dt} = k_i \rho_i^{(3)}(t/\varepsilon_1) h_i\big(u_i^* + \tilde{u}_i + \rho_i^{(3)}(t/\varepsilon_1), u_{-i}^* + \tilde{u}_{-i} + \rho_{-i}^{(3)}(t/\varepsilon_1)\big), \tag{9.19}$$

where $\rho_i^{(3)}(t) = a_i \sin(Z_i(t))$, $\rho_{-i}^{(3)}(t/\varepsilon_1) = [a_1 \sin(Z_1(t/\varepsilon_1)), \ldots, a_{i-1} \sin(Z_{i-1}(t/\varepsilon_1)), a_{i+1} \sin(Z_{i+1}(t/\varepsilon_1)), \ldots, a_N \sin(Z_N(t/\varepsilon_1))]$.

Since $(\chi_i(t), t \geq 0)$ is ergodic and has invariant distribution $\mu_i(dx_i) = \frac{1}{\sqrt{\pi} q_i} \times e^{-x_i^2/q_i^2} dx_i$, by Lemma 8.2, the vector value process $[Z_1(t), \ldots, Z_N(t)]^T$ is also ergodic with invariant distribution $\mu_1 \times \cdots \times \mu_N$. Thus, by (8.9), we have the following average error system

$$\frac{d\tilde{u}_i^{\text{ave}}(t)}{dt} = k_i a_i \int_{\mathbb{R}^N} \sin(x_i) h_i\big(u_i^* + \tilde{u}_i^{\text{ave}} + a_i \sin(x_i), u_{-i}^* + \tilde{u}_{-i}^{\text{ave}} + a_{-i} \sin(x_{-i})\big)$$
$$\times \mu_1(dx_1) \times \cdots \times \mu_N(dx_N), \tag{9.20}$$

where

$$a_{-i} \sin(x_{-i}) = \big[a_1 \sin(x_1), \ldots, a_{i-1} \sin(x_{i-1}), a_{i+1} \sin(x_{i+1}), \ldots, a_N \sin(x_N)\big],$$

and μ_i is the invariant distribution of the process $(\chi_i(t), t \geq 0)$ or $(Z_i(t), t \geq 0)$.

The equilibrium $\tilde{u}^e = [\tilde{u}_1^e, \ldots, \tilde{u}_N^e]$ of (9.20) satisfies

$$0 = \int_{\mathbb{R}^N} \sin(x_i) h_i \left(u_i^* + \tilde{u}_i^e + a_i \sin(x_i), u_{-i}^* + \tilde{u}_{-i}^e + a_{-i} \sin(x_{-i}) \right)$$
$$\times \mu_1(dx_1) \times \cdots \times \mu_N(dx_N) \tag{9.21}$$

for all $i = \{1, \ldots, N\}$.

To calculate the equilibrium of the average error system and analyze its stability, we postulate that $\tilde{u}^e$ has the form

$$\tilde{u}_i^e = \sum_{j=1}^N b_j^i a_j + \sum_{j=1}^N \sum_{k \geq j}^N d_{jk}^i a_j a_k + O\left(\max_i a_i^3\right). \tag{9.22}$$

By expanding h_i about u^* in (9.21) and substituting (9.22), the unknown coefficients b_j^i and d_{jk}^i can be determined.

The Taylor series expansion of h_i about u^* in (9.21) for an N-player game is

$$h_i\left(u^* + v_i, u_{-i}^* + v_{-i}\right) = \sum_{n_1=0}^\infty \cdots \sum_{n_N=0}^\infty \frac{v_1^{n_1} \cdots v_N^{n_N}}{n_1! \cdots n_N!} \left(\frac{\partial^{n_1+\cdots+n_N} h_i}{\partial u_1^{n_1} \cdots \partial u_N^{n_N}}\right)(u^*), \tag{9.23}$$

where $v_i = \tilde{u}_i^e + a_i \sin(x_i)$ and $v_{-i} = \tilde{u}_{-i}^e + a_{-i} \sin(x_{-i})$. Although for any $i = 1, \ldots, N$, h_i may not have its Taylor series expansion only by its smoothness, here we just give the form of Taylor series expansion. In fact, we only need its third order Taylor formula.

Since the invariant distribution $\mu_i(dx_i)$ of the OU process $(\chi_i(t), t \geq 0)$ is $\frac{1}{\sqrt{\pi q_i}} e^{-x_i^2/q_i} dx_i$, we have

$$\int_{\mathbb{R}} \sin^4(x_i) \mu_i(dx_i)$$
$$= \int_{-\infty}^{+\infty} \sin^4(x_i) \frac{1}{\sqrt{\pi q_i}} e^{-x_i^2/q_i^2} dx_i$$
$$= \frac{3}{8} - \frac{1}{2} e^{-q_i^2} + \frac{1}{8} e^{-4q_i^2} \triangleq G_1(q_i), \tag{9.24}$$

$$\int_{\mathbb{R}^2} \sin^3(x_i) \sin(x_j) \mu_i(dx_i) \times \mu_j(dx_j)$$
$$= \int_{-\infty}^{+\infty} \int_{-\infty}^{+\infty} \sin^3(x_i) \sin(x_j) \frac{1}{\sqrt{\pi q_i}} e^{-x_i^2/q_i^2} \frac{1}{\sqrt{\pi q_j}} e^{-x_j^2/q_j^2} dx_i \, dx_j$$
$$= 0, \tag{9.25}$$

$$\int_{\mathbb{R}^2} \sin^2(x_i) \sin^2(x_j) \mu_i(dx_i) \times \mu_j(dx_j)$$
$$= \int_{-\infty}^{+\infty} \int_{-\infty}^{+\infty} \sin^2(x_i) \sin^2(x_j) \frac{1}{\sqrt{\pi q_i}} e^{-x_i^2/q_i^2} \frac{1}{\sqrt{\pi q_j}} e^{-x_j^2/q_j^2} dx_i \, dx_j$$
$$= \frac{1}{4}\left(1 - e^{-q_i^2}\right)\left(1 - e^{-q_j^2}\right) \triangleq G_2(q_i, q_j), \tag{9.26}$$

$$\int_{\mathbb{R}^3} \sin(x_i)\sin^2(x_j)\sin(x_k)\mu_i(dx_i) \times \mu_j(dx_j) \times \mu_k(dx_k)$$

$$= \int_{-\infty}^{+\infty}\int_{-\infty}^{+\infty}\int_{-\infty}^{+\infty} \sin(x_i)\sin^2(x_j)\sin(x_k)$$

$$\times \frac{1}{\sqrt{\pi}q_i}e^{-x_i^2/q_i^2}\frac{1}{\sqrt{\pi}q_j}e^{-x_j^2/q_j^2}\frac{1}{\sqrt{\pi}q_k}e^{-x_k^2/q_k^2}\,dx_i\,dx_j\,dx_k$$

$$= 0. \tag{9.27}$$

Based on the above calculations together with (8.41), (8.42), (8.43), (8.44), (8.45), substituting (9.23) into (9.21) and computing the average of each term gives

$$0 = a_i^2 G_0(q_i)\tilde{u}_i^e\frac{\partial^2 h_i}{\partial u_i^2}(u^*) + a_i^2 G_0(q_i)\sum_{j\neq i}^{N}\tilde{u}_j^e\frac{\partial^2 h_i}{\partial u_i\,\partial u_j}(u^*)$$

$$+ \left(\frac{a_i^2}{2}G_0(q_i)(\tilde{u}_i^e)^2 + \frac{a_i^4}{6}G_1(q_i)\right)\frac{\partial^3 h_i}{\partial u_i^3}(u^*) + a_i^2 G_0(q_i)\tilde{u}_i^e\sum_{j\neq i}^{N}\tilde{u}_j^e\frac{\partial^3 h_i}{\partial u_i^2\,\partial u_j}(u^*)$$

$$+ \sum_{j\neq i}^{N}\left(\frac{a_i^2}{2}G_0(q_i)(\tilde{u}_j^e)^2 + \frac{a_i^2 a_j^2}{2}G_2(q_i,q_j)\right)\frac{\partial^3 h_i}{\partial u_i\,\partial u_j^2}(u^*)$$

$$+ \sum_{j\neq i}^{N}\sum_{k>j,k\neq i}^{N} a_i^2 G_0(q_i)\tilde{u}_j^e\tilde{u}_k^e\frac{\partial^3 h_i}{\partial u_i\,\partial u_j\,\partial u_k}(u^*) + O\left(\max_i a_i^5\right), \tag{9.28}$$

or equivalently,

$$0 = \tilde{u}_i^e\frac{\partial^2 h_i}{\partial u_i^2}(u^*) + \sum_{j\neq i}^{N}\tilde{u}_j^e\frac{\partial^2 h_i}{\partial u_i\,\partial u_j}(u^*) + \left(\frac{1}{2}(\tilde{u}_i^e)^2 + \frac{a_i^2}{6}\frac{G_1(q_i)}{G_0(q_i)}\right)\frac{\partial^3 h_i}{\partial u_i^3}(u^*)$$

$$+ \tilde{u}_i^e\sum_{j\neq i}^{N}\tilde{u}_j^e\frac{\partial^3 h_i}{\partial u_i^2\,\partial u_j}(u^*) + \sum_{j\neq i}^{N}\left(\frac{1}{2}(\tilde{u}_j^e)^2 + \frac{a_j^2}{2}G_0(q_j)\right)\frac{\partial^3 h_i}{\partial u_i\,\partial u_j^2}(u^*)$$

$$+ \sum_{j\neq i}^{N}\sum_{k>j,k\neq i}^{N}\tilde{u}_j^e\tilde{u}_k^e\frac{\partial^3 h_i}{\partial u_i\,\partial u_j\,\partial u_k}(u^*) + O\left(\max_i a_i^3\right). \tag{9.29}$$

Substituting (9.22) into (9.28) and matching first order powers of a_i gives

$$\begin{bmatrix} 0 \\ \vdots \\ 0 \end{bmatrix} = \varXi\begin{bmatrix} b_i^1 \\ \vdots \\ b_i^N \end{bmatrix}, \quad i = 1,\ldots,N, \tag{9.30}$$

which implies that $b^i_j = 0$ for all i, j since Ξ is nonsingular by Assumption 9.2. Similarly, matching second order terms $a_j a_k$ ($j > k$) and a^2_j of a_j, and substituting $b^i_j = 0$ to simplify the resulting expressions, yields

$$\begin{bmatrix} 0 \\ \vdots \\ 0 \end{bmatrix} = \Xi \begin{bmatrix} d^1_{jk} \\ \vdots \\ d^N_{jk} \end{bmatrix}, \quad j = 1, \ldots, N, j > k, \tag{9.31}$$

and

$$\begin{bmatrix} 0 \\ \vdots \\ 0 \end{bmatrix} = \Xi \left(\begin{bmatrix} d^1_{jj} \\ \vdots \\ d^N_{jj} \end{bmatrix} + \begin{bmatrix} \frac{1}{2}G_0(q_j)\frac{\partial^3 h_1}{\partial u_1 \partial u^2_j}(u^*) \\ \vdots \\ \frac{1}{2}G_0(q_j)\frac{\partial^3 h_{j-1}}{\partial u_{j-1} \partial u^2_j}(u^*) \\ \frac{1}{6}\frac{G_1(q_j)}{G_0(q_j)}\frac{\partial^3 h_j}{\partial u^3_j}(u^*) \\ \frac{1}{2}G_0(q_j)\frac{\partial^3 h_{j+1}}{\partial u^2_j \partial u_{j+1}}(u^*) \\ \vdots \\ \frac{1}{2}G_0(q_j)\frac{\partial^3 h_N}{\partial u^2_j \partial u_N}(u^*) \end{bmatrix} \right). \tag{9.32}$$

Thus, $d^i_{jk} = 0$ for all i, j, k when $j \neq k$, and d^i_{jj} is given by (9.14).

Therefore, by (9.22), the equilibrium of the average error system (9.20) is

$$\tilde{u}^e_i = \sum_{j=1}^N d^i_{jj} a^2_j + O\left(\max_i a^3_i\right). \tag{9.33}$$

By the Dominated Convergence Theorem, we obtain that the Jacobian $\Psi^{\text{ave}} = (\psi_{ij})_{N \times N}$ of the average error system (9.20) at $\tilde{u}^e$ has elements given by

$$\psi_{ij} = k_i \int_{\mathbb{R}^N} a_i \sin(x_i) \frac{\partial h_i}{\partial u_j}\left(u^*_i + \tilde{u}^e_i + a_i \sin(x_i), u^*_{-i} + \tilde{u}^e_{-i} + a_{-i}\sin(x_{-i})\right)$$
$$\times \mu_1(dx_1) \times \cdots \times \mu_N(dx_N)$$
$$= k_i a^2_i G_0(q_i)\frac{\partial^2 h_i}{\partial u_i \partial u_j}(u^*) + O\left(\max_i a^3_i\right) \tag{9.34}$$

and is Hurwitz by Assumptions 9.1 and 9.2 for sufficiently small a_i, which implies that the equilibrium (9.33) of the average error system (9.20) is exponentially stable. By the multi-input stochastic averaging theorem given in Theorem 8.1 of Chap. 8, the theorem is proved.

9.4 Numerical Example

We consider two players with payoff functions

$$J_1 = -u_1^3 + 2u_1u_2 + u_1^2 - \frac{3}{4}u_1, \tag{9.35}$$

$$J_2 = 2u_1^2u_2 - u_2^2. \tag{9.36}$$

Since J_1 is not globally concave in u_1, we restrict the action space to $\mathcal{A} = \{u_1 \geq 1/3, u_2 \geq 1/6\}$ in order to avoid the existence of maximizing actions at infinity or Nash equilibria at the boundary of the action space. (However, we do not restrict the extremum seeking algorithm to $\mathcal{A}$. Such a restriction can be imposed using parameter projection, but would complicate our exposition considerably.)

The game (J_1, J_2) yields two Nash equilibria: $(u_1^{*1}, u_2^{*1}) = (0.5, 0.25)$ and $(u_1^{*2}, u_2^{*2}) = (1.5, 2.25)$. The corresponding matrices are $\varXi_1 = \begin{bmatrix} -1 & 2 \\ 2 & -2 \end{bmatrix}$ and $\varXi_2 = \begin{bmatrix} -7 & 2 \\ 6 & -2 \end{bmatrix}$, where $\varXi_1$ is nonsingular but not Hurwitz, while $\varXi_2$ is nonsingular and Hurwitz, and both matrices are not diagonally dominant. From the proof of the algorithm convergence, we know that diagonal dominance is only a sufficient condition for $\varXi$ to be nonsingular and is not required, in general.

The average error system for this game is

$$\frac{d\tilde{u}_1^{\text{ave}}(t)}{dt} = k_1 a_1^2 G_0(q_1)\left(-3\tilde{u}_1^{\text{ave}^2} - 6u_1^* \tilde{u}_1^{\text{ave}} + 2\tilde{u}_2^{\text{ave}} + 2\tilde{u}_1^{\text{ave}}\right)$$
$$- k_1 a_1^4 G_1(q_1), \tag{9.37}$$

$$\frac{d\tilde{u}_2^{\text{ave}}(t)}{dt} = k_2 a_2^2 G_0(q_2)\left(-2\tilde{u}_2^{\text{ave}} + 2\tilde{u}_1^{\text{ave}^2} + 4u_1^* \tilde{u}_1^{\text{ave}}\right)$$
$$+ 2k_2 a_1^2 a_2^2 G_2(q_1, q_2), \tag{9.38}$$

where u_1^* can be u_1^{*1} or u_1^{*2}. The equilibria $(\tilde{u}_1^e, \tilde{u}_2^e)$ of this average system are

$$\tilde{u}_1^e = 1 - u_1^* \pm \sqrt{\left(1 - u_1^*\right)^2 - a_1^2\left(\frac{G_1(q_1)}{G_0(q_1)} - 2G_0(q_1)\right)}, \tag{9.39}$$

$$\tilde{u}_2^e = 2 - 2u_1^* \pm 2\sqrt{\left(1 - u_1^*\right)^2 - a_1^2\left(\frac{G_1(q_1)}{G_0(q_1)} - 2G_0(q_1)\right)}$$
$$- a_1^2\frac{G_1(q_1)}{G_0(q_1)} + 3a_1^2 G_0(q_1), \tag{9.40}$$

and their postulated form is

$$\tilde{u}_1^{e,p} = \frac{1}{2(1 - u_1^*)}\left(\frac{G_1(q_1)}{G_0(q_1)} - 2G_0(q_1)\right)a_1^2 + O\left(\max_i a_i^3\right), \tag{9.41}$$

$$\tilde{u}_2^{e,p} = \left(\frac{u_1^*}{1 - u_1^*}\frac{G_1(q_1)}{G_0(q_1)} + \frac{1 - 3u_1^*}{1 - u_1^*}G_0(q_1)\right)a_1^2 + O\left(\max_i a_i^3\right). \tag{9.42}$$

The corresponding Jacobian matrices are

$$\psi^{\text{ave}} = \begin{bmatrix} (-6\tilde{u}_1^e - 6u_1^* + 2)\gamma_1 & 2\gamma_1 \\ (2\tilde{u}_1^e + 4u_1^*)\gamma_2 & -2\gamma_2 \end{bmatrix}, \tag{9.43}$$

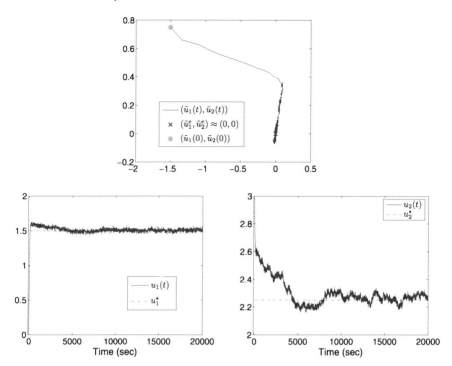

Fig. 9.2 Stochastic Nash equilibrium seeking with an OU process perturbation. *Top*: evolution of the game in the $\tilde{u}$ plane. *Bottom*: two players' actions

where $\gamma_i = k_i a_i^2 G_0(q_i)$, $i = 1, 2$, and their characteristic equation is given by, $\lambda^2 + \alpha_1 \lambda + \alpha_2 = 0$, where

$$\alpha_1 = \left(6\tilde{u}_1^e + 6u_1^* - 2\right)\gamma_1 + 2\gamma_2, \tag{9.44}$$

$$\alpha_2 = \left(2\tilde{u}_1^e + u_1^* - 1\right)4\gamma_1\gamma_2. \tag{9.45}$$

Thus Ψ^{ave} is Hurwitz if and only if α_1 and α_2 are positive. For sufficiently small a_1, which makes $\tilde{u}^e \approx (0, 0)$, α_1 and α_2 are positive for $u_1^* = 1.5$, but for $u_1^* = 0.5$, α_2 is not positive, which is reasonable because $\mathcal{Z}_1$ is not Hurwitz but $\mathcal{Z}_2$ is Hurwitz. Thus, $(u_1^{*1}, u_2^{*1}) = (0.5, 0.25)$ is an unstable Nash equilibrium but $(u_1^{*2}, u_2^{*2}) = (1.5, 2.25)$ is a stable Nash equilibrium. We employ the multi-parameter stochastic extremum seeking algorithm given in Sect. 9.2 to attain this stable equilibrium.

The top picture in Fig. 9.2 depicts the evolution of the game in the $\tilde{u}$ plane, initialized at the point $(u_1(0), u_2(0)) = (0, 3)$, i.e., at $(\tilde{u}_1(0), \tilde{u}_2(0)) = (-1.5, 0.75)$. Note that the initial condition is outside of $\mathcal{A}$. This illustrates the point that the region of attraction of the stable Nash equilibrium under the extremum seeking algorithm is not a subset of $\mathcal{A}$ but a large subset of $\mathbb{R}^2$. The parameters are chosen as $k_1 = 14$, $k_2 = 6$, $a_1 = 0.2$, $a_2 = 0.02$, $\varepsilon_1 = 0.01$, $\varepsilon_2 = 0.8$. The bottom two pictures depict the two players' actions in stochastically seeking the Nash equilibrium

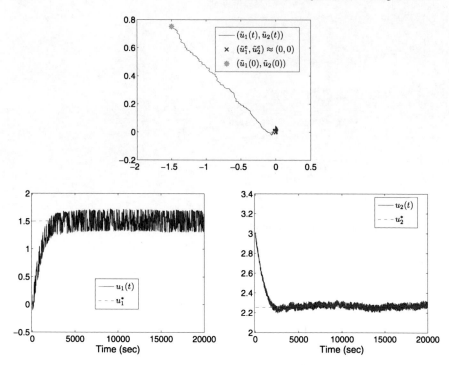

Fig. 9.3 Stochastic Nash equilibrium seeking with Brownian motion on the unit circle as perturbation. *Top*: evolution of the game in the $\tilde{u}$ plane. *Bottom*: two players' actions

$(u_1^*, u_2^*) = (1.5, 2.25)$. From Fig. 9.2, the actions of the players converge to a small neighborhood of the stable Nash equilibrium.

In the algorithm, bounded smooth functions f_i and the excitation processes $(\eta_i(t), t \geq 0)$, $i = 1, \ldots, N$, can be chosen in other forms. We can replace the bounded excitation signal $\sin(\eta_i(t)) = \sin(\chi_i(t/\varepsilon_i))$ with the signal $H^T(\breve{\eta}_i(t/\varepsilon_i))$, where $\breve{\eta}_i(t) = [\cos(W_i(t)), \sin(W_i(t))]^T$ is Brownian motion on the unit circle (see [89]), and $G = [g_1, g_2]^T$ is a constant vector.

Figure 9.3 depicts the evolution of the game in the $\tilde{u}$ plane for games with Brownian motion on the unit circle as perturbation. The initial conditions are the same with the case of the OU process perturbation. The parameters are chosen as $k_1 = 5$, $k_2 = 9$, $a_1 = 0.2$, $a_2 = 0.04$, $\varepsilon_1 = 0.02$, $\varepsilon_2 = 0.02$. From Fig. 9.3, the actions of the players also converge to a small neighborhood of the stable Nash equilibrium.

In these two simulations, possibly different high-pass filter for each player's measurement on the payoff is used to improve the asymptotic performance but is not essential for achieving stability (see [137]), which also can be seen from the stochastic multi-parameter extremum seeking algorithm for static maps in Sect. 8.2.

9.5 Notes and References

Seeking Nash equilibria in continuous games is a difficult problem [74]. Researchers in different fields, including mathematics, computer science, economics, and system engineering, have interest and need for techniques for finding Nash equilibria. Most algorithms designed to achieve convergence to Nash equilibria require modeling information for the game and assume that the players can observe the actions of the other players. An early algorithm is [118], in which a gradient-type algorithm is studied for convex games. Distributed iterative algorithms are designed for the computation of equilibrium in [83] for a general class of non-quadratic convex Nash games. In this algorithm, the agents do not have to know each other's cost functionals and private information, as well as the parameters and subjective probability distributions adopted by the others, but they have to communicate to each other their tentative decisions during each phase of computation. A strategy known as fictitious play is one such strategy that depends on the actions of the other players so that a player can devise a best response. A dynamic version of fictitious play and gradient response is developed in [126]. In [150], a synchronous distributed learning algorithm is designed to the coverage optimization of mobile visual sensor networks. In this algorithm, players remember their own actions and utility values from the previous two times steps, and the algorithm is shown to converge in probability to the set of restricted Nash equilibria. Other diverse engineering applications of game theory include the design of communication networks in [4, 10, 97, 124], integrated structures and controls in [114], and distributed consensus protocols in [12, 99, 125].

Based on the extremum seeking approach with sinusoidal perturbations, in [72], Nash equilibrium seeking is studied for noncooperative games with both finitely and infinitely many players and in [41], Nash equilibrium seeking scheme is supplied for noncooperative games with general payoff functions. In [136], Nash games in mobile sensor networks are solved using extremum seeking.

Compared to the deterministic case, one advantage of stochastic extremum seeking is that there is no need to choose different perturbation frequencies for each player and each player only needs to choose its own perturbation process independently, which may be more realistic in a practical game with adversarial players.

In this chapter which is based on our results in [91], we propose a multi-input stochastic extremum seeking algorithm to solve the problem of seeking Nash equilibria for an N-player nonoperative game. In our algorithm, each player independently employs his seeking strategy using only the value of his own payoff but without any information about the form of his payoff function and other players' actions. Our convergence result is local and the convergence error is in proportion to the third derivatives of the payoff functions and is dependent on the intensity of stochastic perturbation.

Chapter 10
Nash Equilibrium Seeking for Quadratic Games and Applications to Oligopoly Markets and Vehicle Deployment

In this chapter, we consider a special case of Chap. 9: a Nash game with quadratic payoffs. The general case is considered in Sect. 10.1. As applications, we consider an oligopoly market game in Sect. 10.2 and multi-agent deployment in the plane in Sect. 10.3.

10.1 N-Player Games with Quadratic Payoff Functions

10.1.1 General Quadratic Games

We consider static non-cooperative games with N players that wish to maximize their quadratic payoff functions. Specifically, the payoff function of player i is of the form

$$J_i(t) = \frac{1}{2} \sum_{j=1}^{N} \sum_{k=1}^{N} D^i_{jk} u_j(t) u_k(t) + \sum_{j=1}^{N} d^i_j u_j(t) + C_i, \qquad (10.1)$$

where the action of player j is $u_j \in U_j = \mathbb{R}$, D^i_{jk}, d^i_j, and C_i are constants, $D^i_{ii} < 0$, and $D^i_{jk} = D^i_{kj}$.

From Proposition 4.6 in [11], it is known that the N-player game with payoff function (10.1) admits a Nash equilibrium $u^* = [u^*_1, \dots, u^*_N]^T$ if and only if

$$D^i_{ii} u^*_i + \sum_{j \neq i} D^i_{ij} u^*_j + d^i_i = 0, \quad i \in \{1, \dots, N\}, \qquad (10.2)$$

admits a solution. Rewriting (10.2) in matrix form, we have

$$Du^* = -d, \qquad (10.3)$$

S.-J. Liu, M. Krstic, *Stochastic Averaging and Stochastic Extremum Seeking*,
Communications and Control Engineering,
DOI 10.1007/978-1-4471-4087-0_10, © Springer-Verlag London 2012

where

$$D \triangleq \begin{bmatrix} D^1_{11} & D^1_{12} & \cdots & D^1_{1N} \\ D^2_{21} & D^2_{22} & \cdots & D^2_{2N} \\ \vdots & \vdots & \ddots & \vdots \\ D^N_{N1} & D^N_{N2} & \cdots & D^N_{NN} \end{bmatrix}, \qquad d \triangleq \begin{bmatrix} d^1_1 \\ d^2_2 \\ \vdots \\ d^N_N \end{bmatrix}, \tag{10.4}$$

and u^* is unique if D is invertible. We consider the following stronger assumptions about this matrix:

Assumption 10.1 The matrix D given by (10.1) is strictly diagonally dominant, i.e.,

$$\sum_{\substack{j \neq i}}^{N} |D^i_{ij}| < |D^i_{ii}|, \quad i \in \{1, \ldots, N\}. \tag{10.5}$$

By Assumption 10.1, the Nash equilibrium u^* exists and is unique since strictly diagonally dominant matrices are nonsingular by the Levy–Desplanques theorem. We seek a method to attain u^* stably in real time without any modeling information. Let each player employ the stochastic extremum seeking strategy as in (9.5) and (9.6):

$$u_i(t) = \hat{u}(t) + a_i f_i(\eta_i(t)), \tag{10.6}$$

$$\frac{d\hat{u}_i(t)}{dt} = k_i a_i f_i(\eta_i(t)) J_i(t), \tag{10.7}$$

where for any $i = 1, \ldots, N$, $a_i > 0$ is the perturbation amplitude, $k_i > 0$ is the adaptive gain, $J_i(t)$ is the measured payoff value for player i, and f_i is a bounded odd smooth function that player i chooses, e.g., a sine function. The independent ergodic processes $\eta_i(t)$, $i = 1, \ldots, N$, are chosen by player i, e.g., as the Ornstein–Uhlenbeck (OU) processes

$$\eta_i = \frac{\sqrt{\varepsilon_i} q_i}{\varepsilon_i s + 1} [\dot{W}_i], \quad \text{or} \quad \varepsilon_i \, d\eta_i(t) = -\eta_i(t) \, dt + \sqrt{\varepsilon_i} q_i \, dW_i(t), \tag{10.8}$$

where $q_i > 0$, ε_i are small parameters satisfying $0 < \max_i \varepsilon_i < \varepsilon_0$ for fixed $\varepsilon_0 > 0$, and $W_i(t)$, $i = 1, \ldots, N$, are independent 1-dimensional standard Brownian motions on a complete probability space $(\Omega, \mathcal{F}, P)$ with the sample space Ω, σ-field $\mathcal{F}$, and probability measure P.

If the players choose $f_i(x) = \sin x$ for all $i = 1, \ldots, N$, and η_i as OU processes (10.8), we have the following convergence result.

Theorem 10.1 *Consider the system* (10.6)–(10.7) *with* (10.1) *under Assumption* 10.1, *where* $i \in \{1, \ldots, N\}$. *Then there exists a constant* $a^* > 0$ *such that for* $\max_{1 \leq i \leq N} a_i \in (0, a^*)$ *there exist constants* $r > 0$, $c > 0$, $\gamma > 0$, *and a function* $T(\varepsilon_1) : (0, \varepsilon_0) \to \mathbb{N}$ *such that for any initial condition* $|\Lambda^{\varepsilon_1}_1(0)| < r$, *and any* $\delta > 0$,

$$\lim_{\varepsilon_1 \to 0} \inf \left\{ t \geq 0 : \left| \Lambda_1^{\varepsilon_1}(t) \right| > c \left| \Lambda_1^{\varepsilon_1}(0) \right| e^{-\gamma t} + \delta + O\left(\max_i a_i \right) \right\} = \infty \quad a.s. \quad (10.9)$$

and

$$\lim_{\varepsilon_1 \to 0} P \left\{ \left| \Lambda_1^{\varepsilon_1}(t) \right| \leq c \left| \Lambda_1^{\varepsilon_1}(0) \right| e^{-\gamma t} + \delta + O\left(\max_i a_i \right), \right.$$

$$\left. \forall t \in \left[0, T(\varepsilon_1) \right] \right\} = 1 \tag{10.10}$$

with

$$\lim_{\varepsilon_1 \to 0} T(\varepsilon_1) = \infty, \tag{10.11}$$

where

$$\Lambda^{\varepsilon_1}(t) = \left[u_1(t) - u_1^*, \ldots, u_N(t) - u_N^* \right]^T. \tag{10.12}$$

Proof Let $\tilde{u}_i(t) = \hat{u}_i(t) - u_i^*$ denote the error relative to the Nash equilibrium. By substituting (10.1) into (10.6)–(10.7), we obtain the error system

$$\frac{d\tilde{u}_i(t)}{dt} = k_i a_i \sin\left(\eta_i(t) \right) \left(\frac{1}{2} \sum_{j=1}^N \sum_{k=1}^N D_{jk}^i \left(\tilde{u}_j(t) + u_j^* + a_j \sin\left(\eta_j(t) \right) \right) \right.$$

$$\times \left(\tilde{u}_k(t) + u_k^* + a_k \sin\left(\eta_k(t) \right) \right)$$

$$\left. + \sum_{j=1}^N d_j^i \left(\tilde{u}_j(t) + u_j^* + a_j \sin\left(\eta_j(t) \right) \right) + C_i \right). \tag{10.13}$$

Define $\chi_i(t) = \eta_i(\varepsilon_i t)$ and $B_i(t) = \frac{1}{\sqrt{\varepsilon_i}} W_i(\varepsilon_i t)$. Then by (10.8) we have

$$d\chi_i(t) = -\chi_i(t)\, dt + q_i\, dB_i(t), \tag{10.14}$$

where $[B_1(t), \ldots, B_N(t)]^T$ is an N-dimensional standard Brownian motion on the space $(\Omega, \mathcal{F}, P)$. Thus we rewrite the error system (10.13) as

$$\frac{d\tilde{u}_i(t)}{dt} = k_i a_i \sin\left(\chi_i(t/\varepsilon_i) \right) \left(\frac{1}{2} \sum_{j=1}^N \sum_{k=1}^N D_{jk}^i \left(\tilde{u}_j(t) + u_j^* + a_j \sin\left(\chi_j(t/\varepsilon_j) \right) \right) \right.$$

$$\times \left(\tilde{u}_k(t) + u_k^* + a_k \sin\left(\chi_k(t/\varepsilon_k) \right) \right)$$

$$\left. + \sum_{j=1}^N d_j^i \left(\tilde{u}_j(t) + u_j^* + a_j \sin\left(\chi_j(t/\varepsilon_j) \right) \right) + C_i \right). \tag{10.15}$$

Denote

$$\varepsilon_i = \frac{\varepsilon_1}{c_i}, \quad i = 2, \ldots, N, \tag{10.16}$$

for some positive real constants c_i's and consider the change of variable

$$Z_1(t) = \chi_1(t), \qquad Z_2(t) = \chi_2(c_2 t), \qquad \ldots, \qquad Z_N(t) = \chi(c_N t). \qquad (10.17)$$

Then the error system (10.15) can be transformed as one with a single small parameter ε_1:

$$
\frac{d\tilde{u}_i(t)}{dt} = k_i a_i \sin\big(Z_i(t/\varepsilon_1)\big) \bigg(\frac{1}{2} \sum_{j=1}^{N} \sum_{k=1}^{N} D^i_{jk} \big(\tilde{u}_j(t) + u^*_j + a_j \sin\big(Z_j(t/\varepsilon_1)\big) \big)
$$

$$
\times \big(\tilde{u}_k(t) + u^*_k + a_k \sin\big(Z_k(t/\varepsilon_1)\big) \big)
$$

$$
+ \sum_{j=1}^{N} d^i_j \big(\tilde{u}_j(t) + u^*_j + a_j \sin\big(Z_j(t/\varepsilon_1)\big) \big) + C_i \bigg). \qquad (10.18)
$$

Rearranging terms yields

$$
\frac{d\tilde{u}_i(t)}{dt} = \frac{k_i}{2} \sum_{j=1}^{N} \sum_{k=1}^{N} D^i_{jk} \big(\tilde{u}_j(t) + u^*_j \big) \big(\tilde{u}_k(t) + u^*_k \big) a_i \sin\big(Z_i(t/\varepsilon_1)\big)
$$

$$
+ k_i \sum_{j=1}^{N} \sum_{k=1}^{N} D^i_{jk} \big(\tilde{u}_j(t) + u^*_j \big) a_i a_k \sin\big(Z_i(t/\varepsilon_1)\big) \sin\big(Z_k(t/\varepsilon_1)\big)
$$

$$
+ \frac{k_i}{2} \sum_{j=1}^{N} \sum_{k=1}^{N} D^i_{jk} a_i a_j a_k \sin\big(Z_i(t/\varepsilon_1)\big) \sin\big(Z_j(t/\varepsilon_1)\big) \sin\big(Z_k(t/\varepsilon_1)\big)
$$

$$
+ k_i \sum_{j=1}^{N} d^i_j \big(\tilde{u}_j(t) + u^*_j + a_j \sin\big(Z_j(t/\varepsilon_1)\big) \big) a_i \sin\big(Z_i(t/\varepsilon_1)\big)
$$

$$
+ k_i C_i a_i \sin\big(Z_i(t/\varepsilon_1)\big). \qquad (10.19)
$$

By (8.41), (8.42), (8.43), (8.44), and (8.45), together with $Du^* = -d$, we obtain the average system of (10.19):

$$
\frac{d\tilde{u}^{\text{ave}}_i(t)}{dt} = k_i a_i^2 G_0(q_i) \sum_{j=1}^{N} D^i_{ij} \big(\tilde{u}^{\text{ave}}_j(t) + u^*_j \big) + k_i a_i^2 G_0(q_i) d^i_i
$$

$$
= k_i a_i^2 G_0(q_i) \bigg(\sum_{j=1}^{N} D^i_{ij} \tilde{u}^{\text{ave}}_j(t) + \sum_{j=1}^{N} D^i_{ij} u^*_j + d^i_i \bigg)
$$

$$
= k_i a_i^2 G_0(q_i) \sum_{j=1}^{N} D^i_{ij} \tilde{u}^{\text{ave}}_j(t), \qquad \tilde{u}^{\text{ave}}_i(0) = \tilde{u}_i(0), \qquad (10.20)
$$

which in matrix form is

$$\frac{d\tilde{u}^{\text{ave}}(t)}{dt} = A\tilde{u}^{\text{ave}}(t), \tag{10.21}$$

where

$$A = \begin{bmatrix} k_1 a_1^2 G_0(q_1) D_{11}^1 & k_1 a_1^2 G_0(q_1) D_{12}^1 & \cdots & k_1 a_1^2 G_0(q_1) D_{1N}^1 \\ k_2 a_2^2 G_0(q_2) D_{21}^2 & k_2 a_2^2 G_0(q_2) D_{22}^2 & \cdots & k_2 a_2^2 G_0(q_2) D_{2N}^2 \\ \vdots & \vdots & \ddots & \vdots \\ k_N a_N^2 G_0(q_N) D_{N1}^N & k_N a_N^2 G_0(q_N) D_{N2}^N & \cdots & k_N a_N^2 G_0(q_N) D_{NN}^N \end{bmatrix}. \tag{10.22}$$

Now we determine the stability of the average system (10.21). From the Gershgorin Circle Theorem [47, Theorem 7.2.1], we have

$$\lambda(A) \subseteq \bigcup_{i=1}^{N} \rho_i, \tag{10.23}$$

where $\lambda(A)$ denotes the spectrum of A and ρ_i is a Gershgorin disc,

$$\rho_i = k_i a_i^2 G_0(q_i) \left\{ z \in \mathbb{C} \,\middle|\, \left| z - D_{ii}^i \right| < \sum_{j \neq i} \left| D_{ij}^i \right| \right\}. \tag{10.24}$$

Since $D_{ii}^i < 0$ and D is strictly diagonally dominant, the union of the Gershgorin discs lies strictly in the left half of the complex plane, and we conclude that $\text{Re}\{\lambda\} < 0$ for all $\lambda \in \lambda(A)$ and that A is Hurwitz. Thus, there exist positive definite symmetric matrices P and Q that satisfy the Lyapunov equation $PA + A^T P = -Q$. Using $V(t) = (\tilde{u}^{\text{ave}})^T P \tilde{u}^{\text{ave}}$ as a Lyapunov function, we obtain

$$\dot{V} = -\left(\tilde{u}^{\text{ave}}\right)^T Q\tilde{u}^{\text{ave}} \le -\lambda_{\min}(Q)\left|\tilde{u}^{\text{ave}}\right|^2. \tag{10.25}$$

Bounding V and applying the Comparison Lemma [56] gives

$$\left|\tilde{u}^{\text{ave}}(t)\right| \le c e^{-\gamma t}\left|\tilde{u}^{\text{ave}}(0)\right|, \tag{10.26}$$

where

$$c = \sqrt{\frac{\lambda_{\max}(P)}{\lambda_{\min}(P)}}, \tag{10.27}$$

$$\gamma = \frac{\lambda_{\min}(Q)}{2\lambda_{\max}(P)}. \tag{10.28}$$

By the multi-input stochastic averaging theorem given in Theorem 8.1 of Chap. 8, noticing that $u_i(t) - u_i^* = \tilde{u}_i(t) + a_i \sin(\eta_i(t))$ and that $a_i \sin(\eta_i(t))$ is $O(\max_i a_i)$, the proof is completed. $\square$

10.1.2 Symmetric Quadratic Games

If we further restrict the matrix D, we can develop a more precise expression for the convergence rate. Specifically, we now assume the following:

Assumption 10.2

$$D^i_{ij} = D^j_{ji} \quad \text{for all } i, j \in \{1, \ldots, N\}. \tag{10.29}$$

With this additional assumption besides Assumption 10.1, D is a negative definite symmetric matrix.

Corollary 10.1 *Consider the system* (10.6)–(10.7) *with* (10.1) *under Assumptions* 10.1 *and* 10.2, *where* $i \in \{1, \ldots, N\}$. *Then the convergence properties of Theorem* 10.1 *hold with*

$$c = \sqrt{\frac{\max_i\{2k_i a_i^2 G_0(q_i)\}}{\min_i\{2k_i a_i^2 G_0(q_i)\}}} = \sqrt{\frac{\max_i\{k_i a_i^2 G_0(q_i)\}}{\min_i\{k_i a_i^2 G_0(q_i)\}}}, \tag{10.30}$$

$$\gamma = \min_i\{k_i a_i^2 G_0(q_i)\} \min_i\left\{-D^i_{ii} - \sum_{j\neq i}^{N}|D^i_{ij}|\right\}. \tag{10.31}$$

Proof From the proof of Theorem 10.1, there exist positive definite symmetric matrices P and Q that satisfy the Lyapunov equation $PA + A^T P = -Q$ since A, given by (10.22), is Hurwitz. Under Assumption 10.2, we select $Q = -D$ and obtain $P = \text{diag}\left(\frac{1}{2k_1 a_1^2 G_0(q_1)}, \ldots, \frac{1}{2k_N a_N^2 G_0(q_N)}\right)$. Then, we have

$$\lambda_{\max}(P) = \frac{1}{\min_i\{2k_i a_i^2 G_0(q_i)\}}, \tag{10.32}$$

$$\lambda_{\min}(P) = \frac{1}{\max_i\{2k_i a_i^2 G_0(q_i)\}}, \tag{10.33}$$

and using the Gershgorin Circle Theorem [47, Theorem 7.2.1], we can obtain the bound

$$\lambda_{\min}(Q) = \lambda_{\min}(-D) \geq \min_i\left\{-D^i_{ii} - \sum_{j\neq i}^{N}|D^i_{ij}|\right\}, \tag{10.34}$$

where we note that $D^i_{ii} < 0$. From (10.27), (10.28), (10.32), (10.33), and (10.34), we obtain the result. $\qquad\square$

From this corollary, the coefficient c is determined completely by the stochastic extremum seeking parameters k_i, q_i, a_i.

Fig. 10.1 A model of sales s_1, s_2, s_3 in a three-firm oligopoly with price u_1, u_2, u_3 and total sales S. The desirability of product i is proportional to $1/R_i$

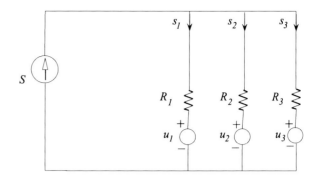

10.2 Oligopoly Price Games

Consider a static non-cooperative game with N firms in an oligopoly market structure that compete to maximize their profits by setting the price u_i of their product. Assume that the profit of the ith firm is

$$J_i(t) = s_i(t)\big(u_i(t) - m_i\big), \tag{10.35}$$

where m_i is the marginal cost of player i, and s_i is its sales volume.

We model the sales volume s_i as

$$s_i(t) = \frac{R_{\|}}{R_i}\left(S - \frac{u_i(t)}{\overline{R}_i} + \sum_{j \neq i}^{N} \frac{u_j(t)}{R_j}\right), \tag{10.36}$$

where S are the total sales of all the firms, $R_i > 0$ for all i, and

$$R_{\|} = \left(\sum_{k=1}^{N} \frac{1}{R_k}\right)^{-1}, \qquad \overline{R}_i = \left(\sum_{k \neq i}^{N} \frac{1}{R_k}\right)^{-1}. \tag{10.37}$$

The sales model (10.36) is motivated by an analogous electric circuit, shown in Fig. 10.1, where S is an ideal current generator, u_i are ideal voltage generators, and most importantly, the resistors R_i represent the "resistance" that consumers have toward buying product i. This resistance may be due to quality or brand image considerations—the most desirable products have the lowest R_i. The sales in (10.36) are inversely proportional to R_i and grow as u_i decreases and as u_j, $j \neq i$, increases. The profit (10.35), in electrical analogy, corresponds to the power absorbed by the $u_i - m_i$ portion of the voltage generator i.

Substituting (10.36) into (10.35) yields quadratic payoff functions of the form

$$J_i(t) = \frac{R_{\|}}{R_i}\left(-\frac{u_i^2}{\overline{R}_i} + u_i \sum_{j \neq i}^{N} \frac{u_j}{R_j} + \left(\frac{m_i}{\overline{R}_i} + S\right)u_i - m_i \sum_{j \neq i}^{N} \frac{u_j}{R_j} - Sm_i\right), \tag{10.38}$$

and the Nash equilibrium u^* satisfies $Du^* = -d$. More specifically, we have

$$
\begin{bmatrix}
-\dfrac{2R_{\|}}{R_1\overline{R}_1} & \dfrac{R_{\|}}{R_1 R_2} & \cdots & \dfrac{R_{\|}}{R_1 R_N} \\[6pt]
\dfrac{R_{\|}}{R_2 R_1} & -\dfrac{2R_{\|}}{R_2 \overline{R}_2} & \cdots & \dfrac{R_{\|}}{R_2 R_N} \\[6pt]
\vdots & \vdots & \ddots & \vdots \\[6pt]
\dfrac{R_{\|}}{R_N R_1} & \dfrac{R_{\|}}{R_N R_2} & \cdots & -\dfrac{2R_{\|}}{R_N \overline{R}_N}
\end{bmatrix}
\underbrace{\phantom{\begin{bmatrix}x\end{bmatrix}}}_{D}
\begin{bmatrix}
u_1^* \\ u_2^* \\ \vdots \\ u_N^*
\end{bmatrix}
= -
\begin{bmatrix}
\dfrac{m_1 R_{\|}}{R_1 \overline{R}_1} + \dfrac{S R_{\|}}{R_1} \\[6pt]
\dfrac{m_2 R_{\|}}{R_2 \overline{R}_2} + \dfrac{S R_{\|}}{R_2} \\[6pt]
\vdots \\[6pt]
\dfrac{m_N R_{\|}}{R_N \overline{R}_N} + \dfrac{S R_{\|}}{R_N}
\end{bmatrix}
\underbrace{\phantom{\begin{bmatrix}x\end{bmatrix}}}_{d}.
$$

$$\tag{10.39}$$

The matrix D has negative diagonal elements and is strictly diagonally dominant, satisfying Assumption 10.1, since

$$
\sum_{\substack{j=1 \\ j \neq i}}^{N} \left| \frac{R_{\|}}{R_i R_j} \right| = \frac{R_{\|}}{R_i \overline{R}_i} < \left| -\frac{2R_{\|}}{R_i \overline{R}_i} \right|, \quad i \in \{1, \ldots, N\}. \tag{10.40}
$$

Thus, the Nash equilibrium of this game exists, is unique, and can be shown to be

$$
u_i^* = \frac{\Pi_1 R_i}{2R_i + \overline{R}_i} \left(\overline{R}_i S + m_i + \sum_{j=1}^{N} \frac{m_j \overline{R}_i - m_i \overline{R}_j}{2R_j + \overline{R}_j} \right), \tag{10.41}
$$

where $\Pi_1^{-1} = 1 - \sum_{j=1}^{N} \frac{\overline{R}_j}{2R_j + \overline{R}_j} > 0$. (The various parameters here are assumed to be selected such that u_i^* is positive for all i.) Moreover, $D_{ij}^i = D_{ji}^j$, so D is a negative definite symmetric matrix, satisfying Assumption 10.2.

Theorem 10.2 *Consider the system* (10.6)–(10.7), *along with* (10.35) *and* (10.36), *where* $i \in \{1, \ldots, N\}$. *Then there exists a constant* $a^* > 0$ *such that for* $\max_{1 \leq i \leq N} a_i \in (0, a^*)$ *there exist constants* $r > 0$, $c > 0$, $\gamma > 0$, *and a function* $T(\varepsilon_1) : (0, \varepsilon_0) \to \mathbb{N}$ *such that for any initial condition* $|\Lambda_2^{\varepsilon_1}(0)| < r$, *and any* $\delta > 0$,

$$
\lim_{\varepsilon_1 \to 0} \inf \left\{ t \geq 0 : \left| \Lambda_2^{\varepsilon_1}(t) \right| > c \left| \Lambda_2^{\varepsilon_1}(0) \right| e^{-\gamma t} + \delta + O\left(\max_i a_i \right) \right\} = \infty \quad a.s. \tag{10.42}
$$

and

$$
\lim_{\varepsilon_1 \to 0} P \left\{ \left| \Lambda_2^{\varepsilon_1}(t) \right| \leq c \left| \Lambda_2^{\varepsilon_1}(0) \right| e^{-\gamma t} + \delta + O\left(\max_i a_i \right), \right.
$$

$$
\left. \forall t \in [0, T(\varepsilon_1)] \right\} = 1 \tag{10.43}
$$

with

$$
\lim_{\varepsilon_1 \to 0} T(\varepsilon_1) = \infty, \tag{10.44}
$$

where

$$\Lambda_2^{\varepsilon_1}(t) = \left[u_1(t) - u_1^*, \ldots, u_N(t) - u_N^*\right]^T, \tag{10.45}$$

$$c = \sqrt{\frac{\max_i\{k_i a_i^2 G_0(q_i)\}}{\min_i\{k_i a_i^2 G_0(q_i)\}}}, \tag{10.46}$$

$$\gamma = \frac{R_{\|} \min_i\{k_i a_i^2 G_0(q_i)\}}{2 \max_i\{R_i \Gamma_i\}}, \tag{10.47}$$

$$\Gamma_i = \min_{j \in \{1,\ldots,N\}, j \neq i} R_j. \tag{10.48}$$

Proof Since Assumptions 10.1 and 10.2 are satisfied for this game, Corollary 10.1 holds. From Corollary 10.1, we obtain the coefficient $c = \sqrt{\frac{\max_i\{k_i a_i^2 G_0(q_i)\}}{\min_i\{k_i a_i^2 G(q_i)\}}}$ and the decay rate $\gamma = \frac{R_{\|} \min_i\{k_i a_i^2 G_0(q_i)\}}{2 \max_i\{R_i \Gamma_i\}}$ since

$$\lambda_{\min}(Q) \geq R_{\|} \min_i \left\{\frac{2}{R_i \overline{R}_i} - \sum_{j \neq i}\left|\frac{1}{R_i R_j}\right|\right\} = \frac{R_{\|}}{\max_i\{R_i \overline{R}_i\}}. \tag{10.49}$$

We further bound this decay rate to obtain γ by noting that $\max_i\{R_i \overline{R}_i\} < \max_i\{R_i \Gamma_i\}$. $\qquad\square$

10.3 Multi-agent Deployment in the Plane

In this section, we consider the problem of deployment of a group of N autonomous fully actuated vehicles (agents) in a non-cooperative manner in a planar signal field using the method of stochastic extremum seeking.

10.3.1 Vehicle Model and Local Agent Cost

We consider vehicles modeled as velocity-actuated point masses,

$$\frac{dx_i}{dt} = V_{xi}, \qquad \frac{dy_i}{dt} = V_{yi}, \tag{10.50}$$

where (x_i, y_i) is the position of the vehicle in the plane, and V_{xi}, V_{yi} are the vehicle velocity inputs. The subscript i is used to denote the ith vehicle.

We assume that the nonlinear map defining the distribution of the signal field is quadratic and takes the form

$$f(x_i, y_i) = f^* + q_x\left(x_i - x^*\right)^2 + q_y\left(y_i - y^*\right)^2, \tag{10.51}$$

where (x^*, y^*) is the minimizer, $f^* = f(x^*, y^*)$ is the minimum, and (q_x, q_y) are unknown positive constants. To account for the interactions between the vehicles, we assume that each vehicle can sense the distance,

$$d(x_i, x_j, y_i, y_j) = \sqrt{(x_i - x_j)^2 + (y_i - y_j)^2}, \tag{10.52}$$

between itself and other vehicles. The cost function

$$J_i(x_i, \ldots, x_N, y_1, \ldots, y_N) = f(x_i, y_i) + \sum_{j \in N} q_{ij} d^2(x_i, x_j, y_i, y_j) \tag{10.53}$$

includes inter-vehicle interactions, where $q_{ij} \geq 0$ is the weighting that vehicle i puts on its distance to vehicle j.

10.3.2 Control Design

To deploy the agents about the source position, we propose a control scheme that utilizes Brownian motion on the unit circle as the excitation signal to perform stochastic extremum seeking.

We propose the following stochastic control algorithm for vehicle i:

$$V_{xi} = -a_{xi}\dot{\eta}_{1i} - c_{xi}\xi_i\eta_{1i} + v_{xi}, \tag{10.54}$$

$$V_{yi} = -a_{yi}\dot{\eta}_{2i} - c_{yi}\xi_i\eta_{2i} + v_{yi}, \tag{10.55}$$

$$\xi_i = \frac{s}{s + h_i}[J_i], \tag{10.56}$$

$$\dot{\eta}_{1i} = -\frac{1}{2\varepsilon_i}\eta_{1i} - \frac{\eta_{2i}}{\sqrt{\varepsilon_i}}\dot{W}_i, \tag{10.57}$$

$$\dot{\eta}_{2i} = -\frac{1}{2\varepsilon_i}\eta_{2i} + \frac{\eta_{1i}}{\sqrt{\varepsilon_i}}\dot{W}_i, \tag{10.58}$$

where ξ_i is the output of the washout filter for the cost J_i, $(\eta_{1i}(t), t \geq 0)$ and $(\eta_{2i}(t), t \geq 0)$ are used as perturbations in the stochastic extremum seeking scheme, $a_{xi}, a_{yi}, c_{xi}, c_{yi}, \varepsilon_i, h_i > 0$ are extremum seeking design parameters, and $v_{xi}, v_{yi} \in \mathbb{R}$. We consider vehicles with $v_{xi}, v_{yi} \neq 0$ to be the anchor agents and those with $v_{xi} = v_{yi} = 0$ to be the follower agents. The signal $\dot{W}_i$ denotes the white noise, and $(W_i(t), t \geq 0)$ is a 1-dimensional Brownian motion which is not necessarily standard in the form $W_i(0) = 0$. The signals $W_1(t), W_2(t), \ldots, W_N(t)$ are independent.

The Eqs. (10.57) and (10.58) are equivalent to

$$d\eta_{1i} = -\frac{1}{2\varepsilon_i}\eta_{1i}\,dt - \frac{\eta_{2i}}{\sqrt{\varepsilon_i}}\,dW_i, \tag{10.59}$$

$$d\eta_{2i} = -\frac{1}{2\varepsilon_i}\eta_{2i}\,dt + \frac{\eta_{1i}}{\sqrt{\varepsilon_i}}\,dW_i, \tag{10.60}$$

which means, by the definition of Ito stochastic differential equation, that

$$\eta_{1i}(t) = \eta_{1i}(0) - \int_0^t \frac{1}{2\varepsilon_i}\eta_{1i}(s)\,ds - \int_0^t \frac{\eta_{2i}(s)}{\sqrt{\varepsilon_i}}\,dW_i(s), \tag{10.61}$$

$$\eta_{2i}(t) = \eta_{2i}(0) - \int_0^t \frac{1}{2\varepsilon_i}\eta_{2i}(s)\,ds + \int_0^t \frac{\eta_{1i}(s)}{\sqrt{\varepsilon_i}}\,dW_i(s). \tag{10.62}$$

Thus it holds that

$$\eta_{1i}(\varepsilon_i t) = \eta_{1i}(0) - \int_0^t \frac{1}{2}\eta_{1i}(\varepsilon_i u)\,du - \int_0^t \frac{\eta_{2i}(s)}{\sqrt{\varepsilon_i}}\,dW_i(\varepsilon_i u), \tag{10.63}$$

$$\eta_{2i}(\varepsilon_i t) = \eta_{2i}(0) - \int_0^t \frac{1}{2}\eta_{2i}(\varepsilon_i u)\,du + \int_0^t \frac{\eta_{1i}(s)}{\sqrt{\varepsilon_i}}\,dW_i(\varepsilon_i u). \tag{10.64}$$

Define

$$B_i(t) = \frac{1}{\sqrt{\varepsilon_i}}W(\varepsilon_i t), \qquad \chi_{1i}(t) = \eta_{1i}(\varepsilon_i t), \qquad \chi_{2i}(t) = \eta_{2i}(\varepsilon_i t). \tag{10.65}$$

Then we have

$$d\chi_{1i} = -\frac{1}{2}\chi_{1i}(t)\,dt - \chi_{2i}(t)\,dB_i(t), \tag{10.66}$$

$$d\chi_{2i} = -\frac{1}{2}\chi_{2i}(t)\,dt + \chi_{1i}(t)\,dB_i(t), \tag{10.67}$$

where $B_i(t)$ is a 1-dimensional Brownian motion which is also not necessarily standard in the form $B_i(0) = 0$.

The solution of stochastic differential equations (10.66) and (10.67) is equivalent to Brownian motion on the unit circle $\chi_i(t) = [\cos(B_i(t)), \sin(B_i(t))]^T$. Thus

$$\eta_i(t) = \left[\eta_{1i}(t), \eta_{2i}(t)\right]^T = \chi_i(t/\varepsilon_i) = \left[\cos\big(B_i(t/\varepsilon_i)\big), \sin\big(B_i(t/\varepsilon_i)\big)\right]^T. \tag{10.68}$$

Hence, the control signals (10.54) and (10.55) become

$$V_{xi} = -\frac{a_{xi}}{2\varepsilon_i}\eta_{1i} - a_{xi}\frac{\eta_{2i}}{\sqrt{\varepsilon_i}}\dot{W}_i - c_{xi}\xi_i\eta_{1i} + v_{xi}, \tag{10.69}$$

$$V_{yi} = -\frac{a_{yi}}{2\varepsilon_i}\eta_{1i} + a_{yi}\frac{\eta_{1i}}{\sqrt{\varepsilon_i}}\dot{W}_i - c_{yi}\xi_i\eta_{2i} + v_{yi}. \tag{10.70}$$

10.3.3 Stability Analysis

In this section, we present and prove the local stability in a specific probabilistic sense for a group of vehicles.

We define an output error variable

$$e_i = \frac{h_i}{s + h_i}[J_i(t)] - f^*, \tag{10.71}$$

where $\frac{h_i}{s+h_i}$ is a low-pass filter applied to the cost J_i, which allows us to express $\xi_i(t)$, the signal from the washout filter, as

$$\xi_i(t) = \frac{s}{s + h_i}[J_i(t)] = J_i(t) - f^* - e_i(t), \tag{10.72}$$

noting also that $\dot{e}_i = h_i \xi_i$.

We have the following stability result for a group of fully actuated vehicles with control laws (10.54)–(10.58).

Theorem 10.3 *Consider the closed-loop system*

$$dx_i = -\frac{a_{xi}}{2\varepsilon_i}\eta_{1i}\,dt - a_{xi}\frac{\eta_{2i}}{\sqrt{\varepsilon_i}}\,dW_i - c_{xi}\xi_i\eta_{1i}\,dt + v_{xi}\,dt, \tag{10.73}$$

$$dy_i = -\frac{a_{yi}}{2\varepsilon_i}\eta_{2i}\,dt + a_{yi}\frac{\eta_{1i}}{\sqrt{\varepsilon_i}}\,dW_i - c_{yi}\xi_i\eta_{2i}\,dt + v_{yi}\,dt, \tag{10.74}$$

$$de_i = h\xi_i\,dt, \tag{10.75}$$

$$d\eta_{1i} = -\frac{1}{2\varepsilon_i}\eta_{1i}\,dt - \frac{\eta_{2i}}{\sqrt{\varepsilon_i}}\,dW_i, \tag{10.76}$$

$$d\eta_{2i} = -\frac{1}{2\varepsilon_i}\eta_{2i}dt + \frac{\eta_{1i}}{\sqrt{\varepsilon_i}}\,dW_i, \tag{10.77}$$

$$\xi_i = q_x(x_i - x^*)^2 + q_y(y_i - y^*)^2$$

$$+ \sum_{j=1}^{N} q_{ij}d^2(x_i, x_j, y_i, y_j) - e_i, \tag{10.78}$$

$\forall i \in \{1, 2, \ldots, N\}$, *with the parameters* $v_x = [v_{x1}, \ldots, v_{xN}]^T$, $v_y = [v_{y1}, \ldots, v_{yN}]^T$, $a_{xi}, a_{yi}, c_{xi}, c_{yi}, h_i, q_x, q_y > 0$, *and* $q_{ij} \geq 0$, $\forall i, j \in \{1, 2, \ldots, N\}$. *If the initial conditions* $x(0)$, $y(0)$, $e(0)$ *are such that the quantities* $|x_i(0) - x^* - \tilde{x}_i^{eq}|$, $|y_i(0) - y^* - \tilde{y}_i^{eq}|$, $|e_i(0) - e_i^{eq}|$, *are sufficiently small, where* (x^*, y^*) *is the minimizer of* (10.51),

$$\tilde{x}^{eq} = (c_x a_x)^{-1} Q_x^{-1} v_x, \tag{10.79}$$

$$\tilde{y}^{eq} = (c_y a_y)^{-1} Q_y^{-1} v_y, \tag{10.80}$$

$$e_i^{eq} = q_x \left(\tilde{x}_i^{eq}\right)^2 + q_y \left(\tilde{y}_i^{eq}\right)^2 + \frac{1}{2}\left(q_x a_{xi}^2 + q_y a_{yi}^2\right)$$
$$+ \sum_{j \in N} q_{ij}\left[\left(\tilde{x}_i^{eq} - x_j^{eq}\right)^2 + \left(\tilde{y}_i^{eq} - y_j^{eq}\right)^2\right]$$
$$+ \sum_{j \in N, j \neq i} q_{ij}\left(\frac{1}{2}\left(a_{xi}^2 + a_{xj}^2\right) + \frac{1}{2}\left(a_{yi}^2 + a_{yj}^2\right)\right), \tag{10.81}$$

and the matrices Q_x and Q_y, given by

$$Q_{xij} = \begin{cases} -q_x - \sum_{k \in N, k \neq i} q_{ik} & \text{if } i = j, \\ q_{ij} & \text{if } i \neq j, \end{cases} \tag{10.82}$$

$$Q_{yij} = \begin{cases} -q_y - \sum_{k \in N, k \neq i} q_{ik} & \text{if } i = j, \\ q_{ij} & \text{if } i \neq j, \end{cases} \tag{10.83}$$

and

$$a_x = \begin{bmatrix} a_{x1} & 0 & \cdots & 0 \\ 0 & a_{x2} & \cdots & 0 \\ \vdots & \vdots & \ddots & \vdots \\ 0 & 0 & \cdots & a_{xN} \end{bmatrix},$$

$$a_y = \begin{bmatrix} a_{y1} & 0 & \cdots & 0 \\ 0 & a_{y2} & \cdots & 0 \\ \vdots & \vdots & \ddots & \vdots \\ 0 & 0 & \cdots & a_{yN} \end{bmatrix}, \quad v_x = \begin{bmatrix} v_{x1} \\ \vdots \\ v_{xN} \end{bmatrix}, \tag{10.84}$$

$$c_x = \begin{bmatrix} c_{x1} & 0 & \cdots & 0 \\ 0 & c_{x2} & \cdots & 0 \\ \vdots & \vdots & \ddots & \vdots \\ 0 & 0 & \cdots & c_{xN} \end{bmatrix},$$

$$c_y = \begin{bmatrix} c_{y1} & 0 & \cdots & 0 \\ 0 & c_{y2} & \cdots & 0 \\ \vdots & \vdots & \ddots & \vdots \\ 0 & 0 & \cdots & c_{yN} \end{bmatrix}, \quad v_y = \begin{bmatrix} v_{y1} \\ \vdots \\ v_{yN} \end{bmatrix}. \tag{10.85}$$

then there exist constants $C_x, C_y, \gamma_x, \gamma_y > 0$, and a function $T(\varepsilon_1) : (0, \varepsilon_0) \to \mathbb{N}$ such that for any $\delta > 0$

$$\lim_{\varepsilon_1 \to 0} \inf\left\{t \geq 0 : \left|x_i(t) - x^* - x_i^{eq}\right| > C_x e^{-\gamma_x t} + \delta + O\left(\|a_x\|\right)\right\}$$
$$= \infty \quad a.s., \tag{10.86}$$

$$\lim_{\varepsilon_1 \to 0} \inf\left\{t \geq 0 : \left|y_i(t) - y^* - y_i^{eq}\right| > C_y e^{-\gamma_y t} + \delta + O\left(\|a_y\|\right)\right\}$$

$$= \infty \quad a.s., \tag{10.87}$$

and

$$\lim_{\varepsilon_1 \to 0} P\left\{\left|x_i(t) - x^* - \tilde{x}_i^{eq}\right| \leq C_x e^{-\gamma_x t} + \delta + O\left(\|a_x\|\right),\right.$$

$$\left. \forall t \in \left[0, T(\varepsilon_1)\right]\right\} = 1, \tag{10.88}$$

$$\lim_{\varepsilon_1 \to 0} P\left\{\left|y_i(t) - y^* - \tilde{y}_i^{eq}\right| \leq C_y e^{-\gamma_y t} + \delta + O\left(\|a_y\|\right),\right.$$

$$\left. \forall t \in \left[0, T(\varepsilon_1)\right]\right\} = 1, \tag{10.89}$$

$\forall i \in \{1, 2, \ldots, N\}$ *with the* $\lim_{\varepsilon_1 \to 0} T(\varepsilon_1) = \infty$. *The constants* C_x, C_y *are dependent on both the initial condition* $(x(0), y(0), e(0))$ *and the parameters* $a_x, c_x, c_y,$ v_x, v_y, h_i $(i = 1, \ldots, N), q_x, q_y$. *The constants* γ_x, γ_y *are dependent on the parameters* $a_x, a_y, c_x, c_y, v_x, v_y, h_i$ $(i = 1, \ldots, N), q_x, q_y$.

Proof We start by defining the error variables

$$\tilde{x}_i = x_i - x^* - a_{xi}\eta_{1i}, \tag{10.90}$$

$$\tilde{y}_i = y_i - y^* - a_{yi}\eta_{2i}. \tag{10.91}$$

Thus

$$d\tilde{x}_i = dx_i - a_{xi}\,d\eta_{1i} = -c_{xi}\xi_i\eta_{1i}dt + v_{xi}\,dt$$

$$= -c_{xi}\xi_i\chi_{1i}(t/\varepsilon_i)\,dt + v_{xi}\,dt, \tag{10.92}$$

$$d\tilde{y}_i = dy_i - a_{yi}\,d\eta_{2i} = -c_{yi}\xi_i\eta_{2i}\,dt + v_{yi}\,dt$$

$$= -c_{yi}\xi_i\chi_{2i}(t/\varepsilon_i)\,dt + v_{yi}\,dt. \tag{10.93}$$

Hence we obtain the following dynamics for the error variables:

$$\frac{d\tilde{x}_i}{dt} = -c_{xi}\xi_i\chi_{1i}(t/\varepsilon_i) + v_{xi}, \tag{10.94}$$

$$\frac{d\tilde{y}_i}{dt} = -c_{yi}\xi_i\chi_{2i}(t/\varepsilon_i) + v_{yi}, \tag{10.95}$$

$$\frac{de_i}{dt} = h_i\xi_i, \tag{10.96}$$

$$\xi_i = q_x\left(\tilde{x}_i + a_{xi}\chi_{1i}(t/\varepsilon_i)\right)^2 + q_y\left(\tilde{y}_i + a_{yi}\chi_{2i}(t/\varepsilon_i)\right)^2$$

$$+ \sum_{j \in N} q_{ij}\left[\left(\tilde{x}_i + a_{xi}\chi_{1i}(t/\varepsilon_i) - \tilde{x}_j - a_{xj}\chi_{1j}(t/\varepsilon_j)\right)^2\right.$$

$$\left. + \left(\tilde{y}_i + a_{yi}\chi_{2i}(t/\varepsilon_i) - \tilde{y}_j - a_{yj}\chi_{2j}(t/\varepsilon_j)\right)^2\right] - e_i, \tag{10.97}$$

$$d\chi_{1i}(t) = -\frac{1}{2}\chi_{1i}(t) - \chi_{2i}(t)\,dB_i(t), \tag{10.98}$$

$$d\chi_{2i}(t) = -\frac{1}{2}\chi_{2i}(t) + \chi_{1i}(t)\,dB_i(t). \tag{10.99}$$

We first calculate the average system of (10.94)–(10.96). Assume that

$$\varepsilon_i = \frac{\varepsilon_1}{c_i}, \quad i = 2, \ldots, N, \tag{10.100}$$

for some positive real constants c_i's. Denote

$$Z_{11}(t) = \chi_{11}(t), \qquad Z_{21}(t) = \chi_{21}(t),$$
$$Z_{1i}(t) = \chi_{1i}(c_i t), \qquad Z_{2i}(t) = \chi_{2i}(c_i t), \quad i = 2, \ldots, N. \tag{10.101}$$

Then (10.94)–(10.96) become

$$\frac{d\tilde{x}_i}{dt} = -c_{xi}\xi_i Z_{1i}(t/\varepsilon_1) + v_{xi}, \tag{10.102}$$

$$\frac{d\tilde{y}_i}{dt} = -c_{yi}\xi_i Z_{2i}(t/\varepsilon_1) + v_{yi}, \tag{10.103}$$

$$\frac{de_i}{dt} = h_i\xi_i, \tag{10.104}$$

$$\xi_i = q_x\big(\tilde{x}_i + a_{xi}Z_{1i}(t/\varepsilon_1)\big)^2 + q_y\big(\tilde{y}_i + a_{yi}Z_{2i}(t/\varepsilon_1)\big)^2$$
$$+ \sum_{j\in N} q_{ij}\big[\big(\tilde{x}_i + a_{xi}Z_{1i}(t/\varepsilon_1) - \tilde{x}_j - a_{xj}Z_{1j}(t/\varepsilon_1)\big)^2$$
$$+ \big(\tilde{y}_i + a_{yi}Z_{2i}(t/\varepsilon_1) - \tilde{y}_j - a_{yj}Z_{2j}(t/\varepsilon_1)\big)^2\big] - e_i. \tag{10.105}$$

The signals Z_{1i} and Z_{2i} are both components of a Brownian motion on a unit circle, which is known to be exponentially ergodic with invariant distribution $\mu(S) = \frac{l(S)}{2\pi}$ for any set $S \subset \mathbb{T} = \{(x, y) \in \mathbb{R}^2 | x^2 + y^2 = 1\}$, where $l(S)$ denotes the length (Lebesgue measure) of S. The integral over the entire space of functions of Brownian motion on a unit circle can be reduced to the integral from 0 to 2π. Since

$$\int_{\mathbb{T}} x^{2k+1}\mu(dx, dy) = \int_0^{2\pi} \cos^{2k+1}(\theta)\frac{1}{2\pi}\,d\theta = 0, \tag{10.106}$$

$$\int_{\mathbb{T}} x^2\mu(dx, dy) = \int_0^{2\pi} \cos^2(\theta)\frac{1}{2\pi}\,d\theta = \frac{1}{2}, \tag{10.107}$$

$$\int_{\mathbb{T}^2} x_1 x_2 \mu(dx_1, dy_1) \times \mu(dx_2, dy_2)$$
$$= \int_0^{2\pi}\int_0^{2\pi} \cos(\theta_1)\cos(\theta_2)\frac{1}{4\pi^2}\,d\theta_1\,d\theta_2 = 0 \tag{10.108}$$

(note that the same applies to the y case) and

$$\int_{\mathbb{T}} xy\mu(dx,dy) = \int_0^{2\pi} \cos(\theta)\sin(\theta)\frac{1}{2\pi}\,d\theta = 0, \tag{10.109}$$

$$\int_{\mathbb{T}} xy^2\mu(dx,dy) = \int_0^{2\pi} \cos(\theta)\sin^2(\theta)\frac{1}{2\pi}\,d\theta = 0, \tag{10.110}$$

$$\int_{\mathbb{T}^2} x_1 y_2^2 \mu(dx_1,dy_1) \times \mu(dx_2,dy_2)$$
$$= \int_0^{2\pi}\int_0^{2\pi} \cos(\theta_1)\sin^2(\theta_2)\frac{1}{4\pi^2}\,d\theta_1\,d\theta_2 = 0, \tag{10.111}$$

$$\int_{\mathbb{T}^2} y_1 x_2^2 \mu(dx_1,dy_1) \times \mu(dx_2,dy_2)$$
$$= \int_0^{2\pi}\int_0^{2\pi} \sin(\theta_1)\cos^2(\theta_2)\frac{1}{4\pi^2}\,d\theta_1\,d\theta_2 = 0, \tag{10.112}$$

$$\int_{\mathbb{T}^2} x_1 x_2^2 \mu(dx_1,dy_1) \times \mu(dx_2,dy_2)$$
$$= \int_0^{2\pi}\int_0^{2\pi} \cos(\theta_1)\cos^2(\theta_2)\frac{1}{4\pi^2}\,d\theta_1\,d\theta_2 = 0, \tag{10.113}$$

we obtain the average system

$$\frac{d\tilde{x}_i^{\text{ave}}}{dt} = a_{xi}c_{xi}\left[-q_x\tilde{x}_i^{\text{ave}} - \sum_{j\in N, j\neq i} q_{ij}\left(\tilde{x}_i^{\text{ave}} - \tilde{x}_j^{\text{ave}}\right)\right] + v_{xi}, \tag{10.114}$$

$$\frac{d\tilde{y}_i^{\text{ave}}}{dt} = a_{yi}c_{xi}\left[-q_y\tilde{y}_i^{\text{ave}} - \sum_{j\in N, j\neq i} q_{ij}\left(\tilde{y}_i^{\text{ave}} - \tilde{y}_j^{\text{ave}}\right)\right] + v_{yi}, \tag{10.115}$$

$$\frac{de_i^{\text{ave}}}{dt} = h_i\left[-e_i^{\text{ave}} + q_x\left(\tilde{x}_i^{\text{ave}}\right)^2 + q_y\left(\tilde{y}_i^{\text{ave}}\right)^2 + \frac{1}{2}\left(q_x a_{xi}^2 + q_y a_{yi}^2\right)\right.$$
$$+ \sum_{j\in N, j\neq i} q_{ij}\left(\left(\tilde{x}_i^{\text{ave}} - \tilde{x}_j^{\text{ave}}\right)^2 + \left(\tilde{y}_i^{\text{ave}} - \tilde{y}_j^{\text{ave}}\right)^2\right.$$
$$\left.\left. + \frac{1}{2}\left(a_{xi}^2 + a_{xj}^2\right) + \frac{1}{2}\left(a_{yi}^2 + a_{yj}^2\right)\right)\right]. \tag{10.116}$$

Rewriting the above systems in the matrix form, we have

$$\frac{d\tilde{x}^{\text{ave}}}{dt} = c_x a_x Q_x \tilde{x}^{\text{ave}} + v_x, \tag{10.117}$$

$$\frac{d\tilde{y}^{\text{ave}}}{dt} = c_y a_y Q_y \tilde{y}^{\text{ave}} + v_y, \tag{10.118}$$

$$\frac{de_i^{\text{ave}}}{dt} = h_i \left(-e_i^{\text{ave}} + q_x \left(\tilde{x}_i^{\text{ave}} \right)^2 + q_y \left(\tilde{y}_i^{\text{ave}} \right)^2 + \frac{1}{2} \left(q_x a_{xi}^2 + q_y a_{yi}^2 \right) \right)$$

$$+ h_i \sum_{j \in N} q_{ij} \left(\left(\tilde{x}_i^{\text{ave}} - \tilde{x}_j^{\text{ave}} \right)^2 + \left(\tilde{y}_i^{\text{ave}} - \tilde{y}_j^{\text{ave}} \right)^2 \right)$$

$$+ h_i \sum_{j \in N, j \neq i} q_{ij} \left(\frac{1}{2} \left(a_{xi}^2 + a_{xj}^2 \right) + \frac{1}{2} \left(a_{yi}^2 + a_{yj}^2 \right) \right). \tag{10.119}$$

The average error system has equilibria (10.79), (10.80), and (10.81) with the Jacobian

$$\Upsilon = \begin{bmatrix} c_x a_x Q_x & 0 & 0 \\ 0 & c_y a_y Q_y & 0 \\ 0 & 0 & -hI \end{bmatrix}, \tag{10.120}$$

where

$$h = \begin{bmatrix} h_1 & 0 & \cdots & 0 \\ 0 & h_2 & \cdots & 0 \\ \vdots & \vdots & \ddots & \vdots \\ 0 & 0 & 0 & h_N \end{bmatrix}. \tag{10.121}$$

Since matrices Q_x and Q_y are given by (10.82) and (10.83), by Gershgorin Circle Theorem [47, Theorem 7.2.1], we know that as long as the constants $q_x, q_y > 0$, the matrices Q_x, Q_y have all of their eigenvalues on the left hand side. Thus Q_x, Q_y are Hurwitz and invertible, which implies that Υ is Hurwitz and that the equilibria (10.79), (10.80), and (10.81) are exponentially stable.

Using the multi-input stochastic averaging theorem given in Theorem 8.1 of Chap. 8, there exist constants $c > 0, r > 0, \gamma > 0$, and a function $T(\varepsilon_1) : (0, \varepsilon_0) \to \mathbb{N}$ such that, for any $\delta > 0$, and any initial conditions $|\Lambda^{\varepsilon_1}(0)| < r$,

$$\lim_{\varepsilon_1 \to 0} \inf \left\{ t \geq 0 : \left| \Lambda^{\varepsilon_1}(t) \right| > c \left| \Lambda^{\varepsilon_1}(0) \right| e^{-\gamma t} + \delta \right\} = \infty \quad \text{a.s.} \tag{10.122}$$

and

$$\lim_{\varepsilon_1 \to 0} P \left\{ \left| \Lambda^{\varepsilon_1}(t) \right| \leq c \left| \Lambda^{\varepsilon_1}(0) \right| e^{-\gamma t} + \delta, t \in \left[0, T(\varepsilon_1) \right) \right\} = 1, \tag{10.123}$$

with $\lim_{\varepsilon_1 \to 0} T(\varepsilon_1) = \infty$, where $\Lambda^{\varepsilon_1}(t) = [\tilde{x} - x^{\text{eq}}, \tilde{y} - y^{\text{eq}}, e - e^{\text{eq}}]^T$.

The results (10.122) and (10.123) state that the norm of the error vector $\Lambda^{\varepsilon_1}(t)$ exponentially converges, both almost surely and in probability, to a point below an arbitrarily small residual value δ over an arbitrarily long time interval, which tends to infinity as ε_1 goes to zero. In particular, each $\tilde{x}_i$-component and $\tilde{y}_i$-component for all $i \in \{1, 2, \ldots, N\}$ of the error vector converges to below δ, which gives us (10.86)–(10.89).   □

Fig. 10.2 Stochastic
extremum seeking of a group
of vehicles. The anchor
agents are denoted by *red
triangles* and the follower
agents are denoted by *blue
dots*. The agents start inside
the *dashed line* and converge
to a circular formation around
the source

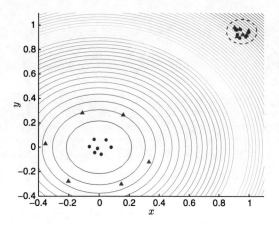

10.3.4 Simulation

In this section, we show numerical results for a group of vehicles with the control
scheme presented in Sect. 10.3.2. For the following simulations, without loss of
generality, we let the unknown location of the signal field be at the origin $(x^*, y^*) = (0, 0)$, and let the unknown signal field parameters be $(q_x, q_y) = (1, 1)$.

In Fig. 10.2, we consider 13 vehicles. We choose the design parameters as
$a = 0.01$, $c_x = c_y = 150$, $h = 10$, and define agents 1 through 6 as the anchor agents
with the forcing terms,

$$(v_{xi}, v_{yi}) = 0.05 \left(\cos\left(\frac{i\pi}{3}\right), \sin\left(\frac{i\pi}{3}\right) \right), \tag{10.124}$$

where $i = 1, \ldots, 6$. In addition to the design parameters, we picked in the interaction
gain q_{ij} such that

$$q_{ij} = \begin{cases} q_{i,i+1} = q_{i+1,i} = 0.5 & \text{if } i \in \{1, \ldots, 12\}, i \neq 6, \\ q_{i,13} = 0.5 & \text{if } i \in \{7, \ldots, 12\}, \\ q_{i,i-6} = q_{i-6,i} = 1 & \text{if } i \in \{7, \ldots, 12\}, \\ q_{i,j} = 0 & \text{otherwise.} \end{cases} \tag{10.125}$$

Figure 10.2 shows the ability of the control algorithm to produce a circular distribution around the source with a higher density of vehicles near the source. In this plot,
the trajectories of the vehicles are not shown, in order to avoid obscuring the final
vehicle formation.

10.4 Notes and References

In this chapter, we specialize the general results of Chap. 9 to games with quadratic
payoffs and illustrate the results with a solution to the problem of seeking a Nash

equilibrium in an economic competition formulated as an oligopoly market, which is treated in the classical game theory [11] and in numerous references in economics. In our solution, companies compete using product pricing and measure only their own profits, while the customers' perceptions of their and the competitors' products, and even the competitors' product pricing are unknown to them. Despite such a lack of model knowledge, the stochastic ES algorithms attain the Nash equilibrium.

In [31, 80, 149], multi-agent deployment is considered as a GPS-enabled game problem where each agent is trying to maximize its own cost function, but in these algorithms the agents also require the cost information of their neighbors. In this chapter, we investigate a stochastic version of non-cooperative source seeking by navigating the autonomous vehicles with the help of a random perturbation. The vehicles have no knowledge of their own position, or the position of the source, and are only required to sense the distances between their neighbors and themselves.

Chapter 11
Newton-Based Stochastic Extremum Seeking

The stochastic extremum seeking algorithms presented in the previous chapters are based on the gradient algorithm. In this chapter, we present a Newton-based stochastic extremum seeking algorithm. The key advantage of the more complicated Newton algorithm relative to the gradient algorithm is that, while the convergence of the gradient algorithm is dictated by the second derivative (Hessian matrix) of the map, which is unknown, rendering the convergence rate unknown to the user, the convergence of the Newton algorithm is independent of the Hessian matrix and can be arbitrarily assigned.

This chapter is organized as follows. Section 11.1 presents the single-parameter stochastic extremum seeking algorithm based on Newton optimization method. Section 11.2 presents the multi-parameter Newton algorithm for static maps. Section 11.3 presents the stochastic extremum seeking Newton algorithm for dynamic systems.

11.1 Single-parameter Newton Algorithm for Static Maps

We consider the following nonlinear static map

$$y = f(\theta), \tag{11.1}$$

where $f(\cdot)$ is not known, but it is known that $f(\cdot)$ has a maximum $y^* = f(\theta^*)$ at $\theta = \theta^*$.

We make the following assumption:

Assumption 11.1 $f(\cdot)$ is twice continuously differentiable and has a unique global maximum, $\theta^* \in \mathbb{R}$ such that

$$\frac{df(\theta)}{d\theta} = 0 \quad \text{if and only if} \quad \theta = \theta^*, \tag{11.2}$$

$$\left.\frac{d^2 f(\theta)}{d\theta^2}\right|_{\theta=\theta^*} \triangleq H < 0. \tag{11.3}$$

S.-J. Liu, M. Krstic, *Stochastic Averaging and Stochastic Extremum Seeking*,
Communications and Control Engineering,
DOI 10.1007/978-1-4471-4087-0_11, © Springer-Verlag London 2012

If $f(\cdot)$ is known, the following Newton optimization algorithm can be used to find θ^*:

$$\frac{d\theta}{dt} = -\left(\frac{d^2 f(\theta)}{d\theta^2}\right)^{-1} \frac{df(\theta)}{d\theta}. \tag{11.4}$$

If $f(\cdot)$ is unknown, then an estimator is needed to approximate $\frac{df(\theta)}{d\theta}$ and $\frac{d^2 f(\theta)}{d\theta^2}$. The purpose of this section is to combine the continuous Newton optimization algorithm (11.4) with estimators of the first and second derivatives to achieve stochastic extremum seeking in such a way that the closed-loop system approximates the behavior of (11.4).

Let $\hat{\theta}$ denote the estimate of θ and let Γ be the estimate of

$$H^{-1} = \left(\frac{d^2 f(\theta)}{d\theta^2}\bigg|_{\theta=\theta^*}\right)^{-1}.$$

We introduce the algorithm

$$\frac{d\hat{\theta}(t)}{dt} = -k\Gamma(t) M\big(\eta(t)\big) y, \quad k > 0, \tag{11.5}$$

$$\frac{d\Gamma(t)}{dt} = h_1 \Gamma(t) - h_1 \Gamma^2(t) N\big(\eta(t)\big) y, \quad \Gamma(0) < 0, \tag{11.6}$$

where $M(\cdot)$ and $N(\cdot)$ are any bounded and odd continuous functions, and $\eta(t)$ is an ergodic stochastic process with an invariant distribution. In the stochastic extremum seeking algorithm (11.5), we use $M(\eta)y$ to estimate the first-order derivative of f. For the estimate Γ of the inverse of the second-order derivative of f, an algebraic division in the form $1/\hat{H}$ would create difficulties when the estimate $\hat{H}$ of $\frac{d^2 f(\theta)}{d\theta^2}|_{\theta=\theta^*}$ is close to zero. To deal with this problem, we employ a dynamic estimator to calculate the inverse of $\hat{H}$ using a Riccati equation. Consider the following filter

$$\frac{d\Xi}{dt} = -h_1 \Xi + h_1 \hat{H}, \tag{11.7}$$

where $h_1 > 0$ is a constant, which guarantees that the state Ξ of the stable filter (11.7) converges to $\hat{H}$. Denote $\Gamma = \Xi^{-1}$. Then

$$\frac{d\Gamma}{dt} = -\Xi^{-2} \frac{d\Xi}{dt}. \tag{11.8}$$

Thus by (11.7) we get the following differential Riccati equation

$$\frac{d\Gamma}{dt} = h_1 \Gamma - h_1 \Gamma^2 \hat{H}, \tag{11.9}$$

which has two equilibria: $\Gamma^* = 0$, $\hat{H}^{-1}$. Since $h_1 > 0$, the equilibrium $\Gamma^* = 0$ is unstable, whereas the equilibrium $\Gamma^* = \hat{H}^{-1}$ is exponentially stable. This shows that after a transient, the Riccati equation (11.9) converges to the actual value of the inverse of H if $\hat{H}$ is a good estimate of H. Comparing (11.6) and (11.9), we use the stochastic excitation signal $N(\eta)$ to generate the estimate $\hat{H} = N(\eta)y$ of H.

Now we perform an illustrative analysis of stability of algorithm (11.5), (11.6). Denote the estimate error $\tilde{\theta} = \hat{\theta} - \theta^*$, $\tilde{\Gamma} = \Gamma - H^{-1}$, and $\theta = \hat{\theta} + a\sin(\eta)$. Then we have the error system

$$\frac{d\tilde{\theta}}{dt} = -k(\tilde{\Gamma} + H^{-1})M(\eta)f(\theta^* + \tilde{\theta} + a\sin(\eta)), \tag{11.10}$$

$$\frac{d\tilde{\Gamma}}{dt} = h_1(\tilde{\Gamma} + H^{-1})$$
$$- h_1(\tilde{\Gamma} + H^{-1})^2 N(\eta)f(\theta^* + \tilde{\theta} + a\sin(\eta)). \tag{11.11}$$

For simplicity and clarity, we consider a quadratic map

$$f(\theta) = f^* + \frac{f''(\theta^*)}{2}(\theta - \theta^*)^2 = f^* + \frac{H}{2}(\theta - \theta^*)^2. \tag{11.12}$$

Then the error system is

$$\frac{d\tilde{\theta}}{dt} = -k(\tilde{\Gamma} + H^{-1})M(\eta)\left(f^* + \frac{H}{2}(\tilde{\theta} + a\sin(\eta))^2\right), \tag{11.13}$$

$$\frac{d\tilde{\Gamma}}{dt} = h_1(\tilde{\Gamma} + H^{-1})$$
$$- h_1(\tilde{\Gamma} + H^{-1})^2 N(\eta)\left(f^* + \frac{H}{2}(\tilde{\theta} + a\sin(\eta))^2\right). \tag{11.14}$$

To obtain an exponentially stable average system, we choose $M(\eta)$, $N(\eta)$ such that

$$\mathrm{Ave}(M(\eta)y) \triangleq \int_{\mathbb{R}} M(x)\left(f^* + \frac{H}{2}(\tilde{\theta} + a\sin(x))^2\right)\mu(dx) = H\tilde{\theta}, \tag{11.15}$$

$$\mathrm{Ave}(N(\eta)y) \triangleq \int_{\mathbb{R}} N(x)\left(f^* + \frac{H}{2}(\tilde{\theta} + a\sin(x))^2\right)\mu(dx) = H, \tag{11.16}$$

where μ is the invariant distribution of the ergodic process $\eta(t)$. We choose the ergodic process as an OU process satisfying

$$\varepsilon\,d\eta = -\eta\,dt + \sqrt{\varepsilon q}\,dW, \tag{11.17}$$

which has invariant distribution $\mu(dx) = \frac{1}{\sqrt{\pi}q}e^{-x^2/q^2}\,dx$. To satisfy (11.15) and (11.16), we choose $M(\eta)$ and $N(\eta)$ that satisfy

$$\left(f^* + \frac{H}{2}\tilde{\theta}^2\right) \times \mathrm{Ave}(M(\eta)) = 0, \tag{11.18}$$

$$H\tilde{\theta}a \times \mathrm{Ave}(M(\eta)\sin(\eta)) = H\tilde{\theta}, \tag{11.19}$$

$$\frac{H}{2}a^2 \times \mathrm{Ave}(M(\eta)\sin^2(\eta)) = 0, \tag{11.20}$$

$$\left(f^* + \frac{H}{2}\tilde{\theta}^2\right) \times \mathrm{Ave}(N(\eta)) = 0, \tag{11.21}$$

$$Ha\tilde{\theta} \times \mathrm{Ave}(N(\eta)\sin(\eta)) = 0, \tag{11.22}$$

$$\frac{H}{2}a^2 \times \mathrm{Ave}(N(\eta)\sin^2(\eta)) = H. \tag{11.23}$$

Since

$$\int_{\mathbb{R}} \sin^{2k+1}(x)\mu(dx) = \int_{-\infty}^{+\infty} \sin^{2k+1}(x)\frac{1}{\sqrt{\pi}q}e^{-x^2/q^2}\,dx = 0, \quad (11.24)$$

$$\int_{\mathbb{R}} \sin^2(x)\mu(dx) = \int_{-\infty}^{+\infty} \sin^2(x)\frac{1}{\sqrt{\pi}q}e^{-x^2/q^2}\,dx$$
$$= \frac{1}{2}\left(1 - e^{-q^2}\right) \triangleq G_0(q), \quad (11.25)$$

$$\int_{\mathbb{R}} \sin^4(x)\mu(dx) = \int_{-\infty}^{+\infty} \sin^4(x)\frac{1}{\sqrt{\pi}q}e^{-x^2/q^2}\,dx$$
$$= \frac{3}{8} - \frac{1}{2}e^{-q^2} + \frac{1}{8}e^{-4q^2} \triangleq G_1(q), \quad (11.26)$$

we choose

$$M(\eta) = \frac{1}{aG_0(q)}\sin(\eta), \quad (11.27)$$

$$N(\eta) = \frac{4}{a^2 G_0^2(\sqrt{2}q)}\left(\sin^2(\eta) - G_0(q)\right), \quad (11.28)$$

where $G_0^2(\sqrt{2}q) = 2(G_1(q) - G_0^2(q))$. Thus we obtain the average system

$$\frac{d\tilde{\theta}^{\text{ave}}}{dt} = -k\tilde{\theta}^{\text{ave}} - k\tilde{\Gamma}^{\text{ave}}H\tilde{\theta}^{\text{ave}}, \quad (11.29)$$

$$\frac{d\tilde{\Gamma}^{\text{ave}}}{dt} = -h_1\tilde{\Gamma}^{\text{ave}} - h_1\left(\tilde{\Gamma}^{\text{ave}}\right)^2 H, \quad (11.30)$$

which has a locally exponentially stable equilibrium at $(\tilde{\theta}^{\text{ave}}, \tilde{\Gamma}^{\text{ave}}) = (0,0)$, as well as an unstable equilibrium at $(0, -1/H)$. Thus, according to the averaging theorem, we have the following result:

Theorem 11.1 *Consider the quadratic map* (11.12) *under the parameter update law* (11.5)–(11.6). *Then there exist constants* $r > 0$, $c > 0$, $\gamma > 0$, *and a function* $T(\varepsilon): (0, \varepsilon_0) \to \mathbb{N}$ *such that, for any initial condition* $|\Lambda^{\varepsilon}(0)| < r$ *and any* $\delta > 0$,

$$\lim_{\varepsilon \to 0}\inf\{t \geq 0: |\Lambda^{\varepsilon}(t)| > c|\Lambda^{\varepsilon}(0)|e^{-\gamma t} + \delta\} = \infty \quad a.s. \quad (11.31)$$

and

$$\lim_{\varepsilon \to 0}P\{|\Lambda^{\varepsilon}(t)| \leq c|\Lambda^{\varepsilon}(0)|e^{-\gamma t} + \delta, \forall t \in [0, T(\varepsilon)]\} = 1 \quad \text{with}$$

$$\lim_{\varepsilon \to 0}T(\varepsilon) = \infty, \quad (11.32)$$

where $\Lambda^{\varepsilon}(t) \triangleq (\tilde{\theta}(t), \tilde{\Gamma}(t))^T$.

Fig. 11.1 Gradient-based
stochastic extremum seeking
scheme for a static map

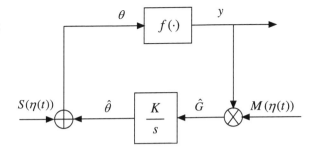

11.2 Multi-parameter Newton Algorithm for Static Maps

11.2.1 Problem Formulation

Consider the static map

$$y = f(\theta), \quad \theta \in \mathbb{R}^n. \tag{11.33}$$

We make the following assumption:

Assumption 11.2 There exist a constant vector $\theta^* \in \mathbb{R}^n$ such that

$$\left. \frac{\partial f(\theta)}{\partial \theta} \right|_{\theta=\theta^*} = 0, \tag{11.34}$$

$$\left. \frac{\partial^2 f(\theta)}{\partial^2 \theta} \right|_{\theta=\theta^*} < 0. \tag{11.35}$$

Assumption 11.2 means that the map (11.33) has a local maximum at θ^*. The cost function is not known in (11.33), but, as usual, we assume that we can measure y and manipulate θ. The gradient-based extremum seeking scheme for this multivariable static map is (shown in Fig. 11.1):

$$\frac{d\hat{\theta}(t)}{dt} = KM(\eta(t))y, \quad \theta(t) = \hat{\theta}(t) + S(\eta(t)), \tag{11.36}$$

where $K = \text{diag}(k_1, \ldots, k_n)$ with $k_i > 0$,

$$S(\eta(t)) = \left[a_1 \sin(\eta_1(t)), \ldots, a_n \sin(\eta_n(t)) \right]^T, \tag{11.37}$$

$$M(\eta(t)) = \left[\frac{1}{a_1 G_0(q_1)} \sin(\eta_1(t)), \ldots, \frac{1}{a_n G_0(q_n)} \sin(\eta_n(t)) \right]^T \tag{11.38}$$

are perturbation signals, and the independent processes $\eta_i(t)$, $i = 1, \ldots, n$, satisfy

$$\varepsilon_i \, d\eta_i = -\eta_i \, dt + \sqrt{\varepsilon_i} q_i \, dW_i, \tag{11.39}$$

where $q_i > 0$, $\varepsilon_i \in (0, \varepsilon_0)$ for fix $\varepsilon_0 > 0$, and $W_i(t)$, $i = 1, \ldots, n$, are independent standard Wiener processes on some complete probability space.

Fig. 11.2 Newton-based
stochastic extremum seeking
scheme for a static map

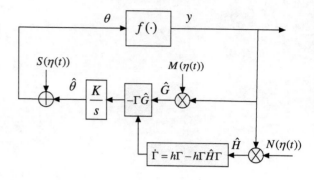

In the parameter error variable $\tilde{\theta} = \hat{\theta} - \theta^*$, the closed-loop system in Fig. 11.1 is
given by

$$\frac{d\tilde{\theta}(t)}{dt} = K M\big(\eta(t)\big) f\big(\theta^* + S\big(\eta(t)\big) + \tilde{\theta}\big). \tag{11.40}$$

For the case of a quadratic static map, $f(\theta) = f^* + \frac{1}{2}(\theta - \theta^*)^T H(\theta - \theta^*)$, the
average system of (11.40) is given by

$$\frac{d\tilde{\theta}^{\text{ave}}(t)}{dt} = K H \tilde{\theta}^{\text{ave}}(t), \tag{11.41}$$

where H is the Hessian matrix of the static map, and it is negative definite. This
observation reveals two things: (i) the gradient-based extremum seeking algorithm
is locally convergent, and (ii) the convergence rate is governed by the unknown
Hessian matrix H. In the next section, we give a stochastic ES algorithm based on
Newton optimization method, which eliminates the dependence of the convergence
rate on the unknown H.

11.2.2 Algorithm Design and Stability Analysis

The Newton-based stochastic extremum seeking algorithm for a static map is shown
in Fig. 11.2, where h is a positive real number. There are two vital parts in the
Newton-based algorithm: the perturbation matrix $N(\eta(t))$, which generates an estimate $\hat{H} = N(\eta)y$ of the Hessian matrix, and the Riccati equation, which generates
an estimate of the inverse of Hessian matrix, even when the estimate of the Hessian
matrix is singular.

The detailed algorithm is as follows:

$$\theta_i = \hat{\theta}_i + a_i \sin(\eta_i), \tag{11.42}$$

$$\frac{d\hat{\theta}}{dt} = -K \Gamma M(\eta) y, \tag{11.43}$$

$$\frac{d\Gamma}{dt} = h\Gamma - h\Gamma N(\eta) y \Gamma, \quad \Gamma(0) < 0, \tag{11.44}$$

where $K = \mathrm{diag}(k_1, \ldots, k_n)$ and $h > 0$ are design parameters, $M(\cdot) \in \mathbb{R}^n$ is given by (11.38), $N(\cdot) \in \mathbb{R}^{n \times n}$ is to be determined, $\Gamma \in \mathbb{R}^{n \times n}$ is used to approximate $(\frac{\partial^2 f(\theta)}{\partial \theta^2})^{-1}|_{\theta = \theta^*} \triangleq (\frac{\partial^2 f(\theta^*)}{\partial \theta^2})^{-1}$, and $\eta_i(t)$, $i = 1, \ldots, n$, are independent ergodic processes.

Denote the estimate error variables $\tilde{\Gamma} = \Gamma - (\frac{\partial^2 f(\theta^*)}{\partial \theta^2})^{-1}$, $\tilde{\theta} = \hat{\theta} - \theta^*$. Then we have the estimate error system

$$\frac{d\tilde{\theta}}{dt} = -K\tilde{\Gamma}M(\eta)y - K\left(\frac{\partial^2 f(\theta^*)}{\partial \theta^2}\right)^{-1} M(\eta)y, \tag{11.45}$$

$$\frac{d\tilde{\Gamma}}{dt} = h\tilde{\Gamma} + h\left(\frac{\partial^2 f(\theta^*)}{\partial \theta^2}\right)^{-1} - h\tilde{\Gamma}N(\eta)y\tilde{\Gamma}$$

$$- h\tilde{\Gamma}N(\eta)y\left(\frac{\partial^2 f(\theta^*)}{\partial \theta^2}\right)^{-1} - h\left(\frac{\partial^2 f(\theta^*)}{\partial \theta^2}\right)^{-1} N(\eta)y\tilde{\Gamma}$$

$$- h\left(\frac{\partial^2 f(\theta^*)}{\partial \theta^2}\right)^{-1} N(\eta)y\left(\frac{\partial^2 f(\theta^*)}{\partial \theta^2}\right)^{-1}. \tag{11.46}$$

For the general map case, the stability analysis is conducted in Sect. 11.3. Here we first give the stability analysis of a quadratic static map.

Consider the quadratic static map,

$$f(\theta) = f^* + \frac{1}{2}(\theta - \theta^*)^T H(\theta - \theta^*), \tag{11.47}$$

where H is negative definite. Then the error system (11.45)–(11.46) becomes

$$\frac{d\tilde{\theta}(t)}{dt} = -K\tilde{\Gamma}M(\eta)\left[f^* + \frac{1}{2}(\tilde{\theta} + a\sin(\eta))^T H(\tilde{\theta} + a\sin(\eta))\right]$$

$$- KH^{-1}M(\eta)\left[f^* + \frac{1}{2}(\tilde{\theta} + a\sin(\eta))^T H(\tilde{\theta} + a\sin(\eta))\right], \tag{11.48}$$

$$\frac{d\tilde{\Gamma}(t)}{dt} = h\tilde{\Gamma} + hH^{-1} - h\tilde{\Gamma}N(\eta)\left[f^* + \frac{1}{2}(\tilde{\theta} + a\sin(\eta))^T H(\tilde{\theta} + a\sin(\eta))\right]\tilde{\Gamma}$$

$$- h\tilde{\Gamma}N(\eta)\left[f^* + \frac{1}{2}(\tilde{\theta} + a\sin(\eta))^T H(\tilde{\theta} + a\sin(\eta))\right]H^{-1}$$

$$- hH^{-1}N(\eta)\left[f^* + \frac{1}{2}(\tilde{\theta} + a\sin(\eta))^T H(\tilde{\theta} + a\sin(\eta))\right]\tilde{\Gamma}$$

$$- hH^{-1}N(\eta)\left[f^* + \frac{1}{2}(\tilde{\theta} + a\sin(\eta))^T H(\tilde{\theta} + a\sin(\eta))\right]H^{-1}. \tag{11.49}$$

Similar to the single parameter case, to make the average system of the error system (11.48)–(11.49) exponentially stable, we choose the matrix function N as

$$(N)_{ii} = \frac{4}{a_i^2 G_0^2(\sqrt{2}q_i)}\left(\sin^2(\eta_i) - G_0(q_i)\right), \tag{11.50}$$

$$(N)_{ij} = \frac{\sin(\eta_i)\sin(\eta_j)}{a_i a_j G_0(q_i)G_0(q_j)}, \quad i \neq j. \tag{11.51}$$

Thus we obtain the average system of the error system (11.48)–(11.49)

$$\frac{d\tilde{\theta}^{\text{ave}}}{dt} = -K\tilde{\theta}^{\text{ave}} - K\tilde{\Gamma}^{\text{ave}}H\tilde{\theta}^{\text{ave}}, \tag{11.52}$$

$$\frac{d\tilde{\Gamma}^{\text{ave}}}{dt} = -h\tilde{\Gamma}^{\text{ave}} - h\tilde{\Gamma}^{\text{ave}}H\tilde{\Gamma}^{\text{ave}}, \tag{11.53}$$

where $K\tilde{\Gamma}^{\text{ave}}H\tilde{\theta}^{\text{ave}}$ is quadratic in $(\tilde{\Gamma}^{\text{ave}}, \tilde{\theta}^{\text{ave}})$, and $h\tilde{\Gamma}^{\text{ave}}H\tilde{\Gamma}^{\text{ave}}$ is quadratic in $\tilde{\Gamma}^{\text{ave}}$. The linearization of this system has all of its eigenvalues at $-K$ and $-h$. Hence, unlike the gradient algorithm, whose convergence is governed by the unknown Hessian matrix H, the convergence rate of the Newton algorithm can be arbitrarily assigned by the designer with an appropriate choice of K and h. By the multi-input stochastic averaging theorem given in Theorem 8.1, we arrive at the following theorem:

Theorem 11.2 *Consider the static map* (11.47) *under the parameter update law* (11.43)–(11.44). *Then there exist constants* $r > 0$, $c > 0$, $\gamma > 0$ *and a function* $T(\varepsilon_1) : (0, \varepsilon_0) \to \mathbb{N}$ *such that for any initial condition* $|\Lambda_1^{\varepsilon_1}(0)| < r$ *and any* $\delta > 0$,

$$\lim_{\varepsilon_1 \to 0} \inf\{t \geq 0 : |\Lambda_1^{\varepsilon_1}(t)| > c|\Lambda_1^{\varepsilon_1}(0)|e^{-\gamma t} + \delta\} = \infty, \quad a.s. \tag{11.54}$$

and

$$\lim_{\varepsilon_1 \to 0} P\{|\Lambda_1^{\varepsilon_1}(t)| \leq c|\Lambda_1^{\varepsilon_1}(0)|e^{-\gamma t} + \delta, \forall t \in [0, T(\varepsilon_1)]\} = 1, \quad with$$

$$\lim_{\varepsilon_1 \to 0} T(\varepsilon_1) = \infty, \tag{11.55}$$

where $\Lambda_1^{\varepsilon_1}(t) \triangleq (\tilde{\theta}^T(t), \mathfrak{Vec}(\tilde{\Gamma}(t)))^T$, $\mathfrak{Vec}(A) \triangleq (A_1^T, \ldots, A_n^T)$, *and* A_i, $i = 1, \ldots, l$, *denote the column vectors of any matrix* $A \in \mathbb{R}^{n \times n}$.

11.3 Newton Algorithm for Dynamic Systems

Consider a general multi-input single-output (MISO) nonlinear model

$$\dot{x} = f(x, u), \tag{11.56}$$

$$y = h(x), \tag{11.57}$$

where $x \in \mathbb{R}^m$ is the state, $u \in \mathbb{R}^n$ is the input, $y \in \mathbb{R}$ is the output, and $f : \mathbb{R}^m \times \mathbb{R}^n \to \mathbb{R}^m$ and $h : \mathbb{R}^m \to \mathbb{R}$ are smooth. Suppose that we know a smooth control law

$$u = \alpha(x, \theta) \tag{11.58}$$

parameterized by a vector parameter $\theta \in \mathbb{R}^n$. Then the closed-loop system

$$\dot{x} = f(x, \alpha(x, \theta)) \tag{11.59}$$

has equilibria parameterized by θ. As in the deterministic case [6], we make the following assumptions about the closed-loop system.

Fig. 11.3 Gradient-based stochastic extremum seeking scheme

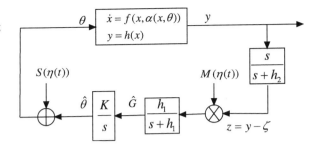

Fig. 11.4 Newton-based stochastic extremum seeking scheme. The initial condition $\Gamma(0)$ should be chosen negative definite and symmetric

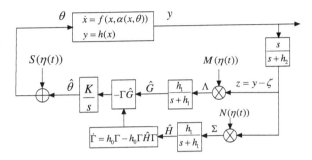

Assumption 11.3 There exists a smooth function $l : \mathbb{R}^n \to \mathbb{R}^m$ such that

$$f\big(x, \alpha(x, \theta)\big) = 0 \quad \text{if and only if} \quad x = l(\theta). \tag{11.60}$$

Assumption 11.4 For each $\theta \in \mathbb{R}^n$, the equilibrium $x = l(\theta)$ of system (11.59) is exponentially stable uniformly in θ.

Assumption 11.5 There exists $\theta^* \in \mathbb{R}^n$ such that

$$\frac{\partial (h \circ l)}{\partial \theta}(\theta^*) = 0, \tag{11.61}$$

$$\frac{\partial^2 (h \circ l)}{\partial \theta^2}(\theta^*) = H < 0, \quad H = H^T. \tag{11.62}$$

Our objective is to develop a feedback mechanism which maximizes the steady-state value of y but without requiring the knowledge of either θ^* or the functions h and l. In Chap. 5, the gradient-based extremum seeking design in the single parameter case achieves this objective. The multi-parameter gradient-based algorithm is shown schematically in Fig. 11.3, whereas Newton-based algorithm is shown in Fig. 11.4.

We introduce error variables

$$\tilde{\theta} = \hat{\theta} - \theta^*, \qquad \theta = \hat{\theta} + S\big(\eta(t)\big), \tag{11.63}$$

$$\tilde{\zeta} = \zeta - h \circ l(\theta^*), \qquad \tilde{\Gamma} = \Gamma - H^{-1}, \tag{11.64}$$

$$\tilde{H} = \hat{H} - H, \tag{11.65}$$

where $S(\eta)$ is given in (11.37). Then we can summarize the system in Fig. 11.4 as

$$
\frac{d}{dt}\begin{bmatrix} x \\ \tilde{\theta} \\ \hat{G} \\ \tilde{\Gamma} \\ \tilde{H} \\ \tilde{\zeta} \end{bmatrix} = \begin{bmatrix} f(x,\alpha(x,\theta^*+\tilde{\theta}+S(\eta(t)))) \\ -K\Gamma\hat{G} \\ -h_1\hat{G}+h_1(y-\zeta)M(\eta(t)) \\ h_0\Gamma-h_0\Gamma\hat{H}\Gamma \\ -h_1\hat{H}+h_1(y-\zeta)N(\eta(t)) \\ -h_2\zeta+h_2y \end{bmatrix}
$$

$$
= \begin{bmatrix} f(x,\alpha(x,\theta^*+\tilde{\theta}+S(\eta(t)))) \\ -K(\tilde{\Gamma}+H^{-1})\hat{G} \\ -h_1\hat{G}+h_1(y-h\circ l(\theta^*)-\tilde{\zeta})M(\eta(t)) \\ h_0(\tilde{\Gamma}+H^{-1})(I-(\tilde{H}+H)(\tilde{\Gamma}+H^{-1})) \\ -h_1\tilde{H}-h_1H+h_1(y-h\circ l(\theta^*)-\tilde{\zeta})N(\eta(t)) \\ -h_2\tilde{\zeta}+h_2(y-h\circ l(\theta^*)) \end{bmatrix}. \quad (11.66)
$$

Denote $\chi_i(\varepsilon_i t)=\eta_i(t)$ and $\chi(t)=[\chi_1(t),\ldots,\chi_n(t)]^T$. Then we change the system (11.66) as

$$
\frac{dx}{dt} = f\big(x,\alpha\big(x,\theta^*+\tilde{\theta}+S\big(\chi(t/\varepsilon)\big)\big)\big), \quad (11.67)
$$

$$
\frac{d}{dt}\begin{bmatrix} \tilde{\theta} \\ \hat{G} \\ \tilde{\Gamma} \\ \tilde{H} \\ \tilde{\zeta} \end{bmatrix}
$$

$$
= \begin{bmatrix} -K(\tilde{\Gamma}+H^{-1})\hat{G} \\ -h_1\hat{G}+h_1(y-h\circ l(\theta^*)-\tilde{\zeta})M(\chi(t/\varepsilon)) \\ h_0(\tilde{\Gamma}+H^{-1})(I-(\tilde{H}+H)(\tilde{\Gamma}+H^{-1})) \\ -h_1\tilde{H}-h_1H+h_1(y-h\circ l(\theta^*)-\tilde{\zeta})N(\chi(t/\varepsilon)) \\ -h_2\tilde{\zeta}+h_2(y-h\circ l(\theta^*)) \end{bmatrix}, \quad (11.68)
$$

where

$$
S\big(\chi(t/\varepsilon)\big)=\big[a_1\sin\big(\chi_1(t/\varepsilon_1)\big),\ldots,a_n\sin\big(\chi_n(t/\varepsilon_n)\big)\big]^T, \quad (11.69)
$$

$$
M\big(\chi(t/\varepsilon)\big)=\Big[\frac{1}{a_1G_0(q_1)}\sin\big(\chi_1(t/\varepsilon_1)\big),\ldots,
$$

$$
\frac{1}{a_nG_0(q_n)}\sin\big(\chi_n(t/\varepsilon_n)\big)\Big]^T, \quad (11.70)
$$

$$
\big(N\big(\chi(t/\varepsilon)\big)\big)_{ii}=\frac{4}{a_i^2G_0^2(\sqrt{2}q_i)}\big(\sin^2\big(\chi_i(t/\varepsilon_i)\big)-G_0(q_i)\big), \quad (11.71)
$$

$$
\big(N\big(\chi(t/\varepsilon)\big)\big)_{ij}=\frac{\sin(\chi_i(t/\varepsilon_i))\sin(\chi_j(t/\varepsilon_j))}{a_ia_jG_0(q_i)G_0(q_j)}, \quad i\neq j. \quad (11.72)
$$

Now, treating ε as large compared to the size of parameters in (11.67), we "freeze" x in (11.67) at its quasisteady-state equilibrium value $x=l(\theta^*+\tilde{\theta}+S(\chi(t/\varepsilon)))$ and substitute it into (11.68), getting the reduced system

$$
\frac{d}{dt}
\begin{bmatrix}
\tilde{\theta}_r \\
\hat{G}_r \\
\tilde{\Gamma}_r \\
\tilde{H}_r \\
\tilde{\zeta}_r
\end{bmatrix}
=
\begin{bmatrix}
-K(\tilde{\Gamma}_r + H^{-1})\hat{G}_r \\
-h_1\hat{G}_r + h_1(v(\tilde{\theta}_r + S(\chi(t/\varepsilon))) - \tilde{\zeta}_r)M(\chi(t/\varepsilon)) \\
h_0(\tilde{\Gamma}_r + H^{-1})(I - (\tilde{H}_r + H)(\tilde{\Gamma}_r + H^{-1})) \\
-h_1\tilde{H}_r - h_1 H + h_1(v(\tilde{\theta}_r + S(\chi(t/\varepsilon))) - \tilde{\zeta}_r)N(\chi(t/\varepsilon)) \\
-h_2\tilde{\zeta}_r + h_2 v(\tilde{\theta}_r + S(\chi(t/\varepsilon)))
\end{bmatrix},
$$

$$(11.73)$$

where $v(z) = h \circ l(\theta^* + z) - h \circ l(\theta^*)$. In view of Assumption 11.5, $v(0) = 0$, $\frac{\partial v}{\partial z}(0) = 0$, and $\frac{\partial^2 v}{\partial z^2}(0) = H < 0$.

Denote $\varepsilon_i = \frac{\varepsilon_1}{c_i}$ for some constants c_i. Then we get the average system of the reduced system (11.73) as

$$
\frac{d}{dt}
\begin{bmatrix}
\tilde{\theta}_r^a \\
\hat{G}_r^a \\
\tilde{\Gamma}_r^a \\
\tilde{H}_r^a \\
\tilde{\zeta}_r^a
\end{bmatrix}
$$

$$
=
\begin{bmatrix}
-K(\tilde{\Gamma}_r^a + H^{-1})\hat{G}_r^a \\
-h_1\hat{G}_r^a + h_1 \int_{\mathbb{R}^n} v(\tilde{\theta}_r + S(\sigma))M(\sigma)\mu_1(d\sigma_1) \times \cdots \times \mu_n(d\sigma_n) \\
h_0(\tilde{\Gamma}_r^a + H^{-1})(I - (\tilde{H}_r^a + H)(\tilde{\Gamma}_r^a + H^{-1})) \\
-h_1\tilde{H}_r^a - h_1 H + h_1 \int_{\mathbb{R}^n} v(\tilde{\theta}_r + S(\sigma))N(\sigma)\mu_1(d\sigma_1) \times \cdots \times \mu_n(d\sigma_n) \\
-h_2\tilde{\zeta}_r^a + h_2 \int_{\mathbb{R}^n} v(\tilde{\theta}_r + S(\sigma))\mu_1(d\sigma_1) \times \cdots \times \mu_n(d\sigma_n)
\end{bmatrix}.
$$

$$(11.74)$$

The equilibrium $(\tilde{\theta}_r^{a,e}, \hat{G}_r^{a,e}, \tilde{\Gamma}_r^{a,e}, \tilde{H}_r^{a,e}, \tilde{\zeta}_r^{a,e})$ of the average reduced system satisfies

$$\hat{G}_r^{a,e} = 0_{n \times 1}, \tag{11.75}$$

$$\int_{\mathbb{R}^n} v(\tilde{\theta}_r^{a,e} + S(\sigma))M(\sigma)\mu_1(d\sigma_1) \times \cdots \times \mu_n(d\sigma_n) = 0_{n \times 1}, \tag{11.76}$$

$$\tilde{\zeta}_r^{a,e} = \int_{\mathbb{R}^n} v(\tilde{\theta}_r^{a,e} + S(\sigma))\mu_1(d\sigma_1) \times \cdots \times \mu_n(d\sigma_n), \tag{11.77}$$

$$\tilde{H}_r^{a,e} + H = \int_{\mathbb{R}^n} v(\tilde{\theta}_r^{a,e} + S(\sigma))N(\sigma)\mu_1(d\sigma_1) \times \cdots \times \mu_n(d\sigma_n), \tag{11.78}$$

$$(\tilde{H}_r^{a,e} + H)(\tilde{\Gamma}_r^{a,e} + H^{-1}) = I. \tag{11.79}$$

By (11.76), for any $p = 1, \dots, n$,

$$\int_{\mathbb{R}^n} v(\tilde{\theta}_r^{a,e} + S(\sigma))\frac{1}{a_p G_0(q_p)} \sin(\sigma_p)\mu_1(d\sigma_1) \times \cdots \times \mu_n(d\sigma_n) = 0. \tag{11.80}$$

By postulating the ith element $\tilde{\theta}_{r,i}^{a,e}$ of $\tilde{\theta}_r^{a,e}$ in the form

$$\tilde{\theta}_{r,i}^{a,e} = \sum_{j=1}^{n} b_j^i a_j + \sum_{j=1}^{n}\sum_{k \geq j}^{n} c_{j,k}^i a_j a_k + O(|a|^3), \tag{11.81}$$

where b_j^i and $c_{j,k}^i$ are real numbers, defining

$$v(z) = \frac{1}{2} \sum_{i=1}^{n} \sum_{j=1}^{n} \frac{\partial^2 v}{\partial z_i \, \partial z_j}(0) z_i z_j$$

$$+ \frac{1}{3!} \sum_{i=1}^{n} \sum_{j=1}^{n} \sum_{k=1}^{n} \frac{\partial^3 v}{\partial z_i \, \partial z_j \, \partial z_k}(0) z_i z_j z_k + O(|z|^4) \qquad (11.82)$$

and substituting (11.82) into (11.80), we have

$$0 = \int_{\mathbb{R}^n} \left[\sum_{i=1}^{n} \sum_{j=1}^{n} \frac{1}{2} \frac{\partial^2 v}{\partial z_i \, \partial z_j}(0) \left(\tilde{\theta}_{r,i}^{a,e} + a_i \sin(\sigma_i) \right) \left(\tilde{\theta}_{r,j}^{a,e} + a_j \sin(\sigma_j) \right) \right.$$

$$+ \sum_{i=1}^{n} \sum_{j=1}^{n} \sum_{k=1}^{n} \frac{1}{3!} \frac{\partial^3 v}{\partial z_i \, \partial z_j \, \partial z_k}(0) \left(\tilde{\theta}_{r,i}^{a,e} + a_i \sin(\sigma_i) \right) \left(\tilde{\theta}_{r,j}^{a,e} + a_j \sin(\sigma_j) \right)$$

$$\left. \times \left(\tilde{\theta}_{r,k}^{a,e} + a_k \sin(\sigma_k) \right) + O(|a|^4) \right]$$

$$\times \frac{1}{a_p G_0(q_p)} \sin(\sigma_p) \mu_1(d\sigma_1) \times \cdots \times \mu_n(d\sigma_n). \qquad (11.83)$$

By calculating the average of each term, we have

$$0 = \tilde{\theta}_{r,p}^{a,e} \frac{\partial^2 v}{\partial z_p^2}(0) + \sum_{j \neq p}^{n} \tilde{\theta}_{r,j}^{a,e} \frac{\partial^2 v}{\partial z_p \, \partial z_j}(0) + \left(\frac{1}{2} (\tilde{\theta}_{r,p}^{a,e})^2 + \frac{1}{3!} a_p^2 \frac{G_1(q_p)}{G_0(q_p)} \right) \frac{\partial^3 v}{\partial z_p^3}(0)$$

$$+ \tilde{\theta}_{r,p}^{a,e} \sum_{j \neq p}^{n} \tilde{\theta}_{r,j}^{a,e} \frac{\partial^3 v}{\partial z_p^2 \, \partial z_j}(0) + \sum_{j \neq p}^{n} \frac{(\tilde{\theta}_{r,j}^{a,e})^2 + a_j^2 G_0(q_j)}{2} \frac{\partial^3 v}{\partial z_p \, \partial z_j^2}(0)$$

$$+ \sum_{j \neq p, k > j}^{n} \sum_{k \neq p}^{n} \tilde{\theta}_{r,j}^{a,e} \tilde{\theta}_{r,k}^{a,e} \frac{\partial^3 v}{\partial z_p \, \partial z_j \, \partial z_k}(0) + O(|a|^3). \qquad (11.84)$$

Substituting (11.81) in (11.84) and matching first order powers of a_i gives

$$\begin{bmatrix} 0 \\ \vdots \\ 0 \end{bmatrix} = H \begin{bmatrix} b_i^1 \\ \vdots \\ b_i^n \end{bmatrix}, \quad i = 1, \ldots, n, \qquad (11.85)$$

which implies that $b_j^i = 0$ for all i, j since H is negative definite (thus nonsingular). Similarly, matching second order term $a_j a_k$ $(j > k)$ and a_j^2 of a_j, and substituting b_j^i to simplify the resulting expressions, yields

$$\begin{bmatrix} 0 \\ \vdots \\ 0 \end{bmatrix} = H \begin{bmatrix} c_{jk}^1 \\ \vdots \\ c_{jk}^n \end{bmatrix}, \quad j = 1, \ldots, n, j > k, \qquad (11.86)$$

and

$$
\begin{pmatrix}
\begin{bmatrix} 0 \\ \vdots \\ 0 \end{bmatrix}
= H
\begin{bmatrix} c_{jj}^{1} \\ \vdots \\ c_{jj}^{n} \end{bmatrix}
+
\begin{bmatrix}
\frac{1}{2} G_0(q_j) \frac{\partial^3 v}{\partial z_1 \partial z_j^2}(0) \\
\vdots \\
\frac{1}{2} G_0(q_j) \frac{\partial^3 v}{\partial z_{j-1} \partial z_j^2}(0) \\
\frac{1}{6} \frac{G_1(q_j)}{G_0(q_j)} \frac{\partial^3 v}{\partial z_j^3}(0) \\
\frac{1}{2} G_0(q_j) \frac{\partial^3 v}{\partial z_j^2 \partial z_{j+1}}(0) \\
\vdots \\
\frac{1}{2} G_0(q_j) \frac{\partial^3 v}{\partial z_j^2 \partial z_n}(0)
\end{bmatrix}
\end{pmatrix}.
\tag{11.87}
$$

Thus $c_{jk}^{i} = 0$ for all i, j, k when $j \neq k$, and c_{jj}^{i} is given by

$$
\begin{bmatrix} c_{jj}^{1} \\ \vdots \\ c_{jj}^{i-1} \\ c_{jj}^{i} \\ c_{jj}^{i+1} \\ \vdots \\ c_{jj}^{n} \end{bmatrix}
= -H^{-1}
\begin{bmatrix}
\frac{1}{2} G_0(q_j) \frac{\partial^3 v}{\partial z_1 \partial z_j^2}(0) \\
\vdots \\
\frac{1}{2} G_0(q_j) \frac{\partial^3 v}{\partial z_{j-1} \partial z_j^2}(0) \\
\frac{1}{6} \frac{G_1(q_j)}{G_0(q_j)} \frac{\partial^3 v}{\partial z_j^3}(0) \\
\frac{1}{2} G_0(q_j) \frac{\partial^3 v}{\partial z_j^2 \partial z_{j+1}}(0) \\
\vdots \\
\frac{1}{2} G_0(q_j) \frac{\partial^3 v}{\partial z_j^2 \partial z_n}(0)
\end{bmatrix}
\qquad \forall i, j \in \{1, 2, \ldots, n\}.
\tag{11.88}
$$

Thus

$$
\tilde{\theta}_{r,i}^{a,e} = \sum_{j=1}^{n} c_{jj}^{i} a_j^2 + O\big(|a|^3\big).
\tag{11.89}
$$

By (11.100), we have

$$
\tilde{\xi}_r^{a,e} = \int_{\mathbb{R}^n} v\big(\tilde{\theta}_r^{a,e} + S(\sigma)\big) \mu_1(d\sigma_1) \times \cdots \times \mu_n(d\sigma_n)
$$

$$
= \int_{\mathbb{R}^n} \Bigg[\sum_{i=1}^{n} \sum_{j=1}^{n} \frac{1}{2} \frac{\partial^2 v}{\partial z_i \, \partial z_j}(0) \big(\tilde{\theta}_{r,i}^{a,e} + a_i \sin(\sigma_i)\big)\big(\tilde{\theta}_{r,j}^{a,e} + a_j \sin(\sigma_j)\big)
$$

$$
+ \sum_{i=1}^{n} \sum_{j=1}^{n} \sum_{k=1}^{n} \frac{1}{3!} \frac{\partial^3 v}{\partial z_i \, \partial z_j \, \partial z_k}(0) \big(\tilde{\theta}_{r,i}^{a,e} + a_i \sin(\sigma_i)\big)\big(\tilde{\theta}_{r,j}^{a,e} + a_j \sin(\sigma_j)\big)
$$

$$
\times \big(\tilde{\theta}_{r,k}^{a,e} + a_k \sin(\sigma_k)\big) + O\big(|a|^4\big) \Bigg] \mu_1(d\sigma_1) \times \cdots \times \mu_n(d\sigma_n)
$$

$$
= \sum_{i=1}^{n} \sum_{j=1}^{n} \frac{1}{2} \frac{\partial^2 v}{\partial z_i \, \partial z_j}(0) \big(\tilde{\theta}_{r,i}^{a,e} \tilde{\theta}_{r,j}^{a,e} + a_i^2 G_0(q_i)\big)
$$

$$+ \sum_{i=1}^{n} \sum_{j=1}^{n} \sum_{k=1}^{n} \frac{1}{3!} \frac{\partial^3 v}{\partial z_i \, \partial z_j \, \partial z_k}(0) \tilde{\theta}_{r,i}^{a,e} \tilde{\theta}_{r,j}^{a,e} \tilde{\theta}_{r,k}^{a,e}$$

$$+ \sum_{i=1}^{n} \sum_{j=k}^{n} \frac{1}{3!} \frac{\partial^3 v}{\partial z_i \, \partial z_j^2}(0) \tilde{\theta}_{r,i}^{a,e} a_j^2 G_0(q_j) + O\left(|a|^4\right). \tag{11.90}$$

This together with (11.89) gives

$$\zeta_r^{a,e} = \frac{1}{2} \sum_{i=1}^{n} H_{ii} a_i^2 G_0(q_i) + O\left(|a|^4\right). \tag{11.91}$$

By (11.101), we have

$$\left(\tilde{H}_r^{a,e}\right)_{pp} = \int_{\mathbb{R}^n} v\left(\tilde{\theta}_r^{a,e} + S(\sigma)\right) \left(N(\sigma)\right)_{pp} \mu_1(d\sigma_1) \times \cdots \times \mu_n(d\sigma_n)$$

$$= \int_{\mathbb{R}^n} \Bigg[\sum_{i=1}^{n} \sum_{j=1}^{n} \frac{1}{2} \frac{\partial^2 v}{\partial z_i \, \partial z_j}(0) \left(\tilde{\theta}_{r,i}^{a,e} + a_i \sin(\sigma_i)\right) \left(\tilde{\theta}_{r,j}^{a,e} + a_j \sin(\sigma_j)\right)$$

$$+ \sum_{i=1}^{n} \sum_{j=1}^{n} \sum_{k=1}^{n} \frac{1}{3!} \frac{\partial^3 v}{\partial z_i \, \partial z_j \, \partial z_k}(0) \left(\tilde{\theta}_{r,i}^{a,e} + a_i \sin(\sigma_i)\right) \left(\tilde{\theta}_{r,j}^{a,e} + a_j \sin(\sigma_j)\right)$$

$$\times \left(\tilde{\theta}_{r,k}^{a,e} + a_k \sin(\sigma_k)\right) + O\left(|a|^4\right) \Bigg] \frac{4}{a_p^2 G_0^2(\sqrt{2}q_p)} \left(\sin^2(\sigma_p) - G_0(q_p)\right)$$

$$\times \mu_1(d\sigma_1) \times \cdots \times \mu_n(d\sigma_n) - (H)_{pp}$$

$$= \int_{\mathbb{R}^n} \frac{1}{2} \frac{\partial^2 v}{\partial z_p^2}(0) a_p^2 \sin^2(\sigma_p) \frac{4}{a_p^2 G_0^2(\sqrt{2}q_p)} \left(\sin^2(\sigma_p) - G_0(q_p)\right)$$

$$+ \int_{\mathbb{R}^n} \sum_{i=p}^{n} \frac{1}{2} \frac{\partial^3 v}{\partial z_p^2}(0) \tilde{\theta}_{r,i}^{a,e} a_p^2 \sin^2(\sigma_p)$$

$$\times \frac{4}{a_p^2 G_0^2(\sqrt{2}q_p)} \left(\sin^2(\sigma_p) - G_0(q_p)\right) - (H)_{pp}$$

$$= (H)_{pp} + \sum_{i=1}^{n} \frac{\partial^3 v}{\partial z_i \, \partial z_p^2} \tilde{\theta}_{r,i}^{a,e} - (H)_{pp}$$

$$= \sum_{i=1}^{n} \frac{\partial^3 v}{\partial z_i \, \partial z_p^2} \tilde{\theta}_{r,i}^{a,e} \tag{11.92}$$

and

$$\left(\tilde{H}_r^{a,e}\right)_{pm} = \int_{\mathbb{R}^n} v\left(\tilde{\theta}_r^{a,e} + S(\sigma)\right) \left(N(\sigma)\right)_{pm} \mu_1(d\sigma_1) \times \cdots \times \mu_n(d\sigma_n)$$

$$= \int_{\mathbb{R}^n} \Bigg[\sum_{i=1}^{n} \sum_{j=1}^{n} \frac{1}{2} \frac{\partial^2 v}{\partial z_i \, \partial z_j}(0) \left(\tilde{\theta}_{r,i}^{a,e} + a_i \sin(\sigma_i)\right) \left(\tilde{\theta}_{r,j}^{a,e} + a_j \sin(\sigma_j)\right)$$

$$+ \sum_{i=1}^{n} \sum_{j=1}^{n} \sum_{k=1}^{n} \frac{1}{3!} \frac{\partial^3 v}{\partial z_i \, \partial z_j \, \partial z_k}(0)\big(\tilde{\theta}_{r,i}^{a,e} + a_i \sin(\sigma_i)\big)\big(\tilde{\theta}_{r,j}^{a,e} + a_j \sin(\sigma_j)\big)$$

$$\times \big(\tilde{\theta}_{r,k}^{a,e} + a_k \sin(\sigma_k)\big) + O\big(|a|^4\big) \bigg]$$

$$\times \frac{\sin(\sigma_p)\sin(\sigma_m)}{a_p a_m \, G_0(q_p) G_0(q_m)} \mu_1(d\sigma_1) \times \cdots \times \mu_n(d\sigma_n) - (H)_{pm}$$

$$= \int_{\mathbb{R}^n} \frac{\partial^2 v}{\partial z_p \, \partial z_m}(0) a_p a_m \sin(\sigma_p)\sin(\sigma_m)$$

$$\times \frac{\sin(\sigma_p)\sin(\sigma_m)}{a_p a_m \, G_0(q_p) G_0(q_m)} \mu_1(d\sigma_1) \times \cdots \times \mu_n(d\sigma_n)$$

$$+ \int_{\mathbb{R}^n} \sum_{i=1}^{n} \frac{1}{3!} \frac{\partial^3 v}{\partial z_i \, \partial z_p \, \partial z_m}(0) \tilde{\theta}_{r,i}^{a,e} a_m a_p \sin(\sigma_m)\sin(\sigma_p)$$

$$\times \frac{\sin(\sigma_p)\sin(\sigma_m)}{a_p a_m \, G_0(q_p) G_0(q_m)} \mu_1(d\sigma_1) \times \cdots \times \mu_n(d\sigma_n) - (H)_{pm}$$

$$= (H)_{pm} + \sum_{i=1}^{n} \frac{\partial^3 v}{\partial z_i \, \partial z_p \, \partial z_m}(0) \tilde{\theta}_{r,i}^{a,e} - (H)_{pm}$$

$$= \sum_{i=1}^{n} \frac{\partial^3 v}{\partial z_i \, \partial z_p \, \partial z_m}(0) \tilde{\theta}_{r,i}^{a,e}. \tag{11.93}$$

This, together with (11.89), gives

$$\tilde{H}_r^{a,e} = \sum_{i=1}^{n} \sum_{j=1}^{n} W^i c_{jj}^i a_j^2 + \big[O\big(|a|^3\big)\big]_{n \times n}, \tag{11.94}$$

where W^i is a $n \times n$ matrix defined by

$$\big(W^i\big)_{j,k} = \frac{\partial^3 v}{\partial z_i \, \partial z_j \, \partial z_k}(0) \quad \forall i, j, \text{ and } k \in \{1, 2, \ldots, n\}. \tag{11.95}$$

By (11.79), we have

$$\tilde{\Gamma}_r^{a,e} = \big(\tilde{H}_r^{a,e} + H\big)^{-1} - H^{-1}$$

$$= \big(H\big(H^{-1}\tilde{H}_r^{a,e} + I\big)\big)^{-1} - H^{-1}$$

$$= \big(H^{-1}\tilde{H}_r^{a,e} + I\big)^{-1} H^{-1} - H^{-1}$$

$$= \big(\big(H^{-1}\tilde{H}_r^{a,e} + I\big)^{-1} - I\big) H^{-1}$$

$$= \big(-H^{-1}\tilde{H}_r^{a,e} + \big(H^{-1}\tilde{H}_r^{a,e}\big)^2 - \big(H^{-1}\tilde{H}_r^{a,e}\big)^3 + \cdots\big) H^{-1}. \tag{11.96}$$

This, together with (11.94), gives that

$$\tilde{\Gamma}_r^{a,e} = -\sum_{i=1}^{n}\sum_{j=1}^{n} H^{-1} W^i H^{-1} c_{jj}^i a_j^2 + \left[O\left(|a|^3\right)\right]_{n\times n}. \tag{11.97}$$

Thus by (11.89), (11.76), (11.97), (11.94), and (11.91), the equilibrium of the average system is

$$\hat{\theta}_{r,i}^{a,e} = \sum_{j=1}^{n} c_{jj}^i a_j^2 + O\left(|a|^3\right), \tag{11.98}$$

$$\hat{G}_r^{a,e} = 0_{n\times 1}, \tag{11.99}$$

$$\tilde{\Gamma}_r^{a,e} = -\sum_{i=1}^{n}\sum_{j=1}^{n} H^{-1} W^i H^{-1} c_{jj}^i a_j^2 + \left[O\left(|a|^3\right)\right]_{n\times n}, \tag{11.100}$$

$$\tilde{H}_r^{a,e} = \sum_{i=1}^{n}\sum_{j=1}^{n} W^i c_{jj}^i a_j^2 + \left[O\left(|a|^3\right)\right]_{n\times n}, \tag{11.101}$$

$$\tilde{\zeta}_r^{a,e} = \frac{1}{2}\sum_{i=1}^{n} H_{ii} a_i^2 G_0(q_i) + O\left(|a|^4\right). \tag{11.102}$$

The Jacobian of the average system (11.74) at the equilibrium is

$$J_r^{a,e} = \begin{bmatrix} A_{2n\times 2n} & 0_{2n\times(2n+1)} \\ B_{(2n+1)\times 2n} & C_{(2n+1)\times(2n+1)} \end{bmatrix}, \tag{11.103}$$

$$A = \begin{bmatrix} 0_{n\times n} & -K(H^{-1}+\tilde{\Gamma}_r^{\tilde{a},e}) \\ h_1 \int_{\mathbb{R}^n} \frac{\partial}{\partial\theta}(vM(\sigma))\mu(d\sigma) & -h_1 I_{n\times n} \end{bmatrix}, \tag{11.104}$$

$$B = \begin{bmatrix} 0_{n\times n} & 0_{n\times n} \\ h_1 \int_{\mathbb{R}^n} \frac{\partial}{\partial\theta}(vN(\sigma))\mu(d\sigma) & 0_{n\times n} \\ h_2 \int_{\mathbb{R}^n} \frac{\partial}{\partial\theta}(v)\mu(d\sigma) & 0_{1\times n} \end{bmatrix}, \tag{11.105}$$

$$C = \begin{bmatrix} -h_0 I_{n\times n} + O_1 & -h_0 H^{-2} + O_2 & 0_{n\times 1} \\ 0_{n\times n} & -h_1 I_{n\times n} & 0_{n\times 1} \\ 0_{1\times n} & 0_{1\times n} & -h_2 \end{bmatrix}, \tag{11.106}$$

$$O_1 = h_0 \sum_{i=1}^{n}\sum_{j=1}^{n} H^{-1} W^i c_{jj}^i a_j^2 + \left[O\left(|a|^3\right)\right]_{n\times n}, \tag{11.107}$$

$$O_2 = h_0 \sum_{i=1}^{n}\sum_{j=1}^{n} H^{-1}\left(W^i H^{-1} - H^{-1} W^i\right) H^{-1} c_{jj}^i a_j^2$$
$$+ \left[O\left(|a|^3\right)\right]_{n\times n}, \tag{11.108}$$

where $\mu(d\sigma) \triangleq \mu_1(d\sigma_1) \times \cdots \times \mu_n(d\sigma_n)$. Since $J_r^{a,e}$ is block-lower-triangular, it is Hurwitz if and only if

$$A_{21} = h_1 \int_{\mathbb{R}^n} M(\sigma)\frac{\partial}{\partial\theta} v(\tilde{\theta}_r^{a,e} + S(\sigma))\mu_1(d\sigma_1) \times \cdots \times \mu_n(d\sigma_n) < 0. \tag{11.109}$$

With a Taylor expansion we get that $A_{21} = h_1 H + [O(|a|)]_{n \times n}$. Hence we have

$$
\begin{aligned}
\det(\lambda I_{2n \times 2n} - A) \\
&= \det\left(\lambda(\lambda + h_1)I_{n \times n} + K\left(H^{-1} + \tilde{\Gamma}_r^{a,e}\right)A_{21}\right) \\
&= \det\left(\left(\lambda^2 + h_1\lambda\right)I_{n \times n} + K\left(H^{-1} + [O(|a|^2)]_{n \times n}\right)\right. \\
&\quad \times \left. \left(h_1 H + [O(|a|)]_{n \times n}\right)\right) \\
&= \det\left(\left(\lambda^2 + h_1\lambda\right)I_{n \times n} + h_1 K + [O(|a|)]_{n \times n}\right),
\end{aligned}
\tag{11.110}
$$

which, in view of $H < 0$, proves that $J_r^{a,e}$ is Hurwitz for a that is sufficiently small in norm. This implies that the equilibrium (11.98)–(11.102) of the average system (11.74) is exponentially stable if all elements of vector a are sufficiently small. Then according to the multi-input stochastic average theorem given in Theorem 8.1, we have the following result.

Theorem 11.3 *Consider the reduced system (11.73). Then there exist $a^* > 0$ such that for all $|a| \in (0, a^*)$, there exist constants $r > 0$, $c > 0$, $\gamma > 0$ and a function $T(\varepsilon_1) : (0, \varepsilon_0) \to \mathbb{N}$ such that for any initial condition $|\Lambda_2^{\varepsilon_1}(0)| < r$ and any $\delta > 0$,*

$$
\lim_{\varepsilon_1 \to 0} \inf\left\{t \geq 0 : \left|\Lambda_2^{\varepsilon_1}(t)\right| > c\left|\Lambda_2^{\varepsilon_1}(0)\right|e^{-\gamma t} + \delta + O(|a|^3)\right\} = \infty, \quad a.s. \tag{11.111}
$$

and

$$
\lim_{\varepsilon_1 \to 0} P\left\{\left|\Lambda_2^{\varepsilon_1}(t)\right| \leq c\left|\Lambda_2^{\varepsilon_1}(0)\right|e^{-\gamma t} + \delta + O(|a|^3), \forall t \in [0, T(\varepsilon_1)]\right\} = 1, \quad with
$$

$$
\lim_{\varepsilon_1 \to 0} T(\varepsilon_1) = \infty, \tag{11.112}
$$

where

$$
\Lambda_2^{\varepsilon_1}(t) \triangleq \left(\tilde{\theta}_r^T(t), \hat{G}_r^T(t), \mathfrak{Vec}\left(\tilde{\Gamma}_r(t)\right), \mathfrak{Vec}\left(\tilde{H}_r(t)\right), \tilde{\zeta}_r(t)\right)^T
$$

$$
- \left(\sum_{j=1}^n c_{jj}a_j^2, 0_{n \times 1}^T, \mathfrak{Vec}\left(-\sum_{i=1}^n \sum_{j=1}^n H^{-1}W^i H^{-1}c_{jj}^i a_j^2\right),\right.
$$

$$
\left.\mathfrak{Vec}\left(\sum_{i=1}^n \sum_{j=1}^n W^i c_{jj}^i a_j^2\right), \frac{1}{2}\sum_{i=1}^n H_{ii}G_0(q_i)a_i^2\right)^T,
$$

$$
\sum_{j=1}^n c_{jj}a_j^2 \triangleq \left(\sum_{j=1}^n c_{jj}^1 a_j^2, \ldots, \sum_{j=1}^n c_{jj}^n a_j^2\right).
$$

11.4 Simulation

To illustrate the results, we consider the static quadratic input-output map:

$$
y = f(\theta) = f^* + \frac{1}{2}\left(\theta - \theta^*\right)^T H\left(\theta - \theta^*\right). \tag{11.113}
$$

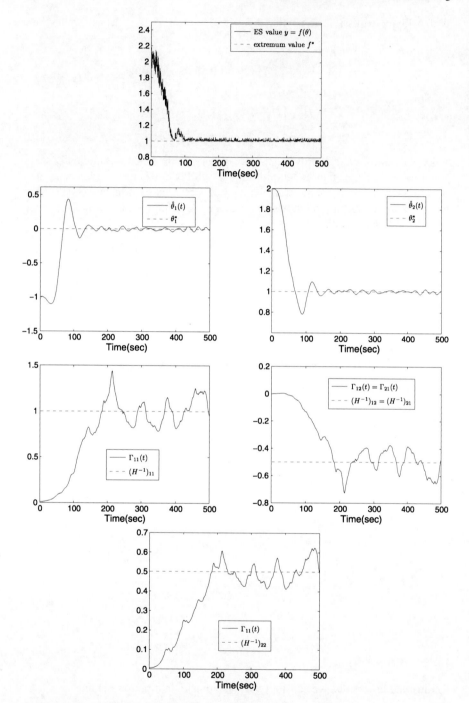

Fig. 11.5 Newton-based stochastic extremum seeking. *Top*: output and extremum values. *Others*: estimate values

Figure 11.5 displays the simulation results with $f^* = 1$, $\theta^* = [0, 1]^T$, $H = \begin{bmatrix} 2 & 2 \\ 2 & 4 \end{bmatrix}$ in the static map (11.47) and $a_1 = 0.1$, $a_2 = 0.1$, $k_1 = 1$, $k_2 = 1$, $h_0 = 0.1$, $h_1 = 0.08$, $h_2 = 0.08$, $q_1 = q_2 = 40$, $\varepsilon_1 = 0.25$, $\varepsilon_2 = 0.01$ in the parameter update law (11.43)–(11.44) and initial condition $\tilde{\theta}_1(0) = 1$, $\tilde{\theta}_2(0) = -1$, $\hat{\theta}_1(0) = -1$, $\hat{\theta}_2(0) = 2$, $\Gamma_{11}(0) = 1/100$, $\Gamma_{22}(0) = 1/200$, $\Gamma_{12}(0) = \Gamma_{21}(0) = 0$.

Comparing Fig. 11.5 with Fig. 8.1, we see that Newton-based stochastic extremum seeking converges faster than gradient-based stochastic extremum seeking by choosing proper design parameters. Note that it was necessary, for the gradient-based simulation in Fig. 8.1, to use gains that are different for the different components of the θ vector (with a gain ratio $k_1/k_2 = 3/4$) to achieve balanced convergence between $\hat{\theta}_1$ and $\hat{\theta}_2$. In Fig. 11.5, the Newton algorithm achieves balanced convergence automatically.

11.5 Notes and References

A Newton-based extremum seeking algorithm was introduced in [103] where, for the single-input case, an estimate of the second derivative of the map was employed in a Newton-like continuous-time algorithm. A generalization employing a different approach than in [103] was presented in [108] where a methodology for generating estimates of higher-order derivatives of the unknown single-input map was introduced for emulating more general continuous-time optimization algorithms, with a Newton algorithm being a special case.

The power of the Newton algorithm is particularly evident in multi-input optimization problems. With the Hessian being a matrix, and with it being typically very different from the identity matrix, the gradient algorithm typically results in different elements of the input vector converging at vastly different speeds. The Newton algorithm, when equipped with a convergent estimator of the Hessian matrix, achieves convergence of all the elements of the input vector at the same, or at arbitrarily assignable, rates.

In this chapter, we generate an estimate of the Hessian matrix by generalizing the idea proposed in [108] for the scalar sinusoid-perturbed case to the multivariable stochastically-perturbed case.

The stochastic continuous-time Newton algorithm that we propose is novel, to our knowledge, even in the case when the cost function being optimized is known. The state-of-the-art continuous-time Newton algorithm in [3] employs a Lyapunov differential equation for estimating the inverse of the Hessian matrix; see (3.2) in [3]. The convergence of this estimator is actually governed by the Hessian matrix itself. This means that the algorithm in [3] removes the difficulty of inverting the estimate of the Hessian, but does not achieve independence of the convergence rate from the Hessian. In contrast, our algorithm's convergence rate is independent of the Hessian and is user-assignable.

This chapter parallels the deterministic Newton-based extremum seeking development in [45].

Appendix A
Some Properties of p-Limit and p-Infinitesimal Operator

Let $\mathcal{F}_t^\varepsilon = \sigma\{X_s^\varepsilon, Y_{s/\varepsilon}, 0 \le s \le t\} = \sigma\{Y_{s/\varepsilon}, 0 \le s \le t\} = \sigma\{Y_s, 0 \le s \le \frac{t}{\varepsilon}\}$, and E_t^ε denote the expectation conditioning on $\mathcal{F}_t^\varepsilon$. Let $\mathcal{M}^\varepsilon$ be the linear space of real-valued processes $f(t, \omega) \triangleq f(t)$ progressively measurable with respect to $\{\mathcal{F}_t^\varepsilon\}$ such that $f(t)$ has a finite expectation for all t, and $\overline{\mathcal{M}}^\varepsilon$ be one subspace of $\mathcal{M}^\varepsilon$ defined by $\overline{\mathcal{M}}^\varepsilon = \{f \in \mathcal{M}^\varepsilon : \sup_{t \ge 0} E|f(t)| < \infty\}$. A function f is said to be p-right continuous (or right continuous in the mean) if for each t,

$$E\big|f(t+\delta) - f(t)\big| \to 0 \quad \text{as } \delta \downarrow 0 \quad \text{and} \tag{A.1}$$

$$\sup_{t \ge 0} E\big|f(t)\big| < \infty. \tag{A.2}$$

Following [76, 116], we define the p-limit and the p-infinitesimal operator $\hat{\mathcal{A}}^\varepsilon$ as follows. Let $f, f^\delta \in \overline{\mathcal{M}}^\varepsilon$ for each $\delta > 0$. Then we say that $f = p\text{-}\lim_{\delta \to 0} f^\delta$ if

$$\sup_{t, \delta} E\big|f^\delta(t)\big| < \infty \quad \text{and} \tag{A.3}$$

$$\lim_{\delta \to 0} E\big|f^\delta(t) - f(t)\big| = 0 \quad \text{for each } t. \tag{A.4}$$

We say that $f \in \mathcal{D}(\hat{\mathcal{A}}^\varepsilon)$, the domain of $\hat{\mathcal{A}}^\varepsilon$, and $\hat{\mathcal{A}}^\varepsilon f = g$ if f and g are in $\overline{\mathcal{M}}^\varepsilon$, and

$$p\text{-}\lim_{\delta \to 0} \frac{E_t^\varepsilon[f(t+\delta)] - f(t)}{\delta} = g(t). \tag{A.5}$$

For our needs, the most useful properties of $\hat{\mathcal{A}}^\varepsilon$ are given by the following theorem:

Theorem A.1 ([76]) *Let* $f(\cdot) \in \mathcal{D}(\hat{\mathcal{A}}^\varepsilon)$. *Then*

$$M_\varepsilon^f(t) = f(t) - f(0) - \int_0^t \hat{\mathcal{A}}^\varepsilon f(u)\,du \tag{A.6}$$

S.-J. Liu, M. Krstic, *Stochastic Averaging and Stochastic Extremum Seeking*, Communications and Control Engineering, DOI 10.1007/978-1-4471-4087-0, © Springer-Verlag London 2012

is a zero-mean martingale with respect to $\{\mathcal{F}_t^\varepsilon\}$, *and*

$$E_t^\varepsilon[f(t+s)] - f(t) = \int_t^{t+s} E_t^\varepsilon[\hat{A}^\varepsilon f(u)] du \quad a.s.$$

Furthermore, if τ *and* σ *are bounded* $\{\mathcal{F}_t^\varepsilon\}$ *stopping times and each takes only countably many values and* $\sigma \geq \tau$, *then*

$$E_\tau^\varepsilon[f(\sigma)] - f(\tau) = E_\tau^\varepsilon\left[\int_\tau^\sigma \hat{A}^\varepsilon f(u) du\right]. \tag{A.7}$$

If $f(\cdot)$ *is right continuous almost surely, we can drop the "countability" requirement.*

Appendix B
Auxiliary Proofs for Section 3.2.2

Lemma B.1 ([77, Lemma 4.4]) *Let $\xi(\cdot)$ be a ϕ-mixing process. Let $\mathcal{F}_0^t = \sigma\{\xi(s) : 0 \le s \le t\}$, $\mathcal{F}_t^\infty = \sigma\{\xi(s) : s \ge t\}$. Suppose that $h(t)$ is bounded with bound $K > 0$, measurable on $\mathcal{F}_t^\infty$. Then*

$$\left| E\left[h(t+s)|\mathcal{F}_0^t\right] - E\left[h(t+s)\right] \right| \le K\phi(s). \tag{B.1}$$

Lemma B.2 $g_\delta^\varepsilon(t) \in \overline{\mathcal{M}_\delta^\varepsilon}$.

Proof By (3.98) and (3.99),

$$\tilde{G}(x, y) = G(x, y) - \bar{G}(x) = \left(\frac{\partial V(x)}{\partial x}\right)^T \left(a(x, y) - \bar{a}(x)\right). \tag{B.2}$$

Then we have that

$$\begin{aligned}
\frac{\partial \tilde{G}(x, y)}{\partial x} &= \left(\frac{\partial^2 V(x)}{\partial x^2}\right)^T \left(a(x, y) - \bar{a}(x)\right) \\
&\quad + \left(\frac{\partial a(x, y)}{\partial x} - \frac{\partial \bar{a}(x)}{\partial x}\right)^T \frac{\partial V(x)}{\partial x}.
\end{aligned} \tag{B.3}$$

By (B.3), (3.27), (3.90), (3.89), (3.23), and (3.26), we get that there exists $C_\delta > 0$ such that for any $x \in D_{\delta+1} = \{x' \in \mathbb{R}^n : |x'| \le \delta + 1\}$ and any $y \in S_Y$,

$$\left| \frac{\partial \tilde{G}(x, y)}{\partial x} \right| \le C_\delta. \tag{B.4}$$

First, we prove that for any $x = [x_1, \ldots, x_n] \in D_\delta$, $t \ge 0$, and $s \ge 0$,

$$\frac{\partial E_t^\varepsilon[\tilde{G}(x, Y_{s/\varepsilon})]}{\partial x} = E_t^\varepsilon\left[\frac{\partial \tilde{G}(x, Y_{s/\varepsilon})}{\partial x}\right]. \tag{B.5}$$

S.-J. Liu, M. Krstic, *Stochastic Averaging and Stochastic Extremum Seeking*, Communications and Control Engineering, DOI 10.1007/978-1-4471-4087-0, © Springer-Verlag London 2012

Without loss of generality, we only need to prove that

$$\frac{\partial E_t^\varepsilon[\tilde{G}(x, Y_{s/\varepsilon})]}{\partial x_1} = E_t^\varepsilon\left[\frac{\partial \tilde{G}(x, Y_{s/\varepsilon})}{\partial x_1}\right]. \tag{B.6}$$

The proofs about the partial derivatives with respect to $x_2, \ldots, x_n$ are similar. By linearity of conditional expectation, the differential mean value theorem, and the dominated convergence theorem for conditional expectation (cf. (B.4)), we obtain

$$\frac{\partial E_t^\varepsilon[\tilde{G}(x, Y_{s/\varepsilon})]}{\partial x_1}$$

$$= \lim_{\Delta x_1 \to 0} \frac{E_t^\varepsilon[\tilde{G}(x_1 + \Delta x_1, x_2, \ldots, x_n, Y_{s/\varepsilon})] - E_t^\varepsilon[\tilde{G}(x_1, x_2, \ldots, x_n, Y_{s/\varepsilon})]}{\Delta x_1}$$

$$= \lim_{\Delta x_1 \to 0} E_t^\varepsilon\left[\frac{\partial \tilde{G}}{\partial x_1}(x_1 + \theta \Delta x_1, x_2, \ldots, x_n, Y_{s/\varepsilon})\right] \quad \text{(where } 0 < \theta < 1)$$

$$= E_t^\varepsilon\left[\lim_{\Delta x_1 \to 0} \frac{\partial \tilde{G}}{\partial x_1}(x_1 + \theta \Delta x_1, x_2, \ldots, x_n, Y_{s/\varepsilon})\right]$$

$$= E_t^\varepsilon\left[\frac{\partial \tilde{G}}{\partial x_1}(x_1, x_2, \ldots, x_n, Y_{s/\varepsilon})\right], \tag{B.7}$$

i.e., $\frac{\partial E_t^\varepsilon[\tilde{G}(x, Y_{s/\varepsilon})]}{\partial x_1} = E_t^\varepsilon[\frac{\partial \tilde{G}(x, Y_{s/\varepsilon})}{\partial x_1}]$ holds. For simplicity, we denote

$$Q(x, y) = \left(\frac{\partial^2 V(x)}{\partial x^2}\right)^T a(x, y) + \left(\frac{\partial a(x, y)}{\partial x}\right)^T \frac{\partial V(x)}{\partial x}. \tag{B.8}$$

Then we have that

$$\int_{S_Y} Q(x, y)\mu(dy) = \left(\frac{\partial^2 V(x)}{\partial x^2}\right)^T \int_{S_Y} a(x, y)\mu(dy)$$

$$+ \left(\int_{S_Y} \frac{\partial a(x, y)}{\partial x}\mu(dy)\right)^T \frac{\partial V(x)}{\partial x}$$

$$= \left(\frac{\partial^2 V(x)}{\partial x^2}\right)^T \bar{a}(x) + \left(\frac{\partial \bar{a}(x)}{\partial x}\right)^T \frac{\partial V(x)}{\partial x}, \tag{B.9}$$

where in the last equality, we used

$$\int_{S_Y} \frac{\partial a(x, y)}{\partial x}\mu(dy) = \frac{\partial}{\partial x}\int_{S_Y} a(x, y)\mu(dy), \tag{B.10}$$

which can be proved by following the deduction in (B.7). By (B.8), (3.27), (3.90), (3.89), and (3.26), we get that for any $x \in \mathbb{R}^n$ with $|x| \leq \delta$, and $y \in S_Y$,

$$|Q(x, y)| \leq (c_3 + c_4)k_\delta|x|. \tag{B.11}$$

By (B.5), (B.3), (B.8), (B.9), the fact that $\mathcal{F}_t^\varepsilon = \mathcal{F}_{t/\varepsilon}^Y$, (B.11), Lemma B.1, (3.90), and (3.91), we obtain that for any $x \in D_\delta$,

$$\left| \int_{\tau_\delta^\varepsilon(t)}^{\tau_\delta^\varepsilon} \left[\frac{\partial E_t^\varepsilon[\tilde{G}(x, Y_{s/\varepsilon})]}{\partial x} \right]^T a(x, Y_{t/\varepsilon}) \, ds \right|$$

$$\leq \int_{\tau_\delta^\varepsilon(t)}^{\tau_\delta^\varepsilon} \left| \left[\frac{\partial E_t^\varepsilon[\tilde{G}(x, Y_{s/\varepsilon})]}{\partial x} \right]^T a(x, Y_{t/\varepsilon}) \right| ds$$

$$= \int_{\tau_\delta^\varepsilon(t)}^{\tau_\delta^\varepsilon} \left| E_t^\varepsilon \left[\frac{\partial \tilde{G}(x, Y_{s/\varepsilon})}{\partial x} \right]^T a(x, Y_{t/\varepsilon}) \right| ds$$

$$= \varepsilon \int_{\tau_\delta^\varepsilon(t)/\varepsilon}^{\tau_\delta^\varepsilon/\varepsilon} \left| E_t^\varepsilon \left[\frac{\partial \tilde{G}(x, Y_u)}{\partial x} \right]^T a(x, Y_{t/\varepsilon}) \right| du \quad \text{(by change of variable)}$$

$$= \varepsilon \int_{\tau_\delta^\varepsilon(t)/\varepsilon}^{\tau_\delta^\varepsilon/\varepsilon} \left| E_t^\varepsilon \left[Q(x, Y_u) - \int_{S_Y} Q(x, y)\mu(dy) \right] \right| |a(x, Y_{t/\varepsilon})| \, du$$

(by (B.3), (B.8), (B.9))

$$= \varepsilon \int_{\tau_\delta^\varepsilon(t)/\varepsilon}^{\tau_\delta^\varepsilon/\varepsilon} \left| E_t^\varepsilon \left[Q(x, Y_u) - \int_{S_Y} Q(x, y)\big(P_u(dy) - P_u(dy) + \mu(dy)\big) \right] \right|$$
$$\times |a(x, Y_{t/\varepsilon})| \, du$$

$$\leq \varepsilon \int_{\tau_\delta^\varepsilon(t)/\varepsilon}^{\tau_\delta^\varepsilon/\varepsilon} \left| E\big[Q(x, Y_u)|\mathcal{F}_t^\varepsilon\big] - E\big[Q(x, Y_u)\big] \right| |a(x, Y_{t/\varepsilon})| \, du$$

$$+ \varepsilon \int_{\tau_\delta^\varepsilon(t)/\varepsilon}^{\tau_\delta^\varepsilon/\varepsilon} \left| \int_{S_Y} Q(x, y)\big(P_u(dy) - \mu(dy)\big) \right| |a(x, Y_{t/\varepsilon})| \, du$$

$$\leq \varepsilon \int_{\tau_\delta^\varepsilon(t)/\varepsilon}^{\tau_\delta^\varepsilon/\varepsilon} \left| E\big[Q(x, Y_u)|\mathcal{F}_{t/\varepsilon}^Y\big] - E\big[Q(x, Y_u)\big] \right| |a(x, Y_{t/\varepsilon})| \, du$$

$$+ \varepsilon \int_{\tau_\delta^\varepsilon(t)/\varepsilon}^{\tau_\delta^\varepsilon/\varepsilon} \left| \int_{S_Y} Q(x, y)\big(P_u(dy) - \mu(dy)\big) \right| |a(x, Y_{t/\varepsilon})| \, du$$

$$\leq \varepsilon(c_3 + c_4)k_\delta|x| \cdot k_\delta|x| \int_{\tau_\delta^\varepsilon(t)/\varepsilon}^{\tau_\delta^\varepsilon/\varepsilon} \phi\left(u - \frac{\tau_\delta^\varepsilon(t)}{\varepsilon} \right) du$$

$$+ \varepsilon\sqrt{2c_5}(c_3 + c_4)k_\delta|x| \cdot k_\delta|x| \int_{\tau_M^\varepsilon(t)/\varepsilon}^{\tau_M^\varepsilon} e^{-(\alpha/2)u} \, du$$

$$\leq \varepsilon C_2(\delta)|x|^2 \quad \text{(see (3.92), (3.93), (3.94), (3.95))}, \tag{B.12}$$

where $C_2(\delta) = \frac{c_6(c_3+c_4)k_\delta^2}{\beta} + \frac{2\sqrt{2c_5}(c_3+c_4)k_\delta^2}{\alpha}$. Hence, by (3.101), (B.12), (3.98), (3.26), (3.90),

$$\sup_{t \geq 0} E\big[|g_\delta^\varepsilon(t)|\big] \leq \sup_{t \geq 0} E\big[I_{\{t < \tau_\delta^\varepsilon\}} \cdot \big(|\bar{G}(X_t^\varepsilon)| + \varepsilon C_2(\delta)|X_t^\varepsilon|^2\big)\big]$$

$$\leq \sup_{t \geq 0} E\left[\sup_{|x| \leq \delta}\left\{\left|\left(\frac{\partial V(x)}{\partial x}\right)^T \bar{a}(x)\right| + \varepsilon C_2(\delta)|x|^2\right\}\right]$$

$$\leq \sup_{|x| \leq \delta}\big\{c_3 k_\delta |x|^2 + \varepsilon C_2(\delta)|x|^2\big\}$$

$$\leq \big(c_3 k_\delta + \varepsilon C_2(\delta)\big)\delta^2$$

$$< \infty, \tag{B.13}$$

and thus $g_\delta^\varepsilon(t) \in \overline{\mathcal{M}}_\delta^\varepsilon$.   □

Lemma B.3

$$p\text{-}\lim_{\delta' \downarrow 0} \frac{E_t^\varepsilon[V^\varepsilon(X_{\tau_\delta^\varepsilon(t+\delta')}^\varepsilon, t+\delta')] - V^\varepsilon(X_{\tau_\delta^\varepsilon(t)}^\varepsilon, t)}{\delta'} = g_\delta^\varepsilon(t).$$

Proof We prove a stronger result

$$\lim_{\delta' \downarrow 0} \frac{E_t^\varepsilon[V^\varepsilon(X_{\tau_\delta^\varepsilon(t+\delta')}^\varepsilon, t+\delta')] - V^\varepsilon(X_{\tau_\delta^\varepsilon(t)}^\varepsilon, t)}{\delta'} = g_\delta^\varepsilon(t) \quad \text{a.s.,} \tag{B.14}$$

from which the statement of the lemma follows. Denote $(\frac{\partial V(x)}{\partial x})^T|_{x=X_{\tau_\delta^\varepsilon(t)}^\varepsilon}$ by $V_x^T(X_{\tau_\delta^\varepsilon(t)}^\varepsilon)$. By (3.87), (3.88), (B.2), and the definition of $V^\varepsilon(X_{\tau_\delta^\varepsilon(t)}^\varepsilon, t)$, the property of conditional expectation, we have that

$$\frac{E_t^\varepsilon[V^\varepsilon(X_{\tau_\delta^\varepsilon(t+\delta')}^\varepsilon, t+\delta')] - V^\varepsilon(X_{\tau_\delta^\varepsilon(t)}^\varepsilon, t)}{\delta'}$$

$$= \frac{1}{\delta'}\left\{E_t^\varepsilon\left[V(X_{\tau_\delta^\varepsilon(t+\delta')}^\varepsilon)\right.\right.$$

$$\left. + \int_{\tau_\delta^\varepsilon(t+\delta')}^{\tau_\delta^\varepsilon} V_x(X_{\tau_\delta^\varepsilon(t+\delta')}^\varepsilon)E_{t+\delta'}^\varepsilon\big[a(X_{\tau_\delta^\varepsilon(t+\delta')}^\varepsilon, Y_{s/\varepsilon}) - \bar{a}(X_{\tau_\delta^\varepsilon(t+\delta')}^\varepsilon)\big]ds\right]$$

$$\left. - \left[V(X_{\tau_\delta^\varepsilon(t)}^\varepsilon) + \int_{\tau_\delta^\varepsilon(t)}^{\tau_\delta^\varepsilon} V_x(X_{\tau_\delta^\varepsilon(t)}^\varepsilon)E_t^\varepsilon\big[a(X_{\tau_\delta^\varepsilon(t)}^\varepsilon, Y_{s/\varepsilon}) - \bar{a}(X_{\tau_\delta^\varepsilon(t)}^\varepsilon)\big]ds\right]\right\}$$

$$= \frac{1}{\delta'}\big\{E_t^\varepsilon\big[V(X_{\tau_\delta^\varepsilon(t+\delta')}^\varepsilon)\big] - V(X_{\tau_\delta^\varepsilon(t)}^\varepsilon)\big\}$$

$$- \frac{1}{\delta'}\int_{\tau_\delta^\varepsilon(t)}^{\tau_\delta^\varepsilon(t+\delta')} V_x(X_{\tau_\delta^\varepsilon(t)}^\varepsilon)E_t^\varepsilon\big[a(X_{\tau_\delta^\varepsilon(t)}^\varepsilon, Y_{s/\varepsilon}) - \bar{a}(X_{\tau_\delta^\varepsilon(t)}^\varepsilon)\big]ds$$

$$+ \frac{1}{\delta'} \int_{\tau_\delta^\varepsilon(t+\delta')}^{\tau_\delta^\varepsilon} \left\{ E_t^\varepsilon \left[V_x\left(X_{\tau_\delta^\varepsilon(t+\delta')}^\varepsilon\right)\left(a\left(X_{\tau_\delta^\varepsilon(t+\delta')}^\varepsilon, Y_{s/\varepsilon}\right) - \bar{a}\left(X_{\tau_\delta^\varepsilon(t+\delta')}^\varepsilon\right)\right) \right. \right.$$

$$\left. \left. - V_x\left(X_{\tau_\delta^\varepsilon(t)}^\varepsilon\right)\left(a\left(X_{\tau_\delta^\varepsilon(t)}^\varepsilon, Y_{s/\varepsilon}\right) - \bar{a}\left(X_{\tau_\delta^\varepsilon(t)}^\varepsilon\right)\right) \right] \right\} ds$$

$$= \frac{1}{\delta'}\left\{ E_t^\varepsilon\left[V\left(X_{\tau_\delta^\varepsilon(t+\delta')}^\varepsilon\right)\right] - V\left(X_{\tau_\delta^\varepsilon(t)}^\varepsilon\right) \right\}$$

$$- \frac{1}{\delta'} \int_{\tau_\delta^\varepsilon(t)}^{\tau_\delta^\varepsilon(t+\delta')} E_t^\varepsilon\left[\tilde{G}\left(X_{\tau_\delta^\varepsilon(t)}^\varepsilon, Y_{s/\varepsilon}\right)\right] ds$$

$$+ \frac{1}{\delta'} \int_{\tau_\delta^\varepsilon(t+\delta')}^{\tau_\delta^\varepsilon} E_t^\varepsilon\left[\tilde{G}\left(X_{\tau_\delta^\varepsilon(t+\delta')}^\varepsilon, Y_{s/\varepsilon}\right) - \tilde{G}\left(X_{\tau_\delta^\varepsilon(t)}^\varepsilon, Y_{s/\varepsilon}\right)\right] ds$$

$$\triangleq g_1^{\varepsilon,\delta}(t, \delta') - g_2^{\varepsilon,\delta}(t, \delta') + g_3^{\varepsilon,\delta}(t, \delta'). \tag{B.15}$$

Following the proof of (B.7), we get

$$\lim_{\delta'\downarrow 0} g_1^{\varepsilon,\delta}(t, \delta')$$

$$= \lim_{\delta'\downarrow 0} \frac{1}{\delta'}\left\{ E_t^\varepsilon\left[V\left(X_{\tau_\delta^\varepsilon(t+\delta')}^\varepsilon\right)\right] - V\left(X_{\tau_\delta^\varepsilon(t)}^\varepsilon\right) \right\}$$

$$= \lim_{\delta'\downarrow 0} E_t^\varepsilon\left[\frac{V_x^T(X_{\tau_\delta^\varepsilon(t)}^\varepsilon + \theta(X_{\tau_\delta^\varepsilon(t+\delta')}^\varepsilon - X_{\tau_\delta^\varepsilon(t)}^\varepsilon))(X_{\tau_\delta^\varepsilon(t+\delta')}^\varepsilon - X_{\tau_\delta^\varepsilon(t)}^\varepsilon)}{\delta'}\right]$$

$$= \lim_{\delta'\downarrow 0} E_t^\varepsilon\left[\frac{V_x^T(X_{\tau_\delta^\varepsilon(t)}^\varepsilon + \theta(X_{\tau_\delta^\varepsilon(t+\delta')}^\varepsilon - X_{\tau_\delta^\varepsilon(t)}^\varepsilon)) \int_{\tau_\delta^\varepsilon(t)}^{\tau_\delta^\varepsilon(t+\delta')} a(X_u^\varepsilon, Y_{u/\varepsilon}) du}{\delta'}\right]$$

$$= \lim_{\delta'\downarrow 0} E_t^\varepsilon\left[\frac{V_x^T(X_{\tau_\delta^\varepsilon(t)}^\varepsilon + \theta(X_{\tau_\delta^\varepsilon(t+\delta')}^\varepsilon - X_{\tau_\delta^\varepsilon(t)}^\varepsilon)) \int_t^{t+\delta'} a(X_u^\varepsilon, Y_{u/\varepsilon})I_{\{u<\tau_\delta^\varepsilon\}} du}{\delta'}\right]$$

$$= V_x^T\left(X_{\tau_\delta^\varepsilon(t)}^\varepsilon\right)a\left(X_t^\varepsilon, Y_{t/\varepsilon}\right) \cdot I_{\{t<\tau_\delta^\varepsilon\}}$$

$$= V_x^T\left(X_t^\varepsilon\right)a\left(X_t^\varepsilon, Y_{t/\varepsilon}\right) \cdot I_{\{t<\tau_\delta^\varepsilon\}} \quad \text{a.s.,} \tag{B.16}$$

$$\lim_{\delta'\downarrow 0} g_2^{\varepsilon,\delta}(t, \delta')$$

$$= \lim_{\delta'\downarrow 0} \frac{1}{\delta'} \int_{\tau_\delta^\varepsilon(t)}^{\tau_\delta^\varepsilon(t+\delta')} E_t^\varepsilon\left[\tilde{G}\left(X_{\tau_\delta^\varepsilon(t)}^\varepsilon, Y_{s/\varepsilon}\right)\right] ds$$

$$= \lim_{\delta'\downarrow 0} \frac{1}{\delta'} \int_{\tau_\delta^\varepsilon \wedge t}^{\tau_\delta^\varepsilon \wedge(t+\delta')} E_t^\varepsilon\left[\tilde{G}\left(X_{\tau_\delta^\varepsilon(t)}^\varepsilon, Y_{s/\varepsilon}\right)\right] ds$$

$$= \lim_{\delta'\downarrow 0} \frac{1}{\delta'} \int_t^{t+\delta'} E_t^\varepsilon\left[\tilde{G}\left(X_{\tau_\delta^\varepsilon(t)}^\varepsilon, Y_{s/\varepsilon}\right)I_{\{s<\tau_\delta^\varepsilon\}}\right] ds$$

$$= \tilde{G}\big(X^\varepsilon_{\tau^\varepsilon_\delta(t)}, Y_{t/\varepsilon}\big)I_{\{t<\tau^\varepsilon_\delta\}}$$

$$= \tilde{G}\big(X^\varepsilon_t, Y_{t/\varepsilon}\big)I_{\{t<\tau^\varepsilon_\delta\}} \quad \text{a.s.} \tag{B.17}$$

Following the proof of (B.16) and by (B.5), we get that

$$\lim_{\delta'\downarrow 0} g_3^{\varepsilon,\delta}(t,\delta)$$

$$= \lim_{\delta'\downarrow 0} \frac{1}{\delta'} \int_{\tau^\varepsilon_\delta(t+\delta')}^{\tau^\varepsilon_\delta} E^\varepsilon_t\Big[\tilde{G}\big(X^\varepsilon_{\tau^\varepsilon_\delta(t+\delta')}, Y_{s/\varepsilon}\big) - \tilde{G}\big(X^\varepsilon_{\tau^\varepsilon_\delta(t)}, Y_{s/\varepsilon}\big)\Big]ds$$

$$= \lim_{\delta'\downarrow 0} \int_{\tau^\varepsilon_\delta(t+\delta')}^{\tau^\varepsilon_\delta} E^\varepsilon_t\left[\frac{\tilde{G}(X^\varepsilon_{\tau^\varepsilon_\delta(t+\delta')}, Y_{s/\varepsilon}) - \tilde{G}(X^\varepsilon_{\tau^\varepsilon_\delta(t)}, Y_{s/\varepsilon})}{\delta'}\right]ds$$

$$= \lim_{\delta'\downarrow 0} \int_{\tau^\varepsilon_\delta(t+\delta')}^{\tau^\varepsilon_\delta} E^\varepsilon_t\Big[\tilde{G}^T_x\big(X^\varepsilon_{\tau^\varepsilon_\delta(t)} + \theta\big(X^\varepsilon_{\tau^\varepsilon_\delta(t+\delta')} - X^\varepsilon_{\tau^\varepsilon_\delta(t)}\big), Y_{s/\varepsilon}\big)$$

$$\times \big(X^\varepsilon_{\tau^\varepsilon_\delta(t+\delta')} - X^\varepsilon_{\tau^\varepsilon_\delta(t)}\big)/\delta'\Big]ds$$

$$= \lim_{\delta'\downarrow 0} \int_{\tau^\varepsilon_\delta(t+\delta')}^{\tau^\varepsilon_\delta} E^\varepsilon_t\left[\frac{\tilde{G}^T_x(X^\varepsilon_{\tau^\varepsilon_\delta(t)} + \theta(X^\varepsilon_{\tau^\varepsilon_\delta(t+\delta')} - X^\varepsilon_{\tau^\varepsilon_\delta(t)}), Y_{s/\varepsilon})}{\delta'}\right.$$

$$\left.\times \int_t^{t+\delta'} a\big(X^\varepsilon_u, Y_{u/\varepsilon}\big)I_{\{u<\tau^\varepsilon_\delta\}}\,du\right]ds$$

$$= \int_{\tau^\varepsilon_\delta(t)}^{\tau^\varepsilon_\delta} E^\varepsilon_t\Big[\tilde{G}^T_x\big(X^\varepsilon_t, Y_{s/\varepsilon}\big)a\big(X^\varepsilon_t, Y_{t/\varepsilon}\big)I_{\{t<\tau^\varepsilon_\delta\}}\Big]ds$$

$$= I_{\{t<\tau^\varepsilon_\delta\}} \int_{\tau^\varepsilon_\delta(t)}^{\tau^\varepsilon_\delta} \left[\frac{\partial E^\varepsilon_t[\tilde{G}(x, Y_{s/\varepsilon})]}{\partial x}\bigg|_{x=X^\varepsilon_t}\right]^T a\big(X^\varepsilon_t, Y_{t/\varepsilon}\big)ds \quad \text{a.s.,} \tag{B.18}$$

which, together with (B.15), (B.17), (3.98), and (3.101), implies that (B.14) holds. □

Lemma B.4 $\hat{A}^\varepsilon_M\big(V^\varepsilon(X^\varepsilon_{\tau^\varepsilon_M(t)}, t) \cdot I_{\{t<\tau^\varepsilon_M\}}\big) = g^\varepsilon_M(t)$, i.e.,

$$p\text{-}\lim_{\delta\downarrow 0} \frac{E^\varepsilon_t[V^\varepsilon(X^\varepsilon_{\tau^\varepsilon_M(t+\delta)}, t+\delta) \cdot I_{\{t+\delta<\tau^\varepsilon_M\}}] - V^\varepsilon(X^\varepsilon_{\tau^\varepsilon_M(t)}, t) \cdot I_{\{t<\tau^\varepsilon_M\}}}{\delta} = g^\varepsilon_M(t).$$

Proof As in the proof of Lemma B.3, we prove

$$\lim_{\delta\downarrow 0} \frac{E^\varepsilon_t[V^\varepsilon(X^\varepsilon_{\tau^\varepsilon_M(t+\delta)}, t+\delta)I_{\{t+\delta<\tau^\varepsilon_M\}}] - V^\varepsilon(X^\varepsilon_{\tau^\varepsilon_M(t)}, t)I_{\{t<\tau^\varepsilon_M\}}}{\delta}$$

$$= g^\varepsilon_M(t) \quad \text{a.s.} \tag{B.19}$$

Denote $(\frac{\partial V(x)}{\partial x})^T|_{x=X^\varepsilon_{\tau^\varepsilon_M(t)}}$ by $V_x^T(X^\varepsilon_{\tau^\varepsilon_M(t)})$. By the definition of $V^\varepsilon(X^\varepsilon_{\tau^\varepsilon_M(t)},t)$, following the proof of Lemma B.3, we get that

$$
\frac{E_t^\varepsilon[V^\varepsilon(X^\varepsilon_{\tau^\varepsilon_M(t+\delta)},t+\delta)I_{\{t+\delta<\tau^\varepsilon_M\}}]-V^\varepsilon(X^\varepsilon_{\tau^\varepsilon_M(t)},t)I_{\{t<\tau^\varepsilon_M\}}}{\delta}
$$

$$
=\frac{1}{\delta}\Bigg\{E_t^\varepsilon\Bigg[V\big(X^\varepsilon_{\tau^\varepsilon_M(t+\delta)}\big)I_{\{t+\delta<\tau^\varepsilon_M\}}
$$

$$
+I_{\{t+\delta<\tau^\varepsilon_M\}}\int_{\tau^\varepsilon_M(t+\delta)}^{\tau^\varepsilon_M}V_x\big(X^\varepsilon_{\tau^\varepsilon_M(t+\delta)}\big)
$$

$$
\times E_{t+\delta}^\varepsilon\big[a\big(X^\varepsilon_{\tau^\varepsilon_M(t+\delta)},Y_{s/\varepsilon}\big)-\bar{a}\big(X^\varepsilon_{\tau^\varepsilon_M(t+\delta)}\big)\big]ds\Bigg]
$$

$$
-\Bigg[V\big(X^\varepsilon_{\tau^\varepsilon_M(t)}\big)I_{\{t<\tau^\varepsilon_M\}}
$$

$$
+I_{\{t<\tau^\varepsilon_M\}}\int_{\tau^\varepsilon_M(t)}^{\tau^\varepsilon_M}V_x\big(X^\varepsilon_{\tau^\varepsilon_M(t)}\big)E_t^\varepsilon\big[a\big(X^\varepsilon_{\tau^\varepsilon_M(t)},Y_{s/\varepsilon}\big)-\bar{a}\big(X^\varepsilon_{\tau^\varepsilon_M(t)}\big)\big]ds\Bigg]\Bigg\}
$$

$$
=\frac{1}{\delta}\Bigg\{E_t^\varepsilon\Bigg[V\big(X^\varepsilon_{\tau^\varepsilon_M(t+\delta)}\big)I_{\{t+\delta<\tau^\varepsilon_M\}}
$$

$$
+\int_{\tau^\varepsilon_M(t+\delta)}^{\tau^\varepsilon_M}V_x\big(X^\varepsilon_{\tau^\varepsilon_M(t+\delta)}\big)E_{t+\delta}^\varepsilon\big[a\big(X^\varepsilon_{\tau^\varepsilon_M(t+\delta)},Y_{s/\varepsilon}\big)-\bar{a}\big(X^\varepsilon_{\tau^\varepsilon_M(t+\delta)}\big)\big]ds\Bigg]
$$

$$
-\Bigg[V\big(X^\varepsilon_{\tau^\varepsilon_M(t)}\big)I_{\{t<\tau^\varepsilon_M\}}
$$

$$
+\int_{\tau^\varepsilon_M(t)}^{\tau^\varepsilon_M}V_x\big(X^\varepsilon_{\tau^\varepsilon_M(t)}\big)E_t^\varepsilon\big[a\big(X^\varepsilon_{\tau^\varepsilon_M(t)},Y_{s/\varepsilon}\big)-\bar{a}\big(X^\varepsilon_{\tau^\varepsilon_M(t)}\big)\big]ds\Bigg]\Bigg\}
$$

$$
=\frac{1}{\delta}\Big\{E_t^\varepsilon\big[V\big(X^\varepsilon_{\tau^\varepsilon_M(t+\delta)}\big)\cdot I_{\{t+\delta<\tau^\varepsilon_M\}}\big]-V\big(X^\varepsilon_{\tau^\varepsilon_M(t)}\big)\cdot I_{\{t<\tau^\varepsilon_M\}}\Big\}
$$

$$
-\frac{1}{\delta}\int_{\tau^\varepsilon_M(t)}^{\tau^\varepsilon_M(t+\delta)}V_x\big(X^\varepsilon_{\tau^\varepsilon_M(t)}\big)E_t^\varepsilon\big[a\big(X^\varepsilon_{\tau^\varepsilon_M(t)},Y_{s/\varepsilon}\big)-\bar{a}\big(X^\varepsilon_{\tau^\varepsilon_M(t)}\big)\big]ds
$$

$$
+\frac{1}{\delta}\int_{\tau^\varepsilon_M(t+\delta)}^{\tau^\varepsilon_M}\Big\{E_t^\varepsilon\big[V_x\big(X^\varepsilon_{\tau^\varepsilon_M(t+\delta)}\big)\big(a\big(X^\varepsilon_{\tau^\varepsilon_M(t+\delta)},Y_{s/\varepsilon}\big)-\bar{a}\big(X^\varepsilon_{\tau^\varepsilon_M(t+\delta)}\big)\big)\big]
$$

$$
-V_x\big(X^\varepsilon_{\tau^\varepsilon_M(t)}\big)\big(a\big(X^\varepsilon_{\tau^\varepsilon_M(t)},Y_{s/\varepsilon}\big)-\bar{a}\big(X^\varepsilon_{\tau^\varepsilon_M(t)}\big)\big)\Big]\Big\}ds
$$

(by the property of conditional expectation)

$$
=\frac{1}{\delta}\Big\{E_t^\varepsilon\big[V\big(X^\varepsilon_{\tau^\varepsilon_M(t+\delta)}\big)\cdot I_{\{t+\delta<\tau^\varepsilon_M\}}\big]-V\big(X^\varepsilon_{\tau^\varepsilon_M(t)}\big)\cdot I_{\{t<\tau^\varepsilon_M\}}\Big\}
$$

$$-\frac{1}{\delta}\int_{\tau_M^\varepsilon(t)}^{\tau_M^\varepsilon(t+\delta)} E_t^\varepsilon\big[\tilde{G}\big(X_{\tau_M^\varepsilon(t)}^\varepsilon, Y_{s/\varepsilon}\big)\big]\,ds$$

$$+\frac{1}{\delta}\int_{\tau_M^\varepsilon(t+\delta)}^{\tau_M^\varepsilon} E_t^\varepsilon\big[\tilde{G}\big(X_{\tau_M^\varepsilon(t+\delta)}^\varepsilon, Y_{s/\varepsilon}\big) - \tilde{G}\big(X_{\tau_M^\varepsilon(t)}^\varepsilon, Y_{s/\varepsilon}\big)\big]\,ds$$

$$\triangleq \bar{g}_1^{\varepsilon,M}(t,\delta) - g_2^{\varepsilon,M}(t,\delta) + g_3^{\varepsilon,M}(t,\delta), \tag{B.20}$$

where the functions $g_2^{\varepsilon,M}(\cdot,\cdot)$ and $g_3^{\varepsilon,M}(\cdot,\cdot)$ are the same as the corresponding ones in (B.15) with δ replaced by M. And so we need only to consider $\bar{g}_1^{\varepsilon,M}(t,\delta)$. Following the proof of (B.16), we get that

$$\lim_{\delta\downarrow 0}\bar{g}_1^{\varepsilon,M}(t,\delta)$$

$$=\lim_{\delta\downarrow 0}\frac{1}{\delta}\big\{E_t^\varepsilon\big[V\big(X_{\tau_M^\varepsilon(t+\delta)}^\varepsilon\big)I_{\{t+\delta<\tau_M^\varepsilon\}}\big] - V\big(X_{\tau_M^\varepsilon(t)}^\varepsilon\big)I_{\{t<\tau_M^\varepsilon\}}\big\}$$

$$=\lim_{\delta\downarrow 0}E_t^\varepsilon\Bigg[\frac{V(X_{\tau_M^\varepsilon(t+\delta)}^\varepsilon)I_{\{t+\delta<\tau_M^\varepsilon\}} - V(X_{\tau_M^\varepsilon(t)}^\varepsilon)I_{\{t<\tau_M^\varepsilon\}}}{\delta}\Bigg]$$

$$=\lim_{\delta\downarrow 0}E_t^\varepsilon\Bigg[\frac{V(X_{\tau_M^\varepsilon(t+\delta)}^\varepsilon)(I_{\{t+\delta<\tau_M^\varepsilon\}} - I_{\{t<\tau_M^\varepsilon\}})}{\delta}\Bigg]$$

$$+\lim_{\delta\downarrow 0}E_t^\varepsilon\Bigg[\frac{(V(X_{\tau_M^\varepsilon(t+\delta)}^\varepsilon) - V(X_{\tau_M^\varepsilon(t)}^\varepsilon))I_{\{t<\tau_M^\varepsilon\}}}{\delta}\Bigg]$$

$$=0+\lim_{\delta\downarrow 0}E_t^\varepsilon\Bigg[\frac{(V(X_{\tau_M^\varepsilon(t+\delta)}^\varepsilon) - V(X_{\tau_M^\varepsilon(t)}^\varepsilon))I_{\{t<\tau_M^\varepsilon\}}}{\delta}\Bigg]$$

$$=\lim_{\delta\downarrow 0}E_t^\varepsilon\Bigg[\frac{V_x^T(X_{\tau_M^\varepsilon(t)}^\varepsilon + \theta(X_{\tau_M^\varepsilon(t+\delta)}^\varepsilon - X_{\tau_M^\varepsilon(t)}^\varepsilon))(X_{\tau_M^\varepsilon(t+\delta)}^\varepsilon - X_{\tau_M^\varepsilon(t)}^\varepsilon)}{\delta}I_{\{t<\tau_M^\varepsilon\}}\Bigg]$$

$$=\lim_{\delta\downarrow 0}E_t^\varepsilon\Bigg[\frac{V_x^T(X_{\tau_M^\varepsilon(t)}^\varepsilon + \theta(X_{\tau_M^\varepsilon(t+\delta)}^\varepsilon - X_{\tau_M^\varepsilon(t)}^\varepsilon))\int_{\tau_M^\varepsilon(t)}^{\tau_M^\varepsilon(t+\delta)} a(X_u^\varepsilon, Y_{u/\varepsilon})\,du}{\delta}I_{\{t<\tau_M^\varepsilon\}}\Bigg]$$

$$=\lim_{\delta\downarrow 0}E_t^\varepsilon\bigg[\bigg(V_x^T\big(X_{\tau_M^\varepsilon(t)}^\varepsilon + \theta\big(X_{\tau_M^\varepsilon(t+\delta)}^\varepsilon - X_{\tau_M^\varepsilon(t)}^\varepsilon\big)\big)$$

$$\times \int_t^{t+\delta} a\big(X_u^\varepsilon, Y_{u/\varepsilon}\big)I_{\{u<\tau_M^\varepsilon\}}\,du\Big/\delta\bigg)I_{\{t<\tau_M^\varepsilon\}}\bigg]$$

$$=V_x^T\big(X_{\tau_M^\varepsilon(t)}^\varepsilon\big)a\big(X_t^\varepsilon, Y_{t/\varepsilon}\big)\cdot I_{\{t<\tau_M^\varepsilon\}}I_{\{t<\tau_M^\varepsilon\}}$$

$$=V_x^T\big(X_t^\varepsilon\big)a\big(X_t^\varepsilon, Y_{t/\varepsilon}\big)I_{\{t<\tau_M^\varepsilon\}}$$

$$=\lim_{\delta\downarrow 0}g_1^{\varepsilon,M}(t,\delta). \tag{B.21}$$

Hence by the proof of Lemma B.3, we get that (B.19) holds. $\square$

Lemma B.5 M_t^ε *is a martingale relative to* $\{\mathcal{F}_t^\varepsilon\}$.

Proof For any $s, t \geq 0$, by (3.139), the property of conditional expectation, and $\hat{\mathcal{A}}_M^\varepsilon(V^\varepsilon(X_{\tau_M^\varepsilon(t)}^\varepsilon, t) \cdot I_{\{t < \tau_M^\varepsilon\}}) = \hat{\mathcal{A}}_M^\varepsilon V^\varepsilon(X_{\tau_M^\varepsilon(t)}^\varepsilon, t)$ (see Lemma B.4), we have that

$$E\left[M_{t+s}^\varepsilon - M_t^\varepsilon | \mathcal{F}_t^\varepsilon\right]$$

$$= E\left[e^{2\hat{\gamma}(t+s)} V^\varepsilon\left(X_{\tau_M^\varepsilon(t+s)}^\varepsilon, t+s\right) I_{\{t+s < \tau_M^\varepsilon\}} - e^{2\hat{\gamma}t} V^\varepsilon\left(X_{\tau_M^\varepsilon(t)}^\varepsilon, t\right) I_{\{t < \tau_M^\varepsilon\}}\right.$$

$$\left. - \int_t^{t+s} e^{2\hat{\gamma}u}\left(\hat{\mathcal{A}}_M^\varepsilon + 2\hat{\gamma}\right)\left(V^\varepsilon\left(X_{\tau_M^\varepsilon(u)}^\varepsilon, u\right) I_{\{u < \tau_M^\varepsilon\}}\right) du \middle| \mathcal{F}_t^\varepsilon\right]$$

$$+ E\left[e^{2\hat{\gamma}\tau_M^\varepsilon} V\left(X_{\tau_M^\varepsilon}^\varepsilon\right) I_{\{\tau_M^\varepsilon \leq t+s\}} - e^{2\hat{\gamma}\tau_M^\varepsilon} V\left(X_{\tau_M^\varepsilon}^\varepsilon\right) I_{\{\tau_M^\varepsilon \leq t\}} | \mathcal{F}_t^\varepsilon\right]$$

$$= E\left[e^{2\hat{\gamma}(t+s)} V^\varepsilon\left(X_{\tau_M^\varepsilon(t+s)}^\varepsilon, t+s\right) \cdot I_{\{t+s < \tau_M^\varepsilon\}} | \mathcal{F}_t^\varepsilon\right]$$

$$- e^{2\hat{\gamma}t} V^\varepsilon\left(X_{\tau_M^\varepsilon(t)}^\varepsilon, t\right) \cdot I_{\{t < \tau_M^\varepsilon\}}$$

$$- \int_t^{t+s} E\left[e^{2\hat{\gamma}u}\left(\hat{\mathcal{A}}_M^\varepsilon + 2\hat{\gamma}\right)\left(V^\varepsilon\left(X_{\tau_M^\varepsilon(u)}^\varepsilon, u\right) \cdot I_{\{u < \tau_M^\varepsilon\}}\right) | \mathcal{F}_t^\varepsilon\right] du$$

$$+ E\left[e^{2\hat{\gamma}\tau_M^\varepsilon} V\left(X_{\tau_M^\varepsilon}^\varepsilon\right) \cdot I_{\{\tau_M^\varepsilon \leq t+s\}} - e^{2\hat{\gamma}\tau_M^\varepsilon} V\left(X_{\tau_M^\varepsilon}^\varepsilon\right) \cdot I_{\{\tau_M^\varepsilon \leq t\}} | \mathcal{F}_t^\varepsilon\right]$$

$$= \left\{E\left[e^{2\hat{\gamma}(t+s)} V^\varepsilon\left(X_{\tau_M^\varepsilon(t+s)}^\varepsilon, t+s\right) | \mathcal{F}_t^\varepsilon\right]\right.$$

$$- e^{2\hat{\gamma}t} V^\varepsilon\left(X_{\tau_M^\varepsilon(t)}^\varepsilon, t\right)$$

$$\left. - \int_t^{t+s} E\left[e^{2\hat{\gamma}u}\left(\hat{\mathcal{A}}_M^\varepsilon + 2\hat{\gamma}\right)\left(V^\varepsilon\left(X_{\tau_M^\varepsilon(u)}^\varepsilon, u\right)\right) | \mathcal{F}_t^\varepsilon\right] du\right\}$$

$$- \left\{E\left[e^{2\hat{\gamma}(t+s)} V^\varepsilon\left(X_{\tau_M^\varepsilon(t+s)}^\varepsilon, t+s\right) I_{\{t+s \geq \tau_M^\varepsilon\}} | \mathcal{F}_t^\varepsilon\right]\right.$$

$$- e^{2\hat{\gamma}t} V^\varepsilon\left(X_{\tau_M^\varepsilon(t)}^\varepsilon, t\right) I_{\{t \geq \tau_M^\varepsilon\}}$$

$$\left. - \int_t^{t+s} E\left[2\hat{\gamma} e^{2\hat{\gamma}u} V^\varepsilon\left(X_{\tau_M^\varepsilon(u)}^\varepsilon, u\right) I_{\{u \geq \tau_M^\varepsilon\}} | \mathcal{F}_t^\varepsilon\right] du\right\}$$

$$+ E\left[e^{2\hat{\gamma}\tau_M^\varepsilon} V\left(X_{\tau_M^\varepsilon}^\varepsilon\right) I_{\{\tau_M^\varepsilon \leq t+s\}} - e^{2\hat{\gamma}\tau_M^\varepsilon} V\left(X_{\tau_M^\varepsilon}^\varepsilon\right) I_{\{\tau_M^\varepsilon \leq t\}} | \mathcal{F}_t^\varepsilon\right]$$

$$\triangleq g_1(t, s, \omega) - g_2(t, s, \omega) + g_3(t, s, \omega). \tag{B.22}$$

For $u \geq t$, define

$$f(u, \omega) = E\left[e^{2\hat{\gamma}(u)} V^\varepsilon\left(X_{\tau_M^\varepsilon(u)}^\varepsilon, u\right) | \mathcal{F}_t^\varepsilon\right](\omega). \tag{B.23}$$

Then for any $u \geq t$, we have (if $u = t$, we consider the right derivative)

$$f'(u, \omega) = \lim_{s \to 0} \frac{f(u+s, \omega) - f(u, \omega)}{s}$$

$$= \lim_{s \to 0} \frac{E[e^{2\hat{\gamma}(u+s)} V^{\varepsilon}(X^{\varepsilon}_{\tau^{\varepsilon}_M(u+s)}, u+s)|\mathcal{F}^{\varepsilon}_t] - E[e^{2\hat{\gamma}(u)} V^{\varepsilon}(X^{\varepsilon}_{\tau^{\varepsilon}_M(u)}, u)|\mathcal{F}^{\varepsilon}_t]}{s}$$

$$= \lim_{s \to 0} E\left[\frac{e^{2\hat{\gamma}(u+s)} V^{\varepsilon}(X^{\varepsilon}_{\tau^{\varepsilon}_M(u+s)}, u+s) - e^{2\hat{\gamma}u} V^{\varepsilon}(X^{\varepsilon}_{\tau^{\varepsilon}_M(u)}, u)}{s}\bigg|\mathcal{F}^{\varepsilon}_t\right]$$

$$= \lim_{s \to 0} E\left[\frac{(e^{2\hat{\gamma}(u+s)} - e^{2\hat{\gamma}u}) V^{\varepsilon}(X^{\varepsilon}_{\tau^{\varepsilon}_M(u+s)}, u+s)}{s}\bigg|\mathcal{F}^{\varepsilon}_t\right]$$

$$+ \lim_{s \to 0} E\left[\frac{e^{2\hat{\gamma}u}(V^{\varepsilon}(X^{\varepsilon}_{\tau^{\varepsilon}_M(u+s)}, u+s) - V^{\varepsilon}(X^{\varepsilon}_{\tau^{\varepsilon}_M(u)}, u))}{s}\bigg|\mathcal{F}^{\varepsilon}_t\right]$$

$$= E[e^{2\hat{\gamma}u}(\hat{A}_M + 2\hat{\gamma})(V^{\varepsilon}(X^{\varepsilon}_{\tau^{\varepsilon}_M(u)}, u))|\mathcal{F}^{\varepsilon}_t], \qquad (B.24)$$

and thus

$$g_1(t, s, \omega) = f(t+s, \omega) - f(t, \omega) - \int_t^{t+s} f'(u, \omega)\, du = 0 \quad \text{a.s.} \quad (B.25)$$

By the definitions of τ^{ε}_M and $V^{\varepsilon}(x, t)$, we have

$$g_2(t, s, \omega) = E\left[e^{2\hat{\gamma}(t+s)} V(X^{\varepsilon}_{\tau^{\varepsilon}_M}) I_{\{t+s \geq \tau^{\varepsilon}_M\}}|\mathcal{F}^{\varepsilon}_t\right] - e^{2\hat{\gamma}t} V(X^{\varepsilon}_{\tau^{\varepsilon}_M}) I_{\{t \geq \tau^{\varepsilon}_M\}}$$

$$- \int_t^{t+s} E\left[2\hat{\gamma} e^{2\hat{\gamma}u} V(X^{\varepsilon}_{\tau^{\varepsilon}_M}) I_{\{u \geq \tau^{\varepsilon}_M\}}|\mathcal{F}^{\varepsilon}_t\right] du$$

$$= E\left[e^{2\hat{\gamma}(t+s)} V(X^{\varepsilon}_{\tau^{\varepsilon}_M}) I_{\{t+s \geq \tau^{\varepsilon}_M\}} - e^{2\hat{\gamma}t} V(X^{\varepsilon}_{\tau^{\varepsilon}_M}) I_{\{t \geq \tau^{\varepsilon}_M\}}\right.$$

$$\left. - \int_t^{t+s} 2\hat{\gamma} e^{2\hat{\gamma}u} V(X^{\varepsilon}_{\tau^{\varepsilon}_M}) I_{\{u \geq \tau^{\varepsilon}_M\}}\, du\bigg|\mathcal{F}^{\varepsilon}_t\right]. \qquad (B.26)$$

Now, we analyze the expression within the conditional expectation on the right-hand side of (B.26). For simplicity, let

$$h(t, s, \omega) = e^{2\hat{\gamma}(t+s)} V(X^{\varepsilon}_{\tau^{\varepsilon}_M}) \cdot I_{\{t+s \geq \tau^{\varepsilon}_M\}} - e^{2\hat{\gamma}t} V(X^{\varepsilon}_{\tau^{\varepsilon}_M}) \cdot I_{\{t \geq \tau^{\varepsilon}_M\}}$$

$$- \int_t^{t+s} 2\hat{\gamma} e^{2\hat{\gamma}u} V(X^{\varepsilon}_{\tau^{\varepsilon}_M}) \cdot I_{\{u \geq \tau^{\varepsilon}_M\}}\, du. \qquad (B.27)$$

Case 1: $t + s < \tau^{\varepsilon}_M(\omega)$. Then $h(t, s, \omega) = 0$.

Case 2: $t \geq \tau_M^\varepsilon(\omega)$. Then we have

$$h(t, s, \omega) = e^{2\hat{\gamma}(t+s)} V\left(X_{\tau_M^\varepsilon}^\varepsilon\right) - e^{2\hat{\gamma}t} V\left(X_{\tau_M^\varepsilon}^\varepsilon\right) - \int_t^{t+s} 2\hat{\gamma} e^{2\hat{\gamma}u} V\left(X_{\tau_M^\varepsilon}^\varepsilon\right) du$$

$$= 0, \tag{B.28}$$

since

$$\frac{d(e^{2\hat{\gamma}u} V(X_{\tau_M^\varepsilon}^\varepsilon))}{du} = 2\hat{\gamma} e^{2\hat{\gamma}u} V\left(X_{\tau_M^\varepsilon}^\varepsilon\right). \tag{B.29}$$

Case 3: $t < \tau_M^\varepsilon(\omega) \leq t + s$. Then by (B.27) and (B.29), we have

$$h(t, s, \omega) = e^{2\hat{\gamma}(t+s)} V\left(X_{\tau_M^\varepsilon}^\varepsilon\right) - \int_{\tau_M^\varepsilon}^{t+s} 2\hat{\gamma} e^{2\hat{\gamma}u} V\left(X_{\tau_M^\varepsilon}^\varepsilon\right) du$$

$$= e^{2\hat{\gamma}\tau_M^\varepsilon} V\left(X_{\tau_M^\varepsilon}^\varepsilon\right). \tag{B.30}$$

Hence we have

$$-g_2(t, s, \omega)$$
$$= -E\left[h(t, s, \omega)|\mathcal{F}_t^\varepsilon\right]$$
$$= -E\left[e^{2\hat{\gamma}\tau_M^\varepsilon} V\left(X_{\tau_M^\varepsilon}^\varepsilon\right) \cdot I_{\{t < \tau_M^\varepsilon \leq t+s\}}|\mathcal{F}_t^\varepsilon\right]$$
$$= -E\left[e^{2\hat{\gamma}\tau_M^\varepsilon} V\left(X_{\tau_M^\varepsilon}^\varepsilon\right) I_{\{\tau_M^\varepsilon \leq t+s\}} - e^{2\hat{\gamma}\tau_M^\varepsilon} V\left(X_{\tau_M^\varepsilon}^\varepsilon\right) I_{\{\tau_M^\varepsilon \leq t\}}|\mathcal{F}_t^\varepsilon\right], \tag{B.31}$$

which implies that

$$-g_2(t, s, \omega) + g_3(t, s, \omega) = E\left[0|\mathcal{F}_t^\varepsilon\right] = 0 \quad \text{a.s.} \tag{B.32}$$

This, together with (B.22) and (B.25), proves that

$$E\left[M_{t+s}^\varepsilon - M_t^\varepsilon|\mathcal{F}_t^\varepsilon\right] = 0 \quad \text{a.s.} \tag{B.33}$$

$\square$

References

1. Adetola V, Guay M (2006) Adaptive output feedback extremum seeking receding horizon control of linear systems. J Process Control 16:521–533
2. Adetola V, Guay M (2007) Guaranteed parameter convergence for extremum-seeking control of nonlinear systems. Automatica 43:105–110
3. Airapetyan R (1999) Continuous Newton method and its modification. Appl Anal 73:463–484
4. Altman E, Başar T, Srikant R (2002) Nash equilibria for combined flow control and routing in networks: asymptotic behavior for a large number of users. IEEE Trans Autom Control 47:917–930
5. Anosov DV (1960) Osrednenie v sistemakh obyknovennykh differentsial'nykh uravenenii s bystro koleblyushchimisya resheniyami. Izv Akad Nauk SSSR, Ser Mat 24(5):721–742
6. Ariyur KB, Krstic M (2003) Real-time optimization by extremum seeking control. Wiley, Hoboken
7. Ariyur K, Krstic M (2004) Slope seeking: a generalization of extremum seeking. Int J Adapt Control Signal Process 18:1–22
8. Aumann RJ (1964) Markets with a continuum of traders. Econometrica 32(1):39–50
9. Banaszuk A, Ariyur KB, Krstic M, Jacobson CA (2004) An adaptive algorithm for control of combustion instability. Automatica 40:1965–1972
10. Başar T (2007) Control and game-theoretic tools for communication networks (overview). Appl Comput Math 6:104–125
11. Başar T, Olsder GJ (1999) Dynamic noncooperative game theory, 2nd edn. SIAM, Philadelphia
12. Bauso D, Giarre L, Pesenti R (2008) Consensus in noncooperative dynamic games: a multi-retailer inventory application. IEEE Trans Autom Control 53:998–1003
13. Baxter JR, Brosamler GA (1976) Energy and the law of the iterated logarithm. Math Scand 38:115–136
14. Becker A, Kumar PR, Wei CZ (1985) Adaptive control with the stochastic approximation algorithm: geometry and convergence. IEEE Trans Autom Control 30:330–338
15. Becker R, King R, Petz W, Nitsche W (2006) Adaptive closed-loop separation control on a high-lift configuration using extremum seeking. AIAA paper 2006-3493
16. Becker R, King R, Petz W, Nitsche W (2007) Adaptive closed-loop separation control on a high-lift configuration using extremum seeking. AIAA J 45:1382–1392
17. Benveniste A, Métivier M, Priouret P (1990) Adaptive algorithms and stochastic approximations. Springer, Berlin
18. Berg H (2003) E. coli in motion. Springer, New York
19. Berg H, Brown DA (1972) Chemotaxis in E. coli analyzed by three-dimensional tracking. Nature 239(5374):500–504

20. Binetti P, Ariyur KB, Krstic M, Bernelli F (2003) Formation flight optimization using extremum seeking feedback. AIAA J Guid Control Dyn 26:132–142
21. Blankenship G, Papanicolaou GC (1978) Stability and control of stochastic systems with wide-band noise disturbances. I. SIAM J Appl Math 34:437–476
22. Bogolyubov NN, Mitropolsky YA (1961) Asymptotic methods in the theory of nonlinear oscillation. Gordon & Breach, New York
23. Bogolyubov NN, Zubarev DN (1955) Metod asimptoticheskogo priblizheniya dlya sistem s vrashchayuschcheisya fazoi i ego primenenie k dvizheniyu zaryazhennykh chastits v magnitnom pole. Ukr Mat Zh 7:5–17
24. Brunn A, Nitsche W, Henning L, King R Application of slope-seeking to a generic car model for active drag control. Preprint
25. Carnevale D, Astolfi A, Centioli C, Podda S, Vitale V, Zaccarian L (2009) A new extremum seeking technique and its application to maximize RF heating on FTU. Fusing Eng Des 84:554–558
26. Cesa-Bianchi N, Lugosi G (2006) Prediction, learning, and games. Cambridge University Press, New York
27. Choi J-Y, Krstic M, Ariyur KB, Lee JS (2002) Extremum seeking control for discrete time systems. IEEE Trans Autom Control 47:318–323
28. Cochran J, Krstic M (2009) Nonholonomic source seeking with tuning of angular velocity. IEEE Trans Autom Control 54(4):717–731
29. Cochran J, Ghods N, Siranosian A, Krstic M (2009) 3D source seeking for underactuated vehicles without position measurement. IEEE Trans Robot 25:117–129
30. Cochran J, Kanso E, Kelly SD, Xiong H, Krstic M (2009) Source seeking for two nonholonomic models of fish locomotion. IEEE Trans Robot 25:1166–1176
31. Cortes J, Martinez S, Karatas T, Bullo F (2004) Coverage control for mobile sensing networks. IEEE Trans Robot Autom 20(2):243–255
32. Creaby J, Li Y, Seem JE (2009) Maximizing wind turbine energy capture using multivariable extremum seeking control. Wind Eng 33:361–387
33. DeHaan D, Guay M (2005) Extremum-seeking control of state-constrained nonlinear systems. Automatica 41:1567–1574
34. Doob JL (1984) Classical potential theory and its probabilistic counterpart. Springer, Berlin
35. Doukhan P (1994) Mixing: properties and examples. Springer, Berlin
36. Favache A, Guay M, Perrier M, Dochain D (2008) Extremum seeking control of retention for a microparticulate system. Can J Chem Eng 86:815–827
37. Foster DP, Young HP (2003) Learning, hypothesis testing, and Nash equilibrium. Games Econ Behav 45:73–96
38. Foster DP, Young HP (2006) Regret testing: learning to play Nash equilibrium without knowing you have an opponent. Theor Econ 1:341–367
39. Freidlin MI, Wentzell AD (1984) Random perturbations of dynamical systems. Springer, Berlin
40. Frihauf P, Krstic M, Başar T (2011) Nash equilibrium seeking with infinitely-many players. In: Proceedings of 2011 American control conference, San Francisco, CA, USA, June 29–July 1, pp 3059–3064
41. Frihauf P, Krstic M, Başar T (2012) Nash equilibrium seeking in noncooperative games. IEEE Trans Autom Control 57:1192–1207
42. Fudenberg D, Levine DK (1998) The theory of learning in games. MIT Press, Cambridge
43. Gelfand SB, Mitter SK (1991) Recursive stochastic algorithms for global optimization in $\mathbb{R}^d$. SIAM J Control Optim 29:999–1018
44. Gelfand SB, Mitter SK (1993) Metropolis-type annealing algorithms for global optimization in $\mathbb{R}^d$. SIAM J Control Optim 31:111–131
45. Ghaffari A, Krstic M, Nešić D (2011) Multivariable Newton-based extremum seeking. In: Proceedings of the 50th IEEE conference on decision and control. Automatica (to appear)
46. Ghods N, Krstic M (2010) Speed regulation in steering-based source seeking. Automatica 46:452–459

47. Golub GH, Van Loan CF (1996) Matrix computations, 3rd edn. Johns Hopkins University Press, Baltimore
48. Green EJ (1984) Continuum and finite-player noncooperative models of competition. Econometrica 52(4):975–993
49. Guay M, Perrier M, Dochain D (2005) Adaptive extremum seeking control of nonisothermal continuous stirred reactors. Chem Eng Sci 60:3671–3681
50. Guay M, Dochain D, Perrier M, Hudon N (2007) Flatness-based extremum-seeking control over periodic orbits. IEEE Trans Autom Control 52:2005–2012
51. Hale JK, Verduyn Lunel SM (1990) Averaging in infinite dimensions. J Integral Equ Appl 2:463–494
52. Hashemi SN, Heunis AJ (1998) Averaging principle for diffusion processes. Stoch Stoch Rep 62:201–216
53. Henning L, Becker R, Feuerbach G, Muminovic R, Brun A, Nitsche W, King R (2008) Extensions of adaptive slope-seeking for active flow control. Proc Inst Mech Eng, Part I, J Syst Control Eng 222:309–322
54. Jafari A, Greenwald A, Gondek D, Ercal G (2001) On no-regret learning, fictitious play, and Nash equilibrium. In: Proceedings of the 18th international conference on machine learning
55. Karatzas I, Shreve SE (2005) Brownian motion and stochastic calculus. Springer, Berlin
56. Khalil HK (2002) Nonlinear systems, 3rd edn. Prentice Hall, Upper Saddle River
57. Khas'minskiĭ RZ (1966) A limit theorem for the solutions of differential equations with random right-hand sides. Theory Probab Appl 11:390–406
58. Khas'minskiĭ RZ (1980) Stochastic stability of differential equations. Sijthoff & Noordhoff, Rockville
59. Khas'minskiĭ RZ, Yin G (2004) On averaging principles: an asymptotic expansion approach. SIAM J Math Anal 35:1534–1560
60. Kifer Y (1992) Averaging in dynamical systems and large deviations. Invent Math 110:337–370
61. Killingsworth NJ, Krstic M (2006) PID tuning using extremum seeking. IEEE Control Syst Mag 26:70–79
62. Killingsworth NJ, Krstic M, Flowers DL, Espinoza-Loza F, Ross T, Aceves SM (2009) HCCI engine combustion timing control: optimizing gains and fuel consumption via extremum seeking. IEEE Trans Control Syst Technol 17:1350–1361
63. Kim K, Kasnakoglu C, Serrani A, Samimy M (2009) Extremum-seeking control of subsonic cavity flow. AIAA J 47:195–205
64. King R, Becker R, Feuerbach G, Henning L, Petz R, Nitsche W, Lemke O, Neise W (2006) Adaptive flow control using slope seeking. In: Proceedings of the 14th IEEE Mediterranean conference on control and automation, June 28–30, pp 1–6
65. Korolyuk VS (1995) Average and stability of dynamical system with rapid stochastic switching. In: Skorokhod AV, Borovskikh YuV (eds) Exploring stochastic laws. Brill, Leiden, pp 219–232
66. Korolyuk VS (1998) Stability of stochastic systems in the scheme of diffusion approximation. Ukr Mat Zh 50:36–47
67. Korolyuk VS, Chabanyuk YM (2002) Stability of dynamical system with semi-Markov switches under conditions of stability of the averaged system. Ukr Mat Zh 54:239–252
68. Korolyuk VS, Swishchuk AV (1995) Evolution systems in random media. CRC Press, Boca Raton
69. Krieger JP, Krstic M (2011) Extremum seeking based on atmospheric turbulence for aircraft endurance. AIAA J Guid Control Dyn 34:1876–1885
70. Krstic M (2000) Performance improvement and limitations in extremum seeking control. Syst Control Lett 39:313–326
71. Krstic M, Wang HH (2000) Stability of extremum seeking feedback for general nonlinear dynamic systems. Automatica 36:595–601
72. Krstic M, Frihauf P, Krieger J, Başar T (2010) Nash eqilibrium seeking with finitely- and infinitely-many players. In: Proceedings of the 8th IFAC symposium on nonlinear control

systems, Bologna, Italy

73. Krylov NM, Bogolyubov NN (1937) Introduction to nonlinear mechanics. Izd AN UkSSR, Kiev (in Russian)

74. Kubica BJ, Wozniak A (2010) An interval method for seeking the Nash equilibrium of non-cooperative games. In: Parallel processing and applied mathematics. Lecture notes in computer science, vol. 6068. Springer, Berlin, pp 446–455

75. Kumar PR, Varaiya P (1986) Stochastic systems: estimation, identification and adaptive control. Prentice Hall, Englewood Cliffs

76. Kurtz TG (1975) Semigroups of conditioned shifts and approximation of Markov processes. Ann Probab 3:618–642

77. Kushner HJ (1984) Approximation and weak convergence method for random processes, with applications to stochastic systems theory. MIT Press, Cambridge

78. Kushner HJ, Ramachandran KM (1988) Nearly optimal singular controls for wideband noise driven systems. SIAM J Control Optim 26:569–591

79. Kushner HJ, Yin G (2003) Stochastic approximation and recursive algorithms and applications, 2nd edn. Springer, Berlin

80. Laventall JCK (2009) Coverage control by multi-robot networks with limited-range anisotropic sensory. Int J Control 82:1113–1121

81. Lei P, Li Y, Chen Q, Seem JE (2010) Extremum seeking control based integration of MPPT and degradation detection for photovoltaic arrays. In: Proceedings of 2010 American control conference, Baltimore, MD, USA, June 30–July 2, pp 3536–3541

82. Li P, Li Y, Seem JE (2009) Extremum seeking control for efficient and reliable operation of air-side economizers. In: Proceedings of 2009 American control conference, St. Louis, MO, USA, June 10–12, pp 20–25

83. Li S, Başar T (1987) Distributed algorithms for the computation of noncooperative equilibria. Automatica 23:523–533

84. Li Y, Rotea MA, Chiu GT-C, Mongeau LG, Paek I-S (2005) Extremum seeking control of a tunable thermoacoustic cooler. IEEE Trans Control Syst Technol 13:527–536

85. Liptser RS, Shiryayev AN (1977) Statistics of random processes 1. Springer, New York

86. Liptser RS, Shiryaev AN (1989) Theory of martingales. Kluwer Academic, Norwell

87. Liptser RS, Stoyanov J (1990) Stochastic version of the averaging principle for diffusion type processes. Stoch Stoch Rep 32:145–163

88. Liu S-J, Krstic M (2010) Continuous-time stochastic averaging on infinite interval for locally Lipschitz systems. SIAM J Control Optim 48:3589–3622

89. Liu S-J, Krstic M (2010) Stochastic averaging in continuous time and its applications to extremum seeking. IEEE Trans Autom Control 55(10):2235–2250

90. Liu S-J, Krstic M (2010) Stochastic source seeking for nonholonomic unicycle. Automatica 46:1443–1453

91. Liu S-J, Krstic M (2011) Stochastic Nash equilibrium seeking for games with general nonlinear payoffs. SIAM J Control Optim 49:1659–1679

92. Ljung L (1977) Analysis of recursive stochastic algorithms. IEEE Trans Autom Control 22:551–575

93. Ljung L (1978) Strong convergence of a stochastic approximation algorithm. Ann Stat 6:680–696

94. Ljung L (2001) Recursive least-squares and accelerated convergence in stochastic approximation schemes. Int J Adapt Control Signal Process 15:169–178

95. Ljung L, Pflug G, Walk H (1992) Stochastic approximation and optimization of random systems. Birkhäuser, Basel

96. Luo L, Schuster E (2009) Mixing enhancement in 2D magnetohydrodynamic channel flow by extremum seeking boundary control. In: Proceedings of the 2009 American control conference, St. Louis, MO, USA, June 10–12, pp 1530–1535

97. MacKenzie AB, Wicker SB (2001) Game theory and the design of self-configuring, adaptive wireless networks. IEEE Commun Mag 39:126–131

98. Manzie C, Krstic M (2009) Extremum seeking with stochastic perturbations. IEEE Trans Autom Control 54:580–585
99. Marden JR, Arslon G, Shamma JS (2009) Cooperative control and potential games. IEEE Trans Syst Man Cybern, Part B, Cybern 39:1393–1407
100. Mesquita AR, Hespanha JP, Åström K (2008) Optimotaxis: a stochastic multi-agent optimization procedure with point measurements. In: Egerstedt M, Mishra B (eds) Hybrid systems: computation and control. Lecture notes in computer science, vol. 4981. Springer, Berlin, pp 358–371
101. Moase WH, Manzie C, Brear MJ (2009) Newton-like extremum-seeking part I: theory. In: Proceedings of the joint 48th IEEE conference on decision and control and 28th Chinese control conference, Shanghai, China, December 16–18, pp 3839–3844
102. Moase WH, Manzie C, Brear MJ (2009) Newton-like extremum-seeking part II: simulation and experiments. In: Proceedings of the joint 48th IEEE conference on decision and control and 28th Chinese control conference, Shanghai, China, December 16–18, pp 3845–3850
103. Moase WH, Manzie C, Brear MJ (2010) Newton-like extremum-seeking for the control of thermoacoustic instability. IEEE Trans Autom Control 55:2094–2105
104. Moeck JP, Bothien MR, Paschereit CO, Gelbert G, King R Two-parameter extremum seeking for control of thermoacoustic instabilities and characterization of linear growth. AIAA paper 2007-1416
105. Murugappan S, Gutmark E, Acharya S, Krstic M (2000) Extremum seeking adaptive controller of swirl-stabilized spray combustion. Proc Combust Inst 28:731–737
106. Neishtadt AI (1975) Averaging in multifrequency systems, I. Sov Phys Dokl 20(7):492–494
107. Neishtadt AI (1976) Averaging in multifrequency systems, II. Sov Phys Dokl 21(2):80–82
108. Nešić D, Tan Y, Moase WH, Manzie C (2010) A unifying approach to extremum seeking: adaptive schemes based on estimation of derivatives. In: Proceedings of the 49th IEEE conference on decision and control, Atlanta, GA, USA, December 15–17, pp 4625–4630
109. Øksendal B (1995) Stochastic differential equations, 4th edn. Springer, Berlin
110. Ou Y, Xu C, Schuster E, Luce TC, Ferron JR, Walker ML, Humphreys DA (2008) Design and simulation of extremum-seeking open-loop optimal control of current profile in the DIII-D tokamak. Plasma Phys Control Fusion 50:115001
111. Papanicolaou GC, Kohler W (1974) Asymptotic theory of mixing stochastic ordinary differential equations. Commun Pure Appl Math 27:641–668
112. Pardoux E, Veretennikov AYu (2001) On the Poisson equation and diffusion approximation. I. Ann Probab 29(3):1061–1085
113. Peterson K, Stefanopoulou A (2004) Extremum seeking control for soft landing of an electromechanical valve actuator. Automatica 29:1063–1069
114. Rao SS, Venkayya VB, Khot NS (1988) Game theory approach for the integrated design of structures and controls. AIAA J 26:463–469
115. Ren B, Frihauf P, Krstic M, Rafac RJ (2012) Laser pulse shaping via extremum seeking. Control Eng Pract 20:678–683
116. Rishel R (1970) Necessary and sufficient dynamic programming conditions for continuous time stochastic optimal control. SIAM J Control 8:559–571
117. Roberts JB, Spanos PD (1986) Stochastic averaging: an approximate method of solving random vibration problems. Int J Non-Linear Mech 21:111–134
118. Rosen JB (1965) Existence and uniqueness of equilibrium points for concave N-person games. Econometrica 33:520–534
119. Rotea MA (2000) Analysis of multivariable extremum seeking algorithms. In: Proceedings of the 2000 American control conference, Chicago, IL, USA, June 28–30, pp 433–437
120. Sanders JA, Verhulst F, Murdock J (2007) Averaging methods in nonlinear dynamical systems, 2nd edn. Springer, Berlin
121. Sastry S, Bodson M (1989) Adaptive control: stability, convergence, and robustness. Prentice Hall, Englewood Cliffs
122. Schuster E, Torres N, Xu C (2006) Extremum seeking adaptive control of beam envelope in particle accelerators. In: Proceedings of the 2006 IEEE conference on control applications,

Munich, Germany, October 4–6, pp 1837–1842

123. Schuster E, Xu C, Torres N, Morinaga E, Allen CK, Krstic M (2007) Beam matching adaptive control via extremum seeking. Nucl Instrum Methods Phys Res, Sect A, Accel Spectrom Detect Assoc Equip 581:799–815

124. Scutari G, Palomar DP, Barbarossa S (2009) The MIMO iterative waterfilling algorithm. IEEE Trans Signal Process 57:1917–1935

125. Semsar-Kazerooni E, Khorasani K (2009) Multi-agent team cooperation: a game theory approach. Automatica 45:2205–2213

126. Shamma JS, Arslan G (2005) Dynamic fictitious play, dynamic gradient play, and distributed convergence to Nash equilibria. IEEE Trans Autom Control 53(3):312–327

127. Sharma R, Gopal M (2010) Synergizing reinforcement learning and game theory—a new direction for control. Appl Soft Comput 10(3):675–688

128. Shitovitz B (1973) Oligopoly in markets with a continuum of traders. Econometrica 41(3):467–501

129. Skorokhod AV (1989) Asymptotic methods in the theory of stochastic differential equations. Translations of mathematical monographs. Amer Math Soc, Providence

130. Skorokhod AV, Hoppensteadt FC, Salehi H (2002) Random perturbation methods with applications in science and engineering. Springer, New York

131. Solo V, Kong X (1994) Adaptive signal processing algorithms: stability and performance. Prentice Hall, New York

132. Spall JC (2003) Introduction to stochastic search and optimization: estimation, simulation, and control. Wiley-Interscience, New York

133. Stanković MS, Stipanović DM (2009) Stochastic extremum seeking with applications to mobile sensor networks. In: Proceedings of the 2009 American control conference, St. Louis, MA, USA, June 10–12, pp 5622–5627

134. Stanković MS, Stipanović DM (2009) Discrete time extremum seeking by autonomous vehicles in a stochastic environment. In: Proceedings of the joint 48th IEEE conference on decision and control and 28th Chinese control conference, Shanghai, China, December 16–18, pp 4541–4546

135. Stanković MS, Stipanović DM (2010) Extremum seeking under stochastic noise and applications to mobile sensors. Automatica 46:1243–1251

136. Stanković MS, Johansson KH, Stipanović DM (2010) Distributed seeking of Nash equilibrium in mobile sensor networks. In: Proceedings of IEEE conference on decision and control, Atlanta, GA, USA, December 15–17, pp 5598–5603

137. Tan Y, Nešić D, Mareels IMY (2006) On non-local stability properties of extremum seeking controllers. Automatica 42:889–903

138. Tao G (2003) Adaptive control design and analysis. Wiley, New York

139. Volosov VM (1962) Averaging in systems of ordinary differential equations. Russ Math Surv 17(6):1–126

140. Wang H-H, Krstic M (2000) Extremum seeking for limit cycle minimization. IEEE Trans Autom Control 45:2432–2437

141. Wang H-H, Krstic M, Bastin G (1999) Optimizing bioreactors by extremum seeking. Int J Adapt Control Signal Process 13:651–669

142. Wang H-H, Yeung S, Krstic M (2000) Experimental application of extremum seeking on an axial-flow compressor. IEEE Trans Control Syst Technol 8:300–309

143. Wehner W, Schuster E (2009) Stabilization of neoclassical tearing modes in tokamak fusion plasmas via extremum seeking. In: Proceedings of the 3rd IEEE multi-conference on systems and control (MSC 2009), Saint Petersburg, Russia, July 8–10

144. Wiederhold O, Neuhaus L, King R, Niese W, Enghardt L, Noack BR, Swoboda M (2009) Extensions of extremum-seeking control to improve the aerodynamic performance of axial turbomachines. In: Proceedings of the 39th AIAA fluid dynamics conference, San Antonio, TX, USA

145. Yin G, Zhang Q (1994) Near optimality of stochastic control in systems with unknown parameter processes. Appl Math Optim 29:263–284

146. Zhang C, Arnold D, Ghods N, Siranosian A, Krstic M (2007) Source seeking with non-holonomic unicycle without position measurement and with tuning of forward velocity. Syst Control Lett 56:245–252

147. Zhang C, Siranosian A, Krstic M (2007) Extremum seeking for moderately unstable systems and for autonomous vehicle target tracking without position measurements. Automatica 43:1832–1839

148. Zhang XT, Dawson DM, Dixon WE, Xian B (2006) Extremum-seeking nonlinear controllers for a human exercise machine. IEEE/ASME Trans Mechatron 11:233–240

149. Zhu M, Martinez S (2009) Distributed coverage games for mobile visual sensors (ii): reaching the set of Nash equilibria. In: Proceedings of the 48th IEEE international conference on decision and control, pp 169–174

150. Zhu M, Martinez S (2010) Distributed coverage games for mobile visual sensor networks. SIAM J Control Optim (submitted). Available at http://arxiv.org/abs/1002.0367

151. Zhu WQ (1988) Stochastic averaging methods in random vibration. Appl Mech Rev 41(5):189–199

152. Zhu WQ, Yang YQ (1997) Stochastic averaging of quasi-non integrable-Hamiltonian systems. J Appl Mech 64:157–164

Index

S.-J. Liu, M. Krstic, *Stochastic Averaging and Stochastic Extremum Seeking*,
Communications and Control Engineering,
DOI 10.1007/978-1-4471-4087-0, © Springer-Verlag London 2012

Practical stability, 57

Q
Quadratic games, 161

R
Riccati equation, 182

S
Sinusoidal excitation signals, 12
Slope seeking, 129, 138
Source seeking, 95, 121
Stability in probability, 40
Stochastic averaging, 1, 7, 129
Stochastic extremum seeking, 12, 79

U
Uniform convergence, 25
Uniform strong ergodic, 22, 26

V
Vehicle deployment, 161

W
Washout filter, 14, 97
Weak attractivity, 60
Weak boundedness, 60
Weak convergence, 7, 25
Weakly asymptotically stable, 60, 62
Weakly exponentially stable, 61, 63
Weakly stable, 60, 62

Printed by Publishers' Graphics LLC
MO20120722